The Secret Geometry of the Dollar

The Secret Geometry and Mathematical Tradition of the 1935 Dollar

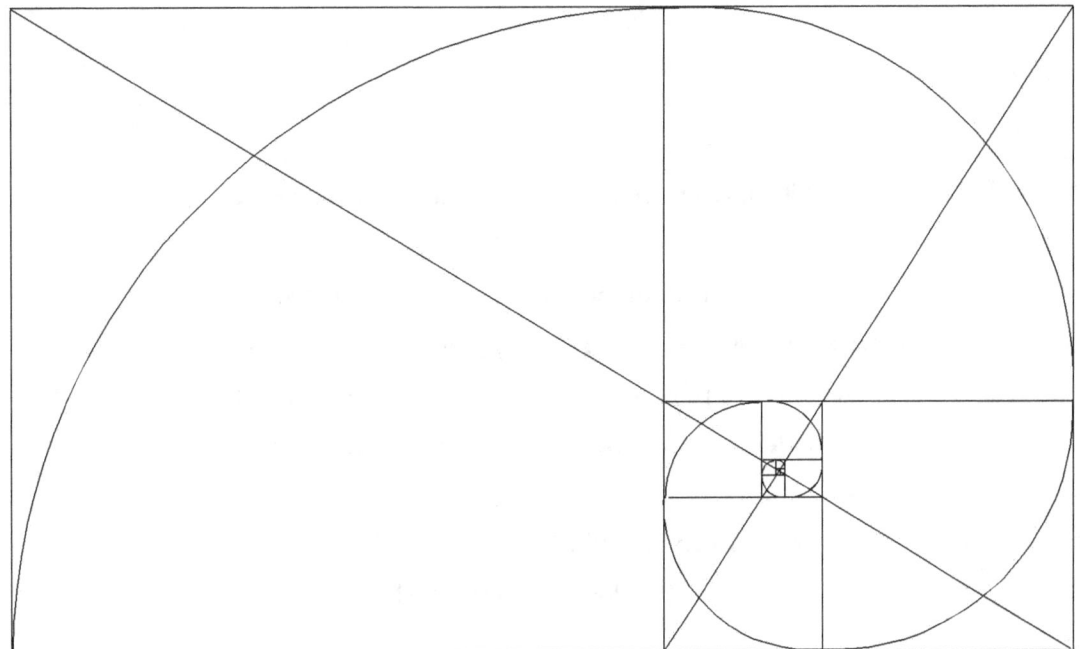

ISBN: 0-7596-1170-X (e-book)
ISBN: 0-7596-1171-8 (Paperback)

This book is printed on acid free paper.

1stBooks – rev. 12/03/02

Dedication

To Judith Ann McGrath, devoted friend and wife.

ACKNOWLEDGMENTS

In the course of sorting out the ideas to write this book, beginning roughly Autumn 1985, no person was more helpful than my wife Judy. On an Osborne CPU, she started typing my handwritten manuscript, editing, spelling and advising, over and over. I started writing shortly after our daughter Elizabeth was born, from notes stretching back to a time in the "gas-crisis"of 1973, when I was laid-off from work for a short time. My notes needed to be put into words, and although I thought I had a good idea of what I wanted to say, she was a good mirror, and she added prudence and promoted maturity in what I wanted to say. Finally I started typing on my own, switching from a manual typewriter, briefly to the original CPM program issue of WordStar, and on through various DOS and Windows releases of WordPerfect. Most of this work was written on an AST 286, some on a IBM 386 and lastly on a Pentium II. Judy has been my primary consultant all along.

Much needed financial assistance was generously provided by Anne W. Russell, my mother, in several tough spots in the writing and publishing steps. Many friends aided me in much needed criticism and advice for readability: first and foremost, traveling men, Wayne Kenaston, Jr., Master Mason, Silvergate Lodge No. 296, (F. and A. M.), and President, Epigraphic Society of So. Calif., San Diego, California, and Vaughn F. Johnson, PM 32°, of Starling J. Hopkins Lodge, No. 88 (P. H. A.), San Diego, California. Great help was rendered by Marshal D. Payn, Epigraphic Society, of Tampa Florida, who provided commentary on my original Entasis essay of 1994, and who located copies of J.H. Cole's nine-page survey report, and John Greaves' *Pyramidographia* written in 1646, both crucial to this work. Julie Gardner-Woodman did early reading and editing. Finally, after the manuscript had expanded dramatically, I hired Daniel Annechrico of ABC Editing, to make an edit of the manuscript, and help make it readable to non-technical readers. Although I can't say that I fully followed all of my editors' recommendations, I hope it is more clear and understandable.

The Temple Tomb doorway picture, on page 27, is from *The Aegean Civilizations* by Peter Warren. In the 1970's, two photographic enlargements used in scans, were made for me by Steve O'Brian of Chicago, Illinois: Chapter 2, Illus. C1, C2, D and E. In recent years, photographic elements were added, beginning on the cover, then page 80 through Illus. Q1 of Chapter 5, the work of Kip Folker of San Diego, California--excepting pictures of an Acanthus on page 146, from a trip I took to the San Francisco area, and Illus Q2 the first "Model," from *The Eagle and Shield*, by Patterson and Dougall. Scanned imagery: page 61 (author), Illus 2, page 245 is from *The Mysteries of the Great Pyramids* by André Pochans; plates from David Davidson; M1 page 112, M6 page 124, M7-9: pages 125-128, M11-12: pages 131-133, and Illus. 1 page 243. The Illus. M2 on page 114 and mapping in M10, page 130 are from *The Secrets of the Great Pyramid,* by Peter Tompkins. The Illustrations on pages 88 and 103 are from The *Egyptian Book of the Dead*, by E.A. Wallis Budge. All line drawings and graphic overlays are by the author. These were created on many different coordinate design program formats: .dxf, .dwg and .dgn, and converted to various kinds of .tif's, .jif's, .jpg's, wpg's and other odd bitmaps and metafiles within CorelDraw.

TABLE OF CONTENTS

PREFACE

This is an unusual book, not belonging to any normal category nor within a regular subject. It is a study of hidden things. It is the result of a detailed study of a small, printed paper surface covering somewhat more than twelve square inches. The surface is the back, or green side of the one dollar bill.

This story is like a detective story of small clues and logical reconstruction. It is a loose chronology of my research, which is an attempt to follow the reasoning behind the many esoteric symbols in the dollar's design. This will touch on much more than the familiar Latin phrases and outward symbols such as the eagle, pyramid, groupings of thirteen, and so on. There are many hidden things within the dollar's design of surprising complexity. The design is known as the *1935 Dollar*, and this book is only concerned with the reverse side of the dollar bearing the Great Seal of the United States of America. Being a unique topic, there will be many peculiar things that the reader will see here that they are not likely to see elsewhere--so I beg the readers' indulgence in the telling of this story.

This is about the gradual uncovering of the hidden work of a brilliant designer, Edward M. Weeks. He was the Superintendent of Engraving of the U.S. Bureau of Printing and Engraving in 1935. This story outlines some of the current dollar bill's enigmatic design: Underneath the familiar green design lies a vast, unknown forest of elegant, geometric constructions with intricate mathematical patterns, some like small "sliding panels," line-edge alignments and other odd things. Under the guise of artwork and symbols, all these puzzles have been waiting silently for years to finally be studied and deciphered. It is now time for others to see what I have seen within the dollar, and in time some adventurous readers may add to a deeper understanding of this design.

The current One Dollar Bill design of 1935--a national treasure of sorts, a comforting image of Americana--is familiar to all Americans and great number of people throughout the world. But the geometric constructions that lie beneath this thoroughly familiar exterior are amazing. The origin of these symbols stem from the ideas of the Founding Fathers at the beginning of the United States of America, at the time of the writing of the U.S. Constitution. And the source of those ideas, in turn, are from a very much older arcane tradition.

Although our record of arcane knowledge during the American Revolutionary period is far from perfectly clear, there is a small amount of tantalizing evidence suggesting that a deeply hidden part of the dollar's mathematical pattern may have been known to some of the founders themselves. These mathematical symbols, I believe, represent an important, invisible aspect of the Founding Fathers thinking. There is solid evidence at a somewhat later period in the 1830's, that there were persons sharing some elements of arcane mathematical traditions within the U.S. elected government *at the time of the legislation that defined the minting of gold and silver coinage.* As you shall see, the real mystery appears in the connection with the ultimate source of this information. And now much later preserved as a quiet tradition, symbolism related to this minting reappears at the time of the mid 1930's, when the current one dollar note was issued. Some of the dollar's mathematical secrets have the appearance of being an updated development from the 1830's period, apparently inspired by the earlier hidden pattern that had been made a part of early U.S. coinage. But certain deeply hidden parts of this message are very complex in character, and would have had their origin more than two hundred years ago, at a time when this sort of mathematical knowledge was unusual. This will, perhaps, be difficult for some to credit. (See Chapter 5.)

Secrets:

Using geometry and certain classical proportions in mathematics, the secret part of the present dollar's design seems to represent a benign philosophical statement. It is, however, *an esoteric statement,* built from the Pyramid of Giza's mathematical of lore, as well as "written" more or less in the dialect of Masonic philosophical symbols. Within the dollar, there is substantial evidence for a very complete knowledge of the measurements, mathematics and theory of the Pyramid of Giza as known to many at the time of the dollar's design. This information is fundamental to the whole dollar design. Its message appears to be more or less non-sectarian Christian in a religious sense, but perhaps mystical, as well as patriotic in character. There are many secrets revealed here that have remained hidden for a very long time, many of which have what may be called a Masonic *character*. But strictly speaking, nothing not already published by Masonry is revealed in this writing, nothing of the sworn secrets of Freemasonry. Yet the dollar's secrets were obviously intended to be openly discovered and puzzled-out by the public. There will be many things here that can only have meaning to students of the Mysteries. This design may contain messages of a prophetic character, or other messages that careful study may in time reveal. I will present the evidence and some of my theories for the evaluation of the reader, who may examine and analyze all these clues as I have, and draw their own conclusions.

Three important elements of the dollar's design:

(1) The most important element of the ornamental design is a well known and beautiful mathematical formula. This is the geometric pattern called *The Golden Ratio.* [1] The ratio is an element of the great arts and a hallowed idea from remotely ancient times. This geometric formula was known to certain classical Greek philosophers, artists and architects. And it was known several thousand years before to earlier peoples in many places. The ratio appears in Greek and Egyptian temples and various other ancient structures of antiquity. Of the various secret proportions used in ancient times, it is the most beautiful and most deeply interesting in terms of nature and mathematical ideas. A history and a practical demonstration of this idea is provided in Chapter 1.

This special mathematical value, like the constant pi, is also an 'irrational number' whose decimal places continue infinitely without repetition. In a similar way to pi (π), it was also assigned a Greek character for its function, the letter φ or phi, (1.618...). In at least one technical reference book [2] phi now takes its place beside pi and the mathematical function e of natural logarithms, where it too has been published to multitudes of decimal places. The unsuspected golden ratio is found hidden in works of art from the Renaissance and in many other places, right up to the present where it is found in the dollar's design.

(2) Beyond this, as a second element, the engravers at the U.S. Treasury also quietly hid *an ancient number motif, within this golden ratio form*. This special motif is the second important element in the dollar's design. Carefully incorporated within the special golden ratio shapes and related shapes, the second element is very much arcane knowledge. It seems surprising that it would appear in the dollar at all. This motif is a mathematical element based on measurements made from an ancient *design offset* found throughout the internal alignment of the architecture of the Great Pyramid of Giza. The design offset was a choice made by the ancient Pyramid builders to place the central shaft alignment in the Pyramid to the East by somewhat more than 23 feet from it's true centerline. This exact, mysterious length was encoded into the dollar's design by a very simple arithmetical method that employed scaled lengths in decimal inch units. This element is David Davidson's "Displacement Factor," or the offset length, of 286.1 Pyramid Inches, which he believed was the basis behind the legendary theme of the "missing capstone of the Pyramid" and its specific dimensions. This number had religious significance to Davidson and many others, and is well known to present-day students of the Great Pyramid of Giza lore.

(3) Together with the two above design elements, is a third important piece that must be taken into consideration. This is a legendary *unit of measure* very nearly the size of an inch, though just slightly larger, called *the Pyramid Inch* which is known throughout Pyramid literature. The presence of this special "inch" length might be likened to an added scale factor over the dollar, serving as another veil in front of the secret design. However, the dollar was so designed in size that it is a veil *that gradually sheds itself, eventually becoming regular inches*, as the paper of the bill ages and shrinks over the life of a given dollar.

At first, there was only the evidence of the golden ratio proportion to go on for a starting point for research. It was this first element of *proportion*, the most important clue left by the designer that I was able to uncover his work.

About This Book

This book is intended as research. It has been written so the average reader will be able to see the fascinating patterns of symbolic mathematics and geometry hidden in the design of the dollar bill. Much of its esoteric symbolism will be shown and analyzed from history of the long train of tradition that led up to the dollar's present design. Although some of these ideas easily lead to wide ranging philosophical speculation, (and I reserve the right to drag out the soap-box occasionally) I will, none the less, try to maintain a neutral or scientific approach to these topics.

Most of this story has been written in the first-person, like a letter to the reader--like a notebook. In order that the reasoning that lead me to these discoveries might be more easily understood, I have tried to show my slow progress and mistakes more or less as they happened, and the gradual development of my thinking as I went along. But to all of this I will add some hindsight, and a certain amount of convenient arrangement of the order of some of the discoveries for clarity. Without this, most of my starting points of investigation and conclusions will not be understandable, and many of my earlier dubious paths can be left unsaid.

Since this curious and strange design is not yet completely known or fully analyzed, this investigation is by no means finished and should be an invitation for more adventurous readers to make their own discoveries. This study is a much larger task than it would appear at first glance. This writing will provide many of the mathematical keys and clues to enable readers to start to investigate on their own, or to demonstrate to themselves the validity of those things shown here. But these are hidden symbols--both philosophical and mathematical--and as such, need to be puzzled out.

****

The Nature of the Dollar's Secrecy

I believe the Designer's motive for this mystical secretiveness and mathematical language, is to make a specialized, classified puzzle--a Mystery, not unlike the Great Pyramid. (A "Mystery" in this usage is a great truth, a deep secret.) The ancient Pyramid's apparent plainness is marked for the careful observer by a large number of sophisticated mathematical "statements" made from subtleties within it's superstructure, most of which are quite out of place within the context of regular stonework of Egyptian pyramids. By analogy, this is similar, in a literary sense, to the occasional odd insertion of French in the torrid Victorian romance novels or strategically placed Latin passages in the medical books of the last century. In the case of the dollar, the dollar's Designer wants to say some things, but wanted only certain people to find them. The difference here is the special language he used to conceal things.

This secrecy by obscure phrase was once commonly found in literature a century or more ago. Political Correctness in speech was once worse than it is now. Latin passages were not generally understood by the unschooled or the dangerously intolerant, who might otherwise raise an outcry at hidden opinions. Such people almost never consult Latin dictionaries or knowledgeable people to figure out the phrases. And, were they to have found out the meaning--accusations are still difficult to prove. Civil authorities charged with censoring such outrages have never known much Latin beyond legal terms, and naturally wouldn't have been comfortable with their ignorance in court. If instead, they *had known* proper Latin, they might then be stuck with the indignity of having to literally reveal the meaning in plain language, openly, in order to specifically accuse or pronounce judgement. A line from another language is very hard to censor.

There has always been a sort of unspoken agreement among the authorities and the educated that foreign passages were exempt from scrutiny. Like science or mathematics, the passages were sophisticated and safely kept from common people by use of special wording. If you *could* read such things, then you must therefore be educated or "certified" to know of such things. And you were obviously of the class that was expected to be able deal with such authors' remarks without resorting to riot and licence. Who knows what was being said? Moreover, for all those of the sort who would loutishly assert, "Its all Greek to me," the matter remained closed and soon happily forgotten.

Secrecy is the beauty of this little classified system. These Latin passages will remain as a kind of a permanent *open secret*, which was the writer's objective. Like the hidden mathematics in the dollar, the odd passages may be seen, as in the case of the medical texts, as a kind of openable door to deeper teachings. Dr. Krafft-Ebing covered his tracks with Latin in his early medical writings on sex. Latin and French graces the pages of fiction such as British crime mysteries of the 1920's, just when the juicy parts came up. Even *Latinized French* graces the pages of The Once and Future King, the classic story by T.H. White, where he hides philosophical ideas.

Encoded ideas are found in Biblical parables that are in common, though oddly worded everyday language. But you do have to learn something of the Latin or French, or closely study Biblical languages to understand their insight. Most Biblical parables have served to avoid the censor, since the censor didn't have any idea what they meant. And so they have survived to our time. The censor reasons if *he* doesn't know what something means, it doesn't mean anything--or, at any rate, you'll never figure it out either. In the case of *mathematical* symbols, I think we might find that the censor couldn't be bothered, even if made aware of them.

Unlike Latin, the clues on the dollar are harder to read, requiring measurement and computation. But in much the same way as the classical epigrams, if you wish to learn something of the dollar's secrets, you must spend at least a small amount of effort studying this special geometry so that you are at least somewhat conversant with it--if you wish to see past the elegant barrier that veils hidden things. The dollar's message was apparently intended for those who would take some time and effort to study and understand. This, like the more easily readable "odd passages" above, is an ancient filtering method used to restrict the special Mysteries to the sincere and interested, while still leaving them mostly invisible to the casual and parochial; something to catch the eye, but not everyone's eye. In the case of the dollar, the "odd passages" are it's special design format, *its geometric form.*

Interestingly, this special form of invisible information extends to cover quite a lot of *plainly* shown things, well known throughout history, containing great secrets. Where the key concept has not been imparted, nothing of importance can be seen at all, even in an object that millions handle every day. This will always be true for the dollar's obvious symbols. Most of the public knows little about, and is unconcerned with the Latin phrases and symbols on the bill as long as it still works well at the bank and grocery store. Others, the very few, who actually go to the trouble to inquire as to the meaning of these things are given standard, more or less satisfactory answers that have always been graciously provided by the U.S. Treasury. Here, the trail seems to end, and would seem to stop. But this is not all there is by any means, as we will see later.

A Delayed Discovery

What if these geometric patterns to have been discovered and published in the Designer's lifetime? He (and the U.S. Treasury Department) probably could have shrugged off any controversy. As will become abundantly clear--most of these cunning patterns are not at all obvious, and might well have been argued by almost anybody in his defense to be purely coincidental, or even complete fantasy. Added to this is the Pyramid inch, *a scale factor* question, where anyone who claimed to have discovered the dollar's secret could be directly refuted, since these lengths are numerically larger than the obviously ideal numbers. Certainly, it wouldn't seem that the Designer is responsible for how people interpreted his design after it shrank over time. (And yet he must be, as you will see.) But as it happened, he didn't have to worry. The Designer died in relative obscurity in the late 'fifties with no need to prevaricate or come up with any sort of explanation (See Appendix B, p. 275).

I have often wondered what explanation he would have given if some clear evidence leading to these questions had come up. If pressed, what would he have said? Maybe he would have something like this totally hypothetical composition, expressed here--perhaps as he might have--in bureaucratese:

> *We chose the classical Golden Mean and some published figures typifying the venerable capstone tradition of the Pyramid as a fitting compliment to the design of the Great Seal now shown on the new issue of the One Dollar Note.*

This of course is not a quote. But would it have answered questions in the time of the New Deal? Or, perhaps another kind of possible answer, such as a claim to have used a purely American motif, such as the date of the Declaration of Independence as a mathematical motif, a subtlety about which we will see more later. (See Chapter 2 and 5.) Could the Designer have claimed that all of this complexity was for anti-counterfeiting? As we will see, this effort would have *far* exceeded any such need. So far as I know he didn't say anything, whatever, about the dollar's design. Until something like a real comment from old correspondence shows up, (which might happen) I think that we are stuck with lame hypothetical guesses as shown above.

It was mostly good luck to have found these patterns. Other than some Latin and a few Masonic symbols there is nothing obvious or unusual that is visible on the dollar bill. Much of the mathematics and geometry intended by the Designer are only hinted through by graphic clues or is only to be inferred by careful measurement and calculation. But there are some openly interesting elements to the design, as well as many more not-so-obvious points of great interest. Although I hope to persuade the reader by reason and evidence, many intelligent people will choose not to accept the evidence, nor believe any of this idea for one reason or other. But I am sure that this selection is as it was supposed to be, as the Designer would have intended.

****

Mathematics

For those who are not mathematically inclined, these ideas will be explained in detail, with extra effort to make them understandable to those who are non-mathematical. To borrow some good advice from the popular British mathematician Roger Penrose, here is what he said at the beginning of his book, The Emperor's New Mind:

> If you are a reader who finds any formula intimidating (and most people do), then I recommend a procedure that I normally adopt myself when such an offending line presents itself. The procedure is, more or less, to ignore that line completely and to skip over to the next actual line of text! Well, not exactly this; one should spare the poor offending formula a perusing, rather than a comprehending glance, and press onwards. After a little, if armed with new confidence, one may return to that neglected formula and try to pick out some salient features. The text itself may be helpful in letting one know what is important and what can be safely ignored about it. If not, then do not be afraid to leave a formula behind altogether.

This is advice that I use myself. In this book, I will try to provide complete explanations, drawings, useful captions, notes and arrows pointing to important things and parts of diagrams. In any event, I hope to simplify and cut out any unnecessary complexity in this odd and detailed subject.

You won't see, as you might read somewhere else, "you can skip over the mathematics and turn to the text on page..." since for the most part the math will be part of the narrative and may be crucial to the ideas being discussed. Of course, many, if not most readers will have to skip over formulas. But I will lead the reader through most of the mathematics and reasoning, and I will always try to paraphrase my reasoning in simple English. It is my belief that if you can't say it in plain language, you probably don't know what you are talking about. Most every equation here is (hopefully) translated into plain English.

To understand this book, *most* of the math can be done on a cheap, scientific-type calculator, which is what I have generally used. Present day electronics stores are full of various kinds of inexpensive calculators. I will show how the math is done, explaining and showing some of the basic keystrokes for calculators and also certain specific routines for some of the types of calculations. You are encouraged to get an inexpensive ten place calculator--one specifically with trigonometry functions--to help follow the ideas and arguments as they are developed. Otherwise, you will have to "take my word for it," receiving whatever I write here on faith alone. I think the interested reader has a sort of obligation to keep authors honest and accountable, as well as to try their own variations of the ideas, or even their own completely new ones. (For simplicity, some math was done on a graphics computer program.) I fully expect many sharp readers to catch some errors and blunders, as well as a good many of the Designer's ideas and nuances I have missed. None of this is very difficult, it's all just multiplication and division.

I do not share the prejudices of the stuffy, academic-type mathematician. I come from a land survey background, and do not share disdain for those not proficient in this art. In my profession, math is simply a *tool*, something easily taught to new-comers to survey work, usually out in the field, informally, with no pompous rigamarole. What will be shown here will mostly require poking buttons on a calculator, not unlike the figuring for setting floor tile, carpentry, or other kind of craft. Mathematics is a practical and beautiful tool. It is a special form of communication, an *etiquette,* a recognized form by which other users can exactly manipulate values that you have measured. And in their own way, with enlightening, reproducible and often surprising results. Philosophically, math has a beauty like frozen music, it is rigorous, clarifying and far-reaching. This will not be the tedious ritual, very hard and dismally endured as most were taught to do. Since most people have few *sweet* uses for this tool, much of mathematics is empty of real interest. We were taught the useful forms, but little if anything at all about it's beauty--such as the golden ratio, for instance.

Since I am more a mathematical cobbler than a purist-type mathematician, my use of the mathematical forms will be a little different from this I hope; less terse, or a more verbal "show your work" style, with visual demonstrations wherever possible, and without attention to formal proofs. This is more of a practical "what was done" and "how to do it" narrative. This work is about measurements and the mystery of the Designer's intended message, leaning more in a physical and empirical direction, rather than one of rigorous schoolbook form. But there still will be some real mathematical surprises for an academic here, at least in a historical sense, and certainly in the subtlety of it's use.

****

Measurement Units

The dimensions in this book are in four principal units: **British Inches**, 1.000" = one inch, as indicated by the double apostrophe symbol; **U.S. Survey Feet** by single apostrophe such as 1.0' = one foot; **Pyramid Inches**, where 1.000 PI = one pyramid inch; and **meters** which I will indicate as 1.000 m being "one meter." If fractional, 0.123 m may be worded "One hundred and twenty-three millimeters."

Most of the dimensions in this book can be thought of as regular "inches" before Chapter 3, since up until that point in my thinking I was not persuaded that I was dealing with any special units such Pyramid inches. The differences between the two kinds of unit are very small. But the so called Pyramid Inch also called: Polar Inch, or, Primitive Inch of David Davidson and Smyth, is clearly present in the dollar, and we will find a few metric unit situations also. (In the Appendix A, we will also touch on various U.S. and British foot and metric standards. These values are for a much larger scale of size, and I will use somewhat different numerical conventions than found in the beginning of this book.)

My tools measure in regular British inches. Any "Pyramid Inches" are therefore a conversion by a multiplier. When it becomes important to the story, we will look at Pyramid Inches, and these can be determined by multiplying an inch dimension by a scale factor. In early books on the subject of the Great Pyramid, the scale multiplier was thought to be 1.0011, but since the turn of the Century this has been

vi

calculated to be somewhere between 1.00106 and 1.00108. The easy way to remember how these units work, is to know that Pyramid Inch units are always a little *bigger,* physically, than British inches. For instance, to convert British to Pyramid units: 10 British inches are multiplied by 1.0011, which equals 10.011 Pyramid inches--a larger number. To convert Pyramid to British units: 10 Pyramid inches are multiplied by the reciprocal 0.9989, equaling 9.989 British inches, which is a smaller number of bigger units. But for the most part, these considerations will not greatly affect this account of ideas leading up to the various discoveries.

Measured Dimensions as opposed to Theory

Since I believe that one-thousandth of an inch is just about the general limit of reliability for measured precision on the paper dollar together with true accuracy, few ***measured dimensions*** will be shown past the third decimal place (0.001", or a thousandth of an inch). In those cases where I venture into the abode of the fourth decimal place (0.0001", or a ten-thousandth of an inch), such *measurements* will have been made with great care over short distances, ordinarily not exceeding two inches by direct measure, or certain larger lengths by means of careful statistical methods, and with some trepidation.

If you see a number in this book showing *more than four places* past the decimal place, you may be certain that this is *a* ***theoretical number only***, that has been derived from (1) some theoretical idea that I or someone else may have had; or, (2) an Ideal Constant--like the irrational number π (pi) or φ (phi); or, (3) some combined use of the above forms of numbers together, and so therefore these become artificial and theoretical.

There are two entirely distinct ideas in use here: the **Theoretical Distance** and the **Measured Distance**. But each is necessarily written in the *same language as the same decimal strings of numbers*. This is a difficulty, so due to their apparent interchangablity, I will at all times try to make the distinction clear to the reader. Those who don't feel sensitive to this distinction, should be clearly advised that a vast gulf of difference lies between them. These can be looked at on the one hand as beliefs, experimentation, imagination, or just ideas, (**theoretical**); as opposed to actual experience with a measuring tool, on the other (**measured**). The problem comes when we write these numbers down somewhere, that they both take on the *same character* in appearance to the reader. Since I will often show theoretical experiments based on measured dimensions, they must all be very distinctly separated.

The measured dimension is offered as something that I think is true, *from experience, from my observations with measuring tools, at least to the stated decimal place*. Whereas the longer, theoretical numbers are offered *as possible answers to what the Designer may have had in mind, or a true mathematical proportion*, by means of the measured shapes and from various formulas that come to mind from them. But then again, the theories are only just that--**Theories**. **Measurement** however, is the reliable and, for the most part, unchanging base of evidence upon which the far less reliable theoretical constructions rest. I will often insert a note in the form of an (**m**) or a (**t**) *before* certain numbers and formulas containing them to show this crucial distinction.

Measurement and Notekeeping

For those who are adventurous, who wish to follow up with their own measurements on the dollar, I strongly suggest that you discipline yourself and make some sort of notes. Measurements then, of *any kind,* even your own check, are absolutely useless unless you write them down or recorded somewhere, such as in a notebook, or even in the margins here if nowhere else. You should keep good, careful notes for yourself in a *notebook.* Don't rely on computer discs of any kind--go to paper. *Date* all your notes, since, although you may not see it now, the sequence of your discoveries and the trail of your thoughts, will be quite important and useful to your later interpretation and understanding of whatever you find.

Those who have computer aided drafting programs (CAD) available may find them very useful in separating geometric patterns and data. But I caution you about the use of the electronic graphics for simulated measurements of small areas. Although I have used this on occasion in this book, few systems will do this accurately, even with correctly setup electronic drafting units. Though it is tempting to try to "measure" small dimensions based on sound computer based data, these should only be used as "drafting notes" and *approximations* of small distances to be used as good clues to further research. (Calculation can however be done with a suitable coordinate geometry program, or COGO, where the units are set to at least ten places past the decimal. But this is not at all the same as a regular CAD program, which cannot be made to work quite the same way.) Instead, distances should be actually calculated mathematically *by hand*, on a hand calculator, computer spread-sheet or with some other reliable mathematical technique. In this way--and *only* in this way--can you be sure of your mathematical discoveries.[3]

****

Dollar Design, United States Government, Currency and Freemasonry

Called in Federal Law the "Great Seal of The United States of America," the One Dollar Design of 1935 was the first open, public display of both sides of our National Seal. Before this time, the complete, dual, appearance of this seal was found only on the face page of internal, official federal documents, diplomat's credentials, and among the notes and publications of certain scholars and philosophical writers where it didn't attract much attention. The "eagle and arrows" side of the Seal had appeared on coins and elsewhere in public view much before this time, but "pyramid and all-seeing eye" side of the Seal had remained mostly hidden before 1935.

The United States is the only country in the world that has a two-sided seal. It is described after the fashion of coins, as having 'Obverse' (eagle and arrows) and a 'Reverse' (pyramid and eye) faces. The fact that it is two-sided, is thought by some to indicate a special intent on the part of the Founders to announce a sort of unique agenda for an ideal human progress. This is supposed to be on a plan very different from that of the design and destiny of other nations. A second, or *hidden* destiny is revealed.

This was to be an ascent of philosophical, intellectual and scientific progress, not known in human history, rising in time in increasing perfection, like the courses of stone of the unfinished pyramid. Inaugurated by the founding of the Republic, somehow especially blessed by God and through Divine Providence. The Latin words 'Annuit Coeptis' above the pyramid on the Seal mean something like "In our doings He is pleased" or, more literally "a nodding assent," and refer to a blessing. Perhaps it reflects a recognized feeling among the Founders at time of the American Revolution.

The present 'New Age' movement with its many unusual and varied forms, is originally from the Seal's quiet and stately Latin motto, "Novus Ordo Seclorum" which is translated as "new order of the ages." As a title for political policies, we have also seen recent Administrations' ominous borrowing of the term 'New Order' out of this motto. And, no doubt we will see others do this at various later times for their own purposes. These people mean something entirely different from the original intent. Wild-eyed idealists and self-serving politicians of dubious motives have finally given a fringe-element twist to the use of the term "New Age." But a genuine New Age *did* and really does now exist: we and our times are the result of it. The tolerant, egalitarian and republic state concept is seen in the motto "E Pluribus Unum" means "out of many, one." This idea now, perhaps more than at any other time in our previous history, might be said to characterize the present demography of the U.S.

****

The Great Seal of the United States of America

In the afternoon, late in the day of July 4, 1776, a special committee of three senators was chosen to design the Great Seal,. John Hancock was president of the Continental Congress, and announced:

"We are now a nation, and I am ready to hear you vote on the question, Resolved that Dr. Franklin, Mr. Thomas Jefferson and Mr. John Adams be a committee to prepare a device for a seal of the United States of North America."

Several trial designs were drafted by these august senators (none of which were well liked), and later several additional developmental ideas were provided by other senators over time. Over several years, and after much work a description rather than a design was finally agreed upon, and voted into law.

This unusual dual seal became official on June 20, 1782, as adopted by the Continental Congress at Philadelphia, about six years before the time of the full ratification of the U.S. Constitution on *September 17, 1789.* This was its final adoption as the "Great Seal of the United States of America." We can see here, that this was among the first acts of the new government. The form as finally adopted consisted of a verbal legal description in heraldic terms for the Obverse and Reverse of the seal, included in a document called "Remarks and Explanation." (See pages 186 and 187.) There was no official graphic design, which allowed for some latitude in later forms of execution in stamps and seal die stamps. The final, verbal form was the creation of William Barton, M.A., a young lawyer from Philadelphia, a private citizen and an authority on heraldry, and Charles Thomson the Secretary of Congress, who is said to have assembled and simplified the design. From this same period of time President George Washington's Masonic Apron shows a pyramid and radiant All-Seeing Eye embroidered on it along with other familiar symbols. There can be no doubt of a common source of inspiration for the final design of the Great Seal.

Much later in history, President Franklin D. Roosevelt (also a Master Mason, as have been many U.S. Presidents), approved a commission to the U.S. Bureau of printing and Engraving to change the shape and design of the dollar, and to put both sides of the Great Seal into its new design. In 1934, Secretary of Agriculture, Henry A. Wallace (a former vice-president) submitted a proposal to Roosevelt to mint a coin showing both the obverse and reverse sides of the Great Seal. Wallace's interest in the Great Seal is said to have been aroused by a book he found by Galliard Hunt, written in 1909 called <u>The History of the Seal of the United States.</u> [4] Roosevelt decided to place the Great Seal on the dollar note rather than a coin. Roosevelt's influence on the execution of the design seems to have been limited to re-ordering the right and left positions of the faces of the Seal after the first design, which evidently must have caused it to be redesigned and its printing plates re-cut.

When released to the public it created a small amount of short-lived controversy. To the public, it must have been quite amazing. The new dollar had an altogether different shape: it was smaller but longer proportionally. It was covered with unusual symbols, such as an unfinished pyramid, an eye in a glowing triangle. And on the other end of the new dollar there were unfamiliar Latin words, in addition to the more familiar eagle, arrows, olive branch, shield with stripes and thirteen stars in a fiery cloud as the public had seen elsewhere. I wonder what the people of our time would have thought were we somehow there at that period, or how it would be accepted by the public if it somehow come out now.

There is also the curious and arcane matter of *the timing* of the new dollar's issue and appearance. For those within the small and quiet world of esoteric philosophy, those familiar with pyramid prophecy lore, this issue seemed appear just when a special, *historic instant in time* called the "Entrance to the Kings Chamber," was scheduled to occur. At the time there were even a few articles about this prophecy in newspapers. This was to occur on *September 16, 1936,* a very special point in a prophetic time-line of chronology. If based on internal measurements and calculations of the Great Pyramid of Giza, (as interpreted by David Davidson, the British Engineer and Pyramid theorist), this is about when the new dollar actually appeared in public. And this date was *also* important to the Designer of the dollar, being clearly evident in several parts of its design.

Were it not for the multiple references to this special pyramid genre, this date might have passed unnoticed. But there is a suspicious coincidence: (1) *this specific date from pyramid lore*; the (2) *same encoded date* appearing in the dollar's design; and (3) *the date of the new dollar's appearance.* This seems startling, and must prompt some speculation. Was this just a fortuitous accident, or was all of this somehow orchestrated by some pre-arrangement? Were there more people involved in this design than just the dollar's Designer and his assistants? Henry A. Wallace, an adviser in Roosevelt's Administration, is the only one known at present to have really had what we might now call "New Age" interests or ideas. (Roosevelt, we should note, was inspired by James Hilton's mystical Lost Horizon, and the *original* name of the present presidential retreat Camp David, was Shangri La, a name taken directly from Hilton's story.)

I was told in grade school (in the sixties) that the dollar's change in shape had been done so that it might "fit in your wallet more easily," although I can't imagine that we couldn't have somehow coped with this problem before. Even if the old dollar was an inconvenient size, who or what prompted this specific shape change? It seems a little surprising that the dollar was issued in any new or unusual form at all--considering, for example, that it was a new currency appearing at an economically and politically difficult time, just before the end of the Great Depression.

More recently during the late 1970's, we have seen the result of what happens when the public totally objects to a new dollar design such as the 'Susan B. Anthony' coin and refuses to accept it. People just quietly objected to them and ignored them. "They were foisted on us. She hadn't been a president," it was said. Then, there were these weird, flat sides along the inside edge of this coin. And, even at that, there were *eleven*, not thirteen sides along the edge--which would have been more in keeping with U.S. tradition. They were much too *small* to be a dollar; insultingly small to many; cashiers had no normal place in their drawers for them. Banks just couldn't pass them off very well as dollars, and bank tellers had to take back in change most of the ones they did manage to force on the public. Then, for many months as we watched, they were relegated to little glass bowls at tellers' windows, quietly returned and rejected as change until their at last their fate was determined by default. The U.S. Treasury Department must have taken back *boxcar loads* of them.

In recent years, the U.S. Post Office stamp vending machines were being used to try to ram these coins down our throats once again, but now with a little warning message at the bottom of the machine saying that "this machine gives Susan B. Anthony dollars in change." In a way, then, they are not really "money" if they have this distinction, if people really don't want them. Now, (in 2000) another effort is afoot with a gold-colored coin of the same size, with another non-presidential image on it. But the brand new paper dollar was well accepted in 1936, even if now the old-timers' phrase "sound as a dollar" has worn a little thin. In those times of yore, before the great swindle of pot-metal coinage of the 1960's, it was an *actual receipt for silver metal*, a Silver Certificate, something very real to it's owner. But that's all on the front face of the dollar anyway--and we are only concerned with the back side.

The 'new symbols' that the people of the 'thirties found on the left side of the dollar didn't come as any surprise to the Masonic Order, or administrators within the U.S. Government. With the printing of the new dollar, it is likely that the some of the politicians and government officials might have been a little uneasy by the public acknowledgment of such peculiar symbols. The Reverse side of the Great Seal *has never been used* in the official stamp of government documents, even though law and tradition require it. And, perhaps some Masons may have been a little surprised with some of the quiet elements their tradition appearing publicly on their currency. Federal Law enacted in 1789, has always required the full presence of the Great Seal on official federal documents. Out of this legal tradition Franklin D. Roosevelt chose to bring this Reverse side out to public view, along with the Obverse. From this new issue, the dollar design has served to openly exhibit the original and fundamental connection between Freemasonry and the U.S. Government.

The symbols on the Great Seal and the philosophical mysteries that they enclose are completely "American"-- every bit as much as Uncle Sam and Apple Pie. After two hundred years, these symbols have now become better known and accepted and are now written about and read about with a greater curiosity than that of any previous time. There is a growing interest for what are now the quaint traditions, mysticism and odd conceits of the Founders of the Republic. These are our philosophical "roots" in the United States of America and there is much of interest beneath the surface.

****

Who were the Masons?

What about the Masonic Order and their history?

What of the esoteric faith or goals of the philosophy of old Freemasonry?

Since we are looking at fundamental secrets, there is something of an inner story to tell in connection with Freemasonry and the origins of the United States and its symbols. The Masonic Fraternity is a very old secret society, dedicated to "brotherly love, relief and truth." They are the inheritors of an ancient Mystery tradition that has its origin in the Mideast. The tradition is taught by symbols and spoken ritual committed to memory. At some point in time, Freemasonry embarked upon a quiet course of moral reform of society from within, which they thought of in a symbolic sense as *the building King Solomon's Temple.* Considering the historical effects of various other events in the past, we were very fortunate to have had their attention and vigorous involvement at that crucial point in our history.

Religious and philosophical symbols are important to Masonic thought. One crucial, symbolic concept within their teachings is the *unfinished process of building of the Temple*, a work always under construction. In the Biblical story of the dedication of the Temple of Solomon, the *mysterious cloud of sacred smoke and fire* associated with the Ark of the Covenant was said to have expanded and completely enveloped the Temple, even driving the priests out by its intensity. This was taken as symbolic of God's approval of the Temple. This was called the *Shekina.* The "Shekina" (a Talmudic word) was the Presence of God, a sacred but dangerous force believed in ancient lore to be a *blue blaze of spirit fire* when seen at night and likened to a *smoke* if seen by day. In the symbol adopted for the Great Seal found on the right side of the dollar, this Shekina of fiery smoke is seen surrounding the Thirteen Stars, or what would be the original Thirteen States, also *at a point of dedication in 1776.* This symbol was obviously employed to show the same approval by God of the United States, and the Shekina occasionally appears elsewhere in Masonic lore and emblems. From the remaining reference in that analogy, the *Thirteen Stars*, we are left to conclude that the original collection of states must have *been thought of as being the temple.* The design of the Field in the American Flag we can see that any added stars are then *enveloped in the sacred blue*, perhaps intended as a "work continually under construction."

Eighteenth Century Masonic thinking in America hoped to promote a balanced, prudent, fully secular state, linked--though separately--to a freely religious citizenry. This was to be through legalized tolerance as well as the trust of the good nature of their fellow man. This trust was to be based on God, or rather, I should say, at least in a *Supreme Being* in Masonic terms, Whose nature was totally impartial and non-partisan to all of mankind. Their idea was a simple one: "the religion to which all men may agree," which is to say, religious and virtuous intent without arguing about the particulars or specific details of any other brother's belief or conscience. This Supreme Being or Providence, was by no means an exclusively Christian type of God--many of the Founders were Deists, various kinds of Christians and whatnot, all holding widely differing views.

In doctrinal specifics, all of the brethren might well have disagreed, but their non-sectarian Deity was only described in very basic terms. The Deity was so thoroughly basic and logical it was also described as the source of pure Geometry. The Colonial Masons linked this Providence, First Cause, or the "Great Architect of the Universe," to a general notion of human progress. They believed in the ideal of self-perfecting individuals and peoples; of a self-representing and self-reliant society having a faith in the common man, devoid of tyranny, devoid of controlling lords, priests and kings. And all in a society where you would profit and progress on your own merit and ability. A society where you were politically *represented* and where your representation *paralleled your neighbors' representation.* This, within an *open and unlicenced political paradise* of civil and religious liberty. A system like their Lodge, with multiple checks and balances to place limits arbitrary authority. Also, the administrators were to be subdued, as *public servants,* having only the rights of regular citizens and were now subject to the same laws, and in open court. The mice had finally put a bell on the cat.

These ideas were--and to whatever extent they have actually been accomplished here today--the natural result and outgrowth of the general ideas of Freemasonry and other developing political ideas current at the time of the Age of Enlightenment. Hidden Masonic society of the time of the 1700's had many interesting and oddly modern characteristics to set them apart from the Old World. Masonry contributed greatly to the Age of Enlightenment, but very much from behind the scenes. In secret, among their own brethren of Master Masons, all those within a lodge were treated as *equals* with no regard to their worldly rank, whether a titled aristocrat, commoner, laborer, merchant, or whoever he may be in the ordinary 18th Century world outside a Lodge of Masons. Rank within the Lodge was determined through ability, by progressing steps of merit, and office was determined through a referendum by Master Masons. They were all equals, special Master Craftsmen, all employed in "the Work." This *Work* was the betterment of one's self and one's local society. Underneath, where it was hard to notice, and gradually over time, their day to day gentle philosophy focused on the social behavior of the everyday man and his dealings. In our country today we still hear people asking if "this is on the level," or speaking of "a square deal" --both references to truth and fairness *straight out of ritual* from Masonic teaching. This moderating thought was also broadly felt in the political action of the time--and Freemasonry's ideas were behind the development of our government.

In hindsight we can see that *it was inevitable* that a government formed on a model of this sort would eventually flower and grow according to the internal logic of its organization--a continually spreading legal process or philosophical political entity. Ultimately, it could not escape the effect of its own rules and discipline, eventually completely extending all these concepts and rights outwardly to all its citizens. And, inevitably in time, these rights passed on even to its slaves, its women finally as citizens, and finally to new naturalized citizens far outside of the original racial and ethnic types. Logically it had to in time, if undisturbed--*brotherhood* was the original idea and practice of the men who carried on its traditions, and it had no other kind of goals.

The period from 1650 to 1717 before the time of the American Republic, mostly brought to a close what is called Operative Masonry: true stone work and building craft training. This began the present expanded phase called Speculative Masonry, which became open to any freeborn male, having vouched for good character by a brother, being at least 21 years of age. There is a small amount of evidence of Masonic activity in North America from this period and even one hundred years or so earlier. After about 1650, perhaps half of the English lodges had a majority of non-tradesmen who had been permitted to join to learn the religious Mysteries involved in the ritual. Quite probably they joined just to find a place allowing un-supervised free speech. Finally divorcing themselves from commitment to any specific type of worship, at this time they began to open their initiation to men of other religions than the dominant local religion near the lodge, which their earlier by-laws had required. In time, the concept developed that the Apprentice was required to be initiated on a book of religious precepts he believed in, such as the Bible, and it was required of him to at least admit of the belief in a Supreme Being.

How far back in history the unusual behavior of equality and social liberty held true for the Brethren we don't really know. But it seems to be true enough to say from various documents, that at the very least as early as the year 1717 in England at the formation of the Grand Lodge, when they wrote their early Constitutions; that they had an already well developed and functional egalitarian philosophy and a sort of internal government. Each brother was assumed to have innate natural rights.

The Masonic concept of basic by-laws and *constitutionality* were important for them as a method of self rule, but also this idea became important *as a model* for the American Republic much later. This was a new idea then, since rulers everywhere ruled by "custom" in those times, and rulers, of course, arbitrarily made up "custom" as they went along. Most forms of Masonry now extant, are constitutionally based, and around the world these origins can traced to the rites and by-laws preserved after the opening of the Grand Lodge in 1717, which were derived from earlier constitutions. In practically any place and in any culture in the world where "a British ship could go to port," these traditions are to be found. Although little is known of the exact form of Masonic rites and by-laws before that time, it is likely that what surfaced during this unifying period--which is still largely preserved today--represents a family of similar rites and very old surviving traditions of several English Masonic lodges as well as from continental Europe. The increasingly popular underground swell of activity as these ideas spread was to greatly affect the general aspirations and sentiment of both the French and American Revolutions as well as those in South America in Bolivar's time. (Bolivar also was a Freemason.)

American Masonic Lodges--long prior to the Revolution--had sophisticated parliamentary procedure and the practice of a secret ballot, making them almost unique as early exponents for a kind of democracy. Their sworn concern for justice and truth set the Freemasons apart from most, if not all social and governmental structures of their day. Politically, they were officially neither Whig nor Tory types of sentiment (those being the political left and right of their day) and they were committed to political neutrality within the lodge. They were also committed to *religious* neutrality. No religious body was connected with the Masonic rites. We can see that this religious neutrality aspect was adopted by the U.S. Government and still continues today. This became the standard for republics all over the world. Freemasonry was fundamentally, though very quietly, influential in the formation of early Colonial Law, the U.S. Constitution, the Bill of Rights, and later U.S. law and governing philosophy. Freemasons themselves were often in important positions in the early U.S. government and still to a large extent for more than a century afterward.

Various elements of the U.S. Constitution and the Bill of Rights were directly influenced by Masonic Constitutions and Masonic by-laws during the colonial period of the Thirteen Colonies. Historians have shown that many of the framers of the Constitution and signers of the Declaration of Independence were Masons. Yet, it is surprising how often one encounters biographical material on George Washington, that completely fails to note his connection with the Masonic order, or belittles it--when you consider that at the time, he was the Master of his lodge, perhaps the most influential lodge in North America, while simultaneously the United States of America's first President. Few scholars have pursued the question of Masonic influence in early America. The Lodges were already doing things internally in their self-government that we now take for granted as civil procedure everywhere.

This then, is a hidden source from where much of our idea and working practice of government, liberty and equality came. This quiet system was almost completely derived from contemporary lodge activity, fortunately shut down or disabled almost all of the Old World traditions of authority here. Do we ever wonder how this happened? Few groups other than the Freemasons and Colonial governments were doing anything like this, anywhere in the world. A solid, pre-built and workable system was quietly grafted onto America in the interim from 1787 to 1789 *with no real awareness on our part as to just how or what had happened.* It is as though the Republic was always "there" with no good historical explanation other than the "Founding Fathers did it at the Constitutional Convention." And we take this all for granted--no fanfare, no credit taken. It was the Delegates *who did it* and haggled out a structure. And this was with the help of non-Masons, but they left a lot out about *how* Delegates did it, and they left no complete record. It was done in *closed session* at the Constitutional Convention. The resulting governmental structure resembles the tripartite Masonic lodge in a number of ways, with elements adopted from other republics known from history and the rejection of a state religion. Immediately thereafter, we began the long, difficult process of adapting the document to our purposes, as well as dumping laws that didn't work, with a surprising amount of success. We are the oldest remaining republic in the world. With regard to our earlier legendary beginnings, our civil system certainly did not come about by way of the "Puritan Pilgrims" as we heard in grade school. The Constitution disabled that kind of cultural control and rule. That tradition--the Puritans were Calvinists--produced a colonial flowering that lead to the Union of South Africa.

It is important to realize that all these good ideas such as "freedom" and "democracy" are a relatively modern addition to our life, and owe little to religious doctrine of any kind. "Truth" and "freedom" do indeed appear in the Scriptures but have been almost completely unsupported by any doctrine or action of the churches throughout Western Civilization. The Scriptures certainly provided the Light, but it is prosperity as a result of economic freedom that must precede the development of representative government in any society. And, these must be internally developed first, by that society, from itself alone. No one gets "empowered" to become a republic, anywhere--it is a cultural development. In our case, we were fortunate enough to have had the seed form of the *working tools* of this life quietly active within a mostly prosperous and homogeneous entrepreneur population in the Thirteen Colonies.

The Constitution was created to *limit and number the powers of government and grants no rights.* We already have these rights as natural rights, and the document is more to enshrine them with regularity. I think the primary historical legacy of these colonial Masons, was their affect on the stable creation of the U.S. Constitution. This document has turned out to be astonishingly long lived and well thought-out. Insofar as legal documents go, it is harmonious and balances the many competing, political forces-- considering that it was determined in secret debate. Our Constitution is one of the few things present in our political life that we still revere without cynicism. The wisdom of the Founders and workable ideas are still revered by the great majority of conservatives and liberals alike, which if you think about it, is one of the odd, unspoken facts of American life. Probably, only a few from either side of the aisle have any clear understanding of how or why it works. Neither side really *disagrees* with the document, but will argue what it specifically means, how it should be interpreted, how it should be applied, or perhaps changed. (It should be left alone.) It should be noted that the first great test of our system, the Civil War, was based on the opposing *interpretations* of constitutional questions about the Northern Tariff and human rights, but *not* a rejection of the Constitution or the idea of *constitutional government.* In this way, strangely enough, it might be said that both sides were patriotic and idealists. In their own way, both were committed to concepts within the Constitution.

When I was growing up, there seemed to be a special feeling among many older people that I talked to, a sort of civic reverence for the Constitution bordering on religious devotion. The physical document of the Declaration of Independence was visited by people on vacations in Washington D.C. almost as though it was a holy relic in a shrine. When present day constitutional scholars and politicians sift the Constitution for the answer to some new question, they are always *reduced to searching out the intent* of these wise Founders. They must search letters, diaries, in the Federalist Papers and so on--whether they like it or not. Even to the extent of reviewing all of their early writings in hopes of finding some lost insight from these old time Statesmen in powdered wigs and frock coats and the Age of Enlightenment. They *have to;* we have largely lost the magic idea of liberty with political balance. These men were the undisputed original source, and no one now has their political legitimacy or brilliance, nor are there any such people on the horizon. And yet, what do we actually have left from that period? We have the Constitution, our system of government, their writings, and *the handful of mysterious symbols and mottos they bequeathed to us through the Great Seal.* This symbolism is what the rest of this book is about.

Masonic Secrecy and Legendary Past

Where did the Masons come from? It is fairly well known at this point in history, that the great Catholic cathedrals of Europe of the preceding several hundred years, were built under the management of various European Freemasons, apparently originating with the Comacine Masters near Lake Como near the Italian-Swiss border. They *closely secured* the building sites and their lodges from non-Masons (called *tiling*), and left their builder's marks and commentary in many places in their stonework. The Mason's Mark was made at the point of a day's limit of an individual's work and used to account for work to be paid. Later adopted by trade unions, their craft science was gradually taught by *degrees,* and these are the origin of our "apprenticeship" and "master" ranking process. Like some unions, they had almost unchallenged control of the construction and skilled craft market. Unlike most unions though, the Masonic Guilds also maintained and controlled a high standard of quality in their work, perhaps resulting from centuries of market evolution. And as idealistic as the Craft may have been ("the Craft" is synonymous with Masonry), Masons were not monastics--they were in it for freedom: they were to be *paid well* and to *travel*. They had the unusual right of *free passage* between countries (*a Journeyman*), and in mediaeval illustrations the Master of a Lodge of Masons stood erect with, and was depicted at equal height, with the aristocrats and clergy--a recognition of their social status in those times.

But long before the time of cathedrals in the much more ancient Roman city of Pompey, certain marks and painted illustrations found in the well preserved houses show that an earlier, very similar building society was at work in that early period. Many identifiable symbols are present with what must have been similar rites as the Masonic ritual has now. Apart from more obscure symbols, the plumb and the square are easily identified here as associated with the skilled arts. Secret builders' marks are found in various places in the very ancient world. There are certain indications of the Craft in the ruins of ancient Greece, Egypt and the Palestine. In many places in ancient times they were simply known as *"the builders"* and such people were spoken of as belonging to an old fraternity.

Masonic legendary tradition claims great antiquity for their fundamental *drama*. This involves the story of the character Hiram Abiff, the Architect of the Temple of Solomon, son of the Biblical King Hyram of Tyre. In a ceremony using a symbolic drama, a fellowcraft mason--a second degree Mason--in the part of Hyram Abiff, "is raised a Master Mason" through the acting out of his persecution and death, where he becomes a third degree Mason. This drama is said to have had its origins in ancient events at the building of the Temple of Solomon, the First Temple of the ancient Hebrews at Jerusalem. Hiram Abiff is symbolized by the *Junior Warden, at the South of the temple*, in Masonic lodges. In various forms, this ancient rebirth drama, always performed in secret, is to be found in a multitude of similar ancient Mystery initiation societies. Mystery initiation societies were well known throughout the ancient Mediterranean world, and some forms greatly predate the time of the Hebrews.

The ancient working Masons may have actually "traveled in foreign countries" all over the ancient world, perhaps after the fashion of the Hiram Abiff, who was originally from the Phoenician City of Tyre. He was a Palestinian and is thought to have been the son of King Hyram of Tyre, the Phoenician architect and building contractor to the Hebrew King Solomon. Architects and other skilled builders, then as now, undoubtedly got around internationally due to demand. Quite apart from their construction and craft science, their hidden theory and use of geometry was held to be very sacred, not just for practical reasons, but because *Geometry* was a tangible Manifestation of God. Masons were the practitioners of this art, and it was their craft that could bring this special, sacred essence into intimate association with materials in their work *by the practical application of ratio and proportion*. Originally, this special perfection was only appropriate for enclosures dedicated to God. Masons operated in most of the detailed technical arts and crafts, generally excluding unskilled workers outside of apprenticeship. The distinctly Masonic arts, especially of planning, design, stone setting and metal work, were a sacred calling, all learned through secret instruction.

Many Masons believe that the tragedy drama of Hyram Abiff, the Architect in the Temple, was like an early forerunner of the drama of Christ with respect to the Temple authorities. Christ makes personal and specific reference to *the builders'* traditions, claiming to be "The stone that the builders rejected." The tradition was that a *perfect stone* was to be the final, crowning addition to the Temple. Brought to the Unfinished Pyramid, this Stone, was then *tragically rejected,* wrongly judged as being imperfect like other imperfect work made for the temple construction. The Temple from Masonic lore, is seen as a symbolic edifice under construction, which represented an developing Society as well as the refining of ones Self, not unlike Christ's symbolism. This was, and still is, a very important element of the internal philosophical teaching of Masonry. This symbolism appears in the small pyramidal form on the left side of the Great Seal on the dollar.

In remotely ancient times, Masons were said to be chiefly concerned with the study of Geometry and the symbolic, pure form and proportion of their work. They were in the very specialized and secret business of the laying-out and building of temples. The unusual care taken in various proportions of temples must have been of intense interest to their clients, the kings and priests of various religions. Special and unlikely building proportions are evident in important temple sites. These proportions are almost like a form of the Mason's Mark, indelible evidence of their knowledge and devotion. These are unmistakable forms which silence the scoffer who would say "they couldn't have known that." The basic "three, four, five" right triangle, for instance, the sacred proportion for obtaining a correct right angle out of three lengths--was one of their primaeval, original secrets. In Egypt we can still see the *characteristic angle* of this triangle, where it serves as a *basis for the vertical slope angle* for the Second Pyramid or Khephren, the second largest of the three pyramids on the Giza plateau. The symbolic implication, once one learns of the unique significance of this slope angle, is that the *sacred square corner* of this triangle-- the one corner of the three angles not shown at the exterior vertical section--is therefore *hidden within* as this pyramid's basis in geometry, *exactly at it's foundation.* Even in this very ancient period of at least 4500 years ago, the builders' special reverence for the idea of the "Square" is evident.

In legend, Masonic rites trace their source to Tubalcain the *original smith* in the Old Testament. Perhaps Tubalcain is also cognate to the Greek and Roman mythical characters of Prometheus and Vulcan/ Hephaistos--both relating to the use of fire or the forge. The Mysteries of Vulcan and related rites go far back into prehistory. Giorgio de Santillana in Hamlet's Mill says that "Prometheus" may be traceable to the ancient term "Pramantha." In Sanskrit it means a "fire drill," the bow-drill of the primitive smith. Vulcan was the legendary blacksmith of the gods, and Greek god Prometheus was the sufferer of a suspiciously familiar symbolic penalty of being *disemboweled for revealing the secret* of fire to mortals.

To some, the medieval secrecy of the Freemasons has always seemed wrong, selfish or clandestine. But it wasn't secret bad that was being done, it was secret good. And without Freemasonry's self interest, their frightening oaths of secrecy, the courage to face the society of their neighbors and oppressive Church--ultimately the system of the United States of America would not exist. If for some reason they had not been able to maintain secrecy, their beneficial action on the world's historical stage would have been brought to a close very early by the European princes and clergy.

The Masonic movement in modern times, though hoary with many centuries of great and mysterious tradition and momentous events throughout history, is now in many places thought of as mearly a social club that is known for little more than charity work. Yet the warm embers of the Mysteries still persist, perhaps to rise phoenix-like out of the ashes of our times to build and create great wonders once again. Our political and religious liberty was largely borrowed from their work, as was the legal philosophy that supported it, and on to the development of the wealthy, workable and pluralistic society that we see today as the United States. Our historical debt to their influence is enormous.

Although much of the story above is far from the beaten path, many parts of this material will surface in one way or another in the following investigation of the dollar. My intent here is to show a fundamental connection to a benign tradition, involving our beginnings, our civil system and the symbols found on our fundamental unit of money--a small defense for something obviously good, but not well understood.

****

It will be very hard for some to accept the nature of these discoveries. It may seem improbable that anyone would have done this within the design of the familiar dollar. Or, that the Founding Fathers might have had some secret philosophy, or that its symbols were imposed on us before we had a chance to object, and so on.[5] Many of these facts will just not fit into some reader's beliefs. Some will be outraged that a public official would have secretly incorporated this sort of thing into U.S. currency--let alone that the Founding Fathers *did exactly the same thing* by secretly building their unusual ideas and mathematics

xvi

into the Great Seal. Many will naturally discount the general idea altogether. But the intricacies of the dollar's design and the mysterious sentiment within it will be unavoidable to the thoughtful sifter of evidence. This design quietly demonstrates a special application of a largely unknown American tradition, a deeper and previously hidden facet of the builder's art.

I hope all of the necessary information for practical method and mathematical precision are provided for as a technical demonstration of these ideas. I greatly encourage the reader's verification and criticism, or for that matter, thorough skepticism and denunciation. I do not expect to make the definitive study of the dollar's design. But it is my hope that this work will open this field of interest to exploration by many adventuresome readers, who will make important contributions to this body of information, and help solve the riddle of the geometry of the Dollar design of 1935.

Notes:

[1] In recent years, quite a lot has been written about this ratio. Probably the most well known popular commentary on it in recent years would be Martin Gardner's occasional articles featuring this subject in his regular Mathematical Games section of Scientific American, during his long career with that magazine. From time to time a book will appear on another topic that mentions the subject of the golden ratio, and occasionally a whole book entirely devoted to the subject appears. I have found a how-to book on furniture design that has a section on it: Cabinetmaking and Millwork, by John Feirer, 1967, see pg 50 . There is a newsletter devoted to it, The Fibonacci Quarterly, that regularly shows new discoveries and applications of the golden ratio, (and many related ratios) a subject that covers all of the sciences, nature, mathematics and the arts.

[2] CRC Handbook of Chemistry and Physics, 72nd ed., pg A-1

[3] Physical Measurement and Evidence: If you are going to measure the dollar, you should get several freshly minted one dollar bills from an bank, and specifically ask for new, un-circulated one dollar bills, that have never been folded or creased, perhaps with sequential serial numbers if possible. Not every bank carries new bills, and it may be necessary to go to a more central bank branch to request them. Any bills that have been folded in a wallet even once or are faded, washed or crumpled in any way, can not be reliable for measurements. Washed and creased bills are *especially bad*, as it turns out, often giving wildly distorted results. I suggest you do not use any old bills, since these will have shrunk since the time of their issue, and perhaps some will not have shrunk evenly. (But it should be noted that my *original* measurements began on bills that I wouldn't use now due to shrinkage, and as we will see later, this occurrence had been helpful for my discoveries in the beginning. Now, I use new bills for all measurements.) Keep your good bills in an envelope, all inside a sealed plastic bag, in a cool, dry place when you are not using them. Never take any less than three measurements of any feature; re-check your work *all the time*.

Write down just what you see, not what you want to believe--or even what I or anyone else said a measurement ought to be. Limit your measurements, or closest estimates of a measurement, to the thousandth of an inch (+/- 0.001"), since that is about the limit of anyone's reliability and credibility. Unless you are using special equipment, with solid statistical techniques, work beyond the one-thousandth of an inch range should be avoided. Make little descriptive sketches what and where exactly you measured. In this way you will have ammunition for logical argument in favor of your thoughts, and you will not have to try to rely on memory what you measured once back at some point in time.

Measuring tools, *for estimation only*, use a twelve inch 100:1" drafting scale, with sharp black lines on a white background with a 5x magnifying glass for guessing an *approximate* one-thousandth of an inch, only for the quick verification. These are available from most drafting equipment suppliers in the U.S. Cheaper steel Machinist's scales in hundredths of an inch are also easily available from machinist's suppliers. But close use of a scale over time will produce a lot of eye strain. Also, using a one hundredth-inch scale will require careful estimation of the thousandths of an inch, which most people aren't very good at doing, and which is not all that reliable even when done well. All measurements made this way *are only approximate, being ten times as small as a one-hundredth of an inch division.*

For *good measurement,* I think most people who are seriously interested in the dollar would be better served by using special equipment such as a machinist's dial caliper, or Vernier caliper, reading directly to thousandths of an inch. This is what I have mostly used, and these are not necessarily very expensive. These also now come in a variety of more easily read tools, such as digital electronic liquid crystal diode (LCD) readout displays that might reduce error and save a lot of time and effort. A cheap, machinist's "1-2-3 block" can be used to check whether one's tools are still good over time, or in need of adjustment.

The *electronic-type caliper* usually rounds to the half thousandth, (0.0005", or 5 ten-thousandths) which is close enough for this sort of work. (This is about the width of a cat hair, which is pretty fine.) The regular mechanical dial caliper being an analog tool is capable of correctly estimating ten-thousandths of an inch with respect to, say, metal parts, where the edges are very clear and hard edged. But I would caution against trying to directly measure to the ten-thousandth in the environment of printed lines on paper such as the dollar. Of course it can be done, over areas of an inch or two where the lines happen to be sharp, but I am inclined to doubt any unqualified measurements in the range of the ten-thousandth of an inch, since the whole length of the paper in the dollar may be able to expand or contract as much as a whole third of a thousandth or more under some conditions (0.0003"). Changing humidity; the observer's pointing error at the caliper jaws; and individual variation and of specific dollar bills are all within this range of error. And this doesn't take into account differences between dollars of different ages, (shrinkage), printings, etc., let alone the same bill under different lighting conditions and so on. Don't bother with any plastic dial calipers, since these will be unreliable for measurement, (they can sometimes vary as much as 0.002", or 2 thousandths between measurements). They are less durable and do not really cost that much less than steel ones. I have found that a 5x magnifying glass which comes with a little swivel stand, such as stamp collectors use, was very helpful, as well as a little clipboard and a bunch of small parallel type clamps and c-clamps, so that things could be held still while I was measuring.

Don't squint. If you find yourself squinting, *you are doing something wrong.* Always use at least a 5x magnifier and in good lighting. Get up and go outside every so often so that you can look at things in the distance. 5x and 10x magnifiers or jeweler's loopes can be found through stamp collectors and are very useful. (This break to look into the distance is *very important* to your eyesight. The lenses in your eyes are flexible and actually need the counter-balancing movement to prevent nearsightedness, which can be the effect of too much close work such as this. Although my sight was probably not worsened by this work, I can easily see how it would be for those that are not careful and spend too much time at it. Safe measuring! Use a magnifying glass.)

If you have a photographic enlarger you can make large scaled copies of the dollar that may be helpful for expanding some areas for the purposes of measurement. But there can be a question of legality. Good scaled enlargements are very tricky to make correctly for a number of technical reasons--and, don't be tempted to make true scale copies of the dollar on paper even for fun, since the U.S. Treasury Department takes *a very dim view* of people making any copies of U.S. currency that might be confused for the real thing. The U.S. Treasury requires that magnifications in prints of the dollar be: *at least 1.5x or, one and a half times original, or 0.75x, in three-quarters reduction.* In this book, no image of the dollar has been made to true scale, but hopefully correct *proportion* has been preserved where possible. All of my illustrations are in conformance with the U.S. Federal Code 18 USC 504. *Please read* Federal Code 18 USC Sec. 504 and 18 USC 474, (These may be found on the Internet under http://www.law2.house.gov/uscode-cgi/f~ or thereabouts. See the page called "Thomas"). This Code does not appear to have been strictly enforced in practice, such as seen in newspaper ads. Perhaps the U.S. Treasury is very busy and more concerned with actual counterfeiters.

[4.] See America's Secret Destiny by Robert Hieronimus, Ph.D., 1989. This is an excellent reference source book for arcane information and mystical lore surrounding the American Revolutionary War period.

[5.] Many are interested in the Illuminati connection to Freemasonry and the American Revolution. Several authors have tarred all of Masonry with a connection to the Bavarian Illuminati, a secret society that began the same year as the Declaration of Independence, on *May 1, 1776.* This is a relatively recent movement with different motives than that of ancient or present Freemasonry. The Illuminati were devoted to *"dictatorship, and the destruction of the existing order of the world: Monarchies, Church, morality and property."* This is the source of Jacobinism and the weird horrors of the French Revolution. American Masonic influence in the U.S. Constitution was always to promote private property, human rights, individual political freedoms and *limited government.* The Illuminati movement infiltrated European Masonry and the French Revolutionaries and even tried subversion of the post-

Revolution United States. This was the basis of the so-called "Whisky Rebellion" which was put down by General George Washington, which was directed by the French Ambassador Edmond Genêt who arrived here in 1793. The Illuminati were roundly denounced by Washington, Adams and many others in letters and public statements. It is interesting to note that their founding was *May 1st:* the same date celebrated by the Socialist Internationale: the date of the spring cross-quarter day being the Celtic *Beltane.* Traditionally, this was always a day of lawless riot and Dionysian licence in ancient Europe. (The United States Declaration of Independence is on *July 4, 1776,* near the astronomical point called the Aphelion or *on the axial line* in the Earth's orbit most distant from the Sun. This point marks the Anomalistic Year.) The profane and power hungry Jacobinism concept in various mutating forms (collectivism, socialism, national socialism, fascism, communism, welfare states, "bread and circus" legislation, etc.) over the last three centuries has been the chief adversary of freedom, private property and the limitation of civil authorities. Its addictive central idea seems to be: "Let's you and him share." As Margaret Thatcher said (approximately): "How come when we share, I always end-up with less?" Its more *about power,* and keeping others busy--by having them steal from, and interfere with, everyone else around them.

CHAPTER 1: THE GOLDEN SECTION

I became interested in the golden section while in my teens, after reading a magazine article that touched on the subject. In those days there was hardly anything to be found on this subject. It was an esoteric idea, a topic found only as an undercurrent in the classics and architecture. This proportion, as it turns out much later, held the fundamental key to the dollar's hidden design.

I think it was the winter of 1965, and this article was in a picture magazine that came to my house.[1] There were a few similar articles going around the magazine circuit at the time, and some of these stories talked about the beauty in mathematics. I think one of these articles must have specifically mentioned the golden section. Perhaps it was a side issue, but it was described in radiant terms. This magazine article gave few facts and no useful details, but the gist of it, was that there was a deep artistic beauty--somehow strangely imbedded in the precise geometric proportions of a special mathematical formula. Staring at an illustration of this shape, it seemed to be familiar and special.

This idea really carried me away. My interests ran toward both art and science, and this idea of connecting the vagaries of artistic beauty, with the sharp clarity of mathematics totally fascinated me. The author said that there was something about this formula that could give beauty to artwork, and that this something-or-other ratio had to do with "classical proportions" in design. The writer believed that there was something uniquely special about this mathematical idea, and I wanted to know all about it.

The article was an in a coffee table magazine, back in the sixties, extolling the virtues of a very obscure mathematical idea, to an unfamiliar audience--and not really saying much about it, or any numbers. The author may have understood the mathematics, but it is likely that he was discouraged from giving any formulas. Perhaps they were edited out of the article. It's even possible that a formula was there, but I don't recall any mathematics in the article--and at this point, I don't remember all that much about it.

I'm sure if I'd have seen a formula, I would have clung to it as though it was magic. I remember thinking that I must have missed something, and I may well have. I had no idea at that point what it might be that I should be looking for. The author described the golden ratio's beauty in an enormously engaging, but frustratingly obscure fashion. This was all the more disappointing because he spoke of "proportions" as key to this idea. I thought I knew what proportions were, and re-readings of the article left me feeling that I was at least onto something important, if I could just figure out what he was talking about.

The author said that great artists had used this special shape. Then, I think there was something said about a "perfect rectangle, not too long and not too wide." It seems to me that there were some line illustrations and pastel colored rectangles, and an superimposed picture of the Parthenon that were all vaguely mentioned in the text. I also thought I knew what a ratio was, but on this point I was not completely certain. I think the article might have used the term "the Golden Rectangle" to describe this shape. I soon lost the article among old newspapers, and it was gone.

Totally fascinated, I sat down and tried my hand at making "a perfect rectangle, not too long and not too wide." How would one do that? I wondered. After trying many ideas, I got into a school mimeograph room that had a large paper cutter. I figured that a paper cutter would produce neater square-edged shapes than had been my luck with scissors. In multitudes of experiments, I cut up several sheets of white paper and leftover scraps into big and little rectangles. This collection of shapes ranged from longish strips to almost perfect squares, but most lay somewhere in between. I looked at these shapes both horizontally and vertically on a dark colored tray and made a groups of the ones I liked. Though I didn't fully understand the idea of ratios, I could see that most of the ones that I liked had a character or shape that was more or less the same no matter how large or small they happened to be. To find a ratio or proportion you simply divide the two sides, but I didn't know that at the time. But I did have a vague idea of fractional proportion and this was more or less what I thought about the subject then:

The first thing I concluded was that any "one by two" (1:2) rectangle was "too long" for artistic magic.

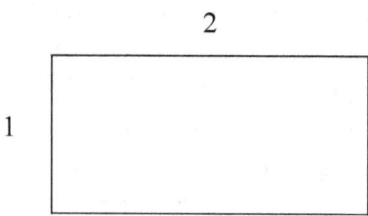

and seemed simple, flat or dull to my eye, as did any "one by one" (1:1) or plain square, which seemed "too short:"

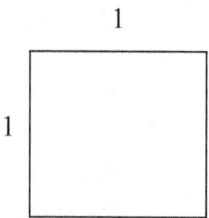

And second, I concluded that some--or maybe any extra length, when added to a square seemed to give it something special at least initially. "Longness" was the question. This extra longness appeared to improve it's appearance to make it a more desirable shape, at least better than a square:

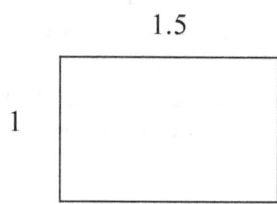

"One by one and a half," (1:1.5) looked nice enough. But the more I looked at it, it seemed as though this *wasn't quite it*--this couldn't have been what the author was talking about, since this would have been very easy to describe. The shape I was looking for needed a "formula." This was still dull or box-like, not much of a departure, conceptually, from "one to one" or "one by two." Certain shapes less than one and a half long seemed interesting also, like "one by three-quarters," (Or: 1 : 0.75):

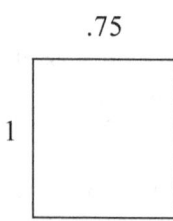

I did settle on a general shape I liked, but couldn't think of a way to define it, or any way to write down what I did, so that I could reliably make one again. I don't know if my shape was even near the elusive golden ratio. But I thought that there probably was somewhere, or should be, an ideal rectangle if I just knew what I was doing. It seems amazing to me now that I went to this much trouble.

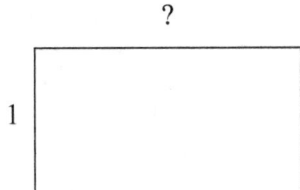

When I read the magazine article, I didn't know if the illustrations were completely uniform with regard to the supposed ideal rectangle (though I couldn't be sure) and I wondered for a short time whether this was one of those "unattainable ideal" type of things. Maybe you had to find it yourself, I thought. For a moment I wondered that maybe it was like the blindfolded Zen archer, who found his target by intuition. Maybe you had to happen upon this exact something-or-other by accident, at the very moment that your intuitive artistic skill surfaced in a blaze of artistic creation, and then you would arrive at this special, mysterious proportion. Or, maybe it was a statistical thing that appeared in artwork.

There had to be some *exact, regular rule* for doing this, if any of what the article said was true. After the paper cutter, I tried drawing some rectangles by compass and straight-edge to see if any moved me, but I didn't find any solid rule for making good-looking shapes.

Sometime later, I found the book <u>Mathematics</u> in the Life Science Library. In it was a small section on the golden ratio spanning a few pages, which at least confirmed the idea that other writers had similar romantic notions about this proportion. This section even had an illustration that showed an exact method to construct this rectangle. This was far better to find than a formula that I could not have understood. Considering the prestigious build up preceding it, it wasn't even difficult to draw. It was a magical shape, and quietly attractive. Not too long and not too short, as advertised. There are those who can't fathom this idea, but I was quite sensitive to it. (This ratio is approximately 8:5 or, about 1:1.6.)

The next step for me was finding the translation of Le Corbusier's book on architecture, <u>The Modulor</u>[2] in a library. The French architect "Le Corbusier," (Charles-Edouard Jeanneret), was a enthusiastic user and promoter of the golden section proportion in his design work--and I tried my hand at mimicking some of his ideas and also tried some of my own. (This book, I should add, was published in a peculiar *square* cover and page format, perhaps for contrast.)

****

I. Where did the Golden Ratio come from?

II. What is the "Golden Ratio" and why is it important?

III. How do you construct it?

(**I.**) The golden ratio in modern times goes by several names in many languages. It is probably now known by more people in the present generation than the total of all others in recorded history. But at this time it is still a more or less unknown even to the majority of the well educated. As a graphic ratio, it seems to have always been with us, but we don't really know where it originally comes from.

It is a subtle, glittering thing from the darkness and distance of remote antiquity, an astonishing element of proportion that appears here and there amidst much that is dull and uninspired. And, this comes from times much earlier than people were supposed to be thinking about such things. It mysteriously appears on occasion in the layout of sacred sites, ancient artwork, and even in pottery styles with no explanation, ordinarily unnoticed by archaeologists. Here lies a world almost unexplored, even to this day: the vast field of sacred proportions in archeology and ancient architecture.

The golden ratio first surfaced from obscurity in the writings of western civilization by way of the ancient Greeks. Even then there are only a few exact references about the ratio in their writings. The Greeks must have gotten it from the Egyptians. To the Greeks and all other ancients, these were holy secrets, mathematics and geometry were *religion*. Any teacher in ancient Greece would actually risk death from zealots were they to reveal anything about geometry to those not initiated within the temples. This, unfortunately went for writing these ideas down--since they might then be read by the unwashed and profane.

Wondrous mathematical ideas were conveyed to temple students by word of mouth only, much in the way Masonic Mysteries are still taught today. Perhaps taught by diagrams sketched in the dust, such as Socrates is said to have done. More solid evidence of the Greeks' knowledge of this ratio can be learned from measuring remaining stone ruins than from reading their literature. For the most part their books will refer to it only through veiled comments and praise for the beauty of this proportion. As we will see later, this ratio is easily generated on paper by means of simple geometry. But its vastly unusual character would have required a great amount of time for thinkers to become acquainted with it in a primitive time, and much investigation for any culture to develop an awareness of it.

The first use of the term "Golden Section" was found in the literature of relatively recent times. At the end of the Middle Ages it appears in book De Divina Proportione, a translation of the Roman work of Campanus by an Italian, Luca Paccioli, in 1509. This work, oddly enough, was illustrated by Leonardo da Vinci, a personal friend of Paccioli. It contains many examples of the use of the golden ratio in two and three dimensions. Paccioli and others of his time attributed various mystical or supernatural properties to this proportion, in addition to believing that it contained the *very essence of beauty*. It is now variously called the "golden section," the "divine proportion," "golden ratio," "golden cut" or "golden mean," all of which are inter-changeable terms for the general idea in present day usage.

The term "*golden mean*" might be said to tie the geometric idea to Greek moral philosophy, such as Aristotle's concept of ideal human virtues, which he said lay as an ideal balance between two extremes in one avenue of moral behavior. For the ideal of "courage" to be at the *mean*, for instance, it must lie at some special point of balance somewhere between being recklessly bold and it's complete absence-- which would be not being bold enough. But not just half-way in between. This idea is said to be similar by analogy to the golden ratio, by this ratio's being a sort of elegant, dynamic balance point on a line between two ultimate extremes, or in a similar way within a geometric rectangle. Ancient Greek culture, especially the Athenian, prized a kind of personal balance and a refinement of virtue. A variety of dissimilar talents and abilities were to be hoped for all in one person, balanced and "polished to a hair." The ability to speak logically, compose poetry, perform well on the battlefield, compete in sports, perform music and oration, for example, were all required of a well rounded individual. Physical beauty and efficient, exact function were thought to go together. This concept was a place that geometry, man, and Divine Essence come together in ancient thought. This was a sort of an over-all aesthetic ideal under one general heading.

Their whole idea of *personal virtue* was based upon a concept that is fundamental to ideas like geometry, mathematics and logic. It was called *"arete"*, meaning more or less what we mean today by "virtue." But to be technically correct, it was a term that conveyed excellence, exactness, precision and nobility all together as one idea.[3] The ancient Greek, Egyptian and Hebrew cultures shared this idea in common; the Egyptian word was "maat", meaning very much the same thing, and in Hebrew, "sidek", or "righteousness." These cultures were concerned with exact measurement as well as careful speech. For these peoples, mathematics and exactness was *a sacred idea*, which allowed them to be ceremonially pure in matters relating to divinity. This sort of exactitude was present in almost all important writings of antiquity, especially those of the Biblical writers, whose attention to precise detail is quite elaborate. This was also true of the sacred architecture of ancient times, almost everywhere in the world.

This concept of *arete* may allow us to look back to a distant time, to many peoples of the bygone ages, who, as a rule, spoke more carefully, wrote more exactly, and gave the works of others a much closer reading than is true of us, even today. It is fascinating somehow, to think of a whole world of religious and intellectual thought requiring excellence and spoken exactitude. This was generally true in their preservation of oral traditions, logical argument and works of art. Our current idea of "classical form" was present even then, with an equal or even greater reverence among the ancients.

Although the term "golden mean" from philosophy may actually be quite ancient, there is little evidence that the geometric golden proportion was actually called by that title in ancient times.[4] Whatever name it had, it would have likely been a secret name, with religious connotations among those of the inner court of the temple. Most likely it was not spoken aloud, but by mouth to ear like the deep secrets of many old traditions. Until Euclid, we have no record of this proportion being called anything at all. He called it "the extreme and mean ratio."

One of the names this proportion has been given is a Greek letter, φ or, "phi," (the Greek letter "f" pronounced "fee".) This seems to have appeared in literature in Victorian times, (1860's) as a symbol to

The Temple Tomb near Knossos, Crete.

denote the function in mathematics. The Letter "phi" is thought to be taken from the name of Phidias, the classical Greek artist and architect who, tradition has it, used the golden ratio at the building of the Parthenon in Athens in the fifth century B.C. About Phidias, Plutarch the first century Greek writer wrote:

"He was made the Superintendent of Architecture by Pericles during the grand building period in Athens that produced the Temples on the Acropolis, and that all of the temple architects were answerable to him." [Plutarch, Pericles]

This is the classical connection suggesting that Phidias must have been among those master artificers who were trusted and knowledgeable with respect to the secret geometric arts. The Parthenon is now known to include the golden ratio among other mathematical concepts. The ratio can be technically demonstrated such as we moderns have now been able to uncover in the plan of the ruined Parthenon. If he was in fact the master architect, he must have been well aware of the golden ratio.

In its article on Greek architecture, the Penguin Dictionary of Architecture quotes Plutarch from the same book commenting on the Buildings at the Acropolis, and adds its own, almost religious commentary:

...they were created in a short time for all time. Each in its fineness was even then ages old; but in the freshness of its vigor it is, even to the present day, recent and newly wrought.' No better description of the aims and achievements of the Greek architects has ever been written. Their ambition was to discover eternally valid rules of form and proportion; to erect buildings human in scale yet suited to the divinity of the gods; to create in other words classically ideal architecture. Their success may be measured by the fact that their works have been copied on and off for some 2,500 years and have never been superseded. Though severely damaged, the Parthenon remains the most nearly perfect building ever erected. Its influence stretches from the immediate followers of its architects to Le Corbusier."

While this may not be an utterly objective statement, it is definitely the way many writers have described the Parthenon down through the ages. Also, though the golden ratio is curiously unmentioned here, we can see this writer's evident awareness of it's use by the direct reference to Le Corbusier, who is mostly known for its popularization. I think it is true, though, as it was said in the passage above, that the Greeks desired "...eternally valid rules of form and proportion; to erect buildings human in scale yet suited to the divinity of the gods..." Forever new, and of ancient form at the same time. It was not only the Greeks who felt this way about their sacred architecture.

Some good examples of ancient architecture that use the golden ratio are: the Great Pyramid of Giza, specifically the height-base ratios of north-south faces, (as opposed to the east-west and over-all base relationship based on pi), The Parthenon North face façade; the ground plan composed of two reciprocal golden rectangles; the façade in a harmonic relationship to another golden rectangle at the outline of the inner chamber; and the Stonehenge-Heel Stone relationship in addition to several other internal relationships indicating a knowledge of sacred geometry.[5] Shown here is a pre-Mycenaean (Minoan) tomb doorway on the Greek island of Crete[6] that appears to show the use of the golden ratio, at the outline of its entrance way. This is called the Temple Tomb, found near the ancient ruined city of Knossos, and is thought to have been built around 1700-1600 BC, or about a thousand years after the Great Pyramid in Egypt. *(Scaling the door way from the photo reveals a ratio of about 1 : 1.6.)*

Euclid seems to have been the first to call it "the extreme and mean ratio," (circa 300 BC), which is the first name we know it by, and it is described by him without fanfare. He lived in Alexandria, Egypt, an unusual Greek colonial area, mostly outside of control of either Greek or Egyptian temple authorities. Free thought, science and even rudimentary steam technology had begun to explode in Alexandria in that period, so it is unlikely that any regular authority of either system was strongly present there as a speed bump to cultural development. Euclid is now thought to be mainly a publisher or editor of that period's mathematical ideas, rather than the actual author of them. He was probably more of a careful, logical compiler of what was openly known of geometry in the Greek speaking world. The mathematical aspect of the golden proportion is originally known to us from only three places in the thirteen books of Euclid's Elements,[7] and it is also found in a portion of Euclid's Stoicheia, which is actually concerned with its geometric construction.

The discussion in Plato's Timaeus which is traditionally thought to be about the golden section, is apparently about *another* related mathematical concept used in architecture, sometimes called the "sacred cut."[8] This is another ancient mathematical pattern that was also involved in ancient temple layout. Plato's description of the idea shares some common mathematical features with the golden ratio, such as the square root function. Though this sacred cut was sometimes incorporated together with the golden ratio in temple layout, (such as the Parthenon) it is a quite a different idea, based on the ratio of one to the square root of 2. Both of these ratios appear here and there in various buildings in a recent survey of the architecture of the Roman villa of the Emperor Hadrian, located to the east of Rome, Italy.[9]

Kepler, the sixteenth century astronomer, was a golden ratio enthusiast. He too, called it the "divine proportion" in most of his writings. He said:

> "Geometry has two great treasures: one is the theorem of Pythagoras; the other, the division of a line into extreme and mean ratio. The first we may compare to a measure of gold; the second we may name a precious jewel."

As Europe progressed from Roman times, newer architecture showed fewer examples of the rare golden ratio, until the time of Christopher Wren, an English architect who is said to have used it in some of his designs in London, around the late 1660's.

Since the publication of Der Goldene Schnitt by Adolf Zeising in 1884, and The Curves of Life, by T.A. Cook in 1914, which covered mostly biological spiral forms and the golden ratio, the field became more open to the modern reader. J. Hambridge wrote Dynamic Symmetry, in 1920, on the mathematics of proportion in the art of Greek vases and temples; F.M. Lund, Ad Quadratum, 1921, on medieval cathedrals; Matila Ghyka, <u>A Practical Handbook of Geometrical Composition and Design</u>, 1952; and <u>le Modulor</u>, in 1954, by Le Corbusier. These were a few of the important modern works on the subject. After Le Corbusier, the use and interest by mainstream readers was greatly expanded and academic research has appeared on the subject. (Also see: Funk-Hallet, in connection with the analysis of Renaissance art showing the golden ratio; and Jean-Lauer and especially Schwaller de Lubicz on the presence of golden ratio and pi in the religious symbolism of ancient Egypt.)

****

The Mathematics and Related Discoveries of the Golden Ratio

(II.) From a strictly abstract mathematical direction, the now famous "Fibonacci Series," or a golden series of numbers was discovered, (or perhaps was rediscovered) by the great Medieval mathematician Fibonacci. Filius Bonacci, sometimes called Leonardo de Pisa, is now more commonly referred to as Fibonacci. (This is the shortened form of the name meaning "son of Bonaccio." Leonardo de Pisa is the "other" important Leonardo of Medieval times.) Fibonacci's father was an official in an Italian mercantile factory in Bougie, a city in the country of Algeria in or around 1170. As a boy, he received mathematical training from Arab tutors, where he learned the use of the ghobar mathematical notation. This is the Hindu-Arabic number system, which he later brought to Europe and which we use today, which we now call "Arabic Numerals."

His book, <u>Liber Abaci</u>, was Europe's first introduction to the Arabic Numerals and algebraic forms for which he is mostly known. In his book there was a trivial problem involving the growth of rabbit populations, that was the fundamental example of the golden ratio based series:

"Suppose, Leonardo wrote, a male-female pair of adult rabbits is placed inside of an enclosure to breed. Assume that rabbits start to bear young two months after their own birth, producing only a single male-female pair, and that they have one such pair at the end of each subsequent month. If none of the rabbits die, how many pairs of rabbits will there be inside the enclosure at the end of one year?"[10]

This is where the famous golden series called the Fibonacci Series has its origin. This is an abstract, linear series with a common basis to the geometric form of the golden ratio. So, at a constant rate, and at regular intervals, the lines of rabbit descendants, (and lines of lines of descendants, etc.) proceed to pile up, fanning out into a tree shape.

```
(month)          (Production of bunnies per month)     (No. of total male/female pairs per month)

0                                        m-f
1                               m-f                                                            2
2                                                     m-f                                      3
3                                        m-f          m-f                                      5
4                     m-f                 m-f          m-f                                      8
5               m-f m-f     m-f           m-f               m-f                               13
6        m-f        m-f m-f m-f m-f m-f m-f      m-f m-f          m-f                          21
7        m-f m-f m-f m-f m-f m-f m-f      m-f m-f m-f m-f      m-f m-f                         34
8   m-f= 21                                                                                    55
9   m-f=34                                                                                     89
10  m-f=55                                                                                    144
11  m-f=89                                                                                    233
12  m-f=144                                                                                   377
etc.
```

Notice from the columns that the quantities of new bunnies per month strangely increases *at the same rate as the totals*, but on a different step of months. The procedure to produce the correct sum is a series, possibly the simplest of all series, where you add the last sum to the previous sum:

1 plus 1 = 2; 2 plus 1 = 3; 3 plus 2 = 5; 5 plus 3 = 8 etc.

1,1,2,3,5,8,13,21,34,55,89,144...

A French number theorist Edouard Lucas, who wrote a book on recreational mathematics, attached Fibonacci's name to this series about 600 years later. There is no evidence that Fibonacci ever studied this series very deeply, and we don't know if he came up with this problem himself, or whether he discovered the Rabbit Problem somewhere in the lore of the Algerian Moslem world. Although he may well have thought it up himself, there is some evidence that this series was known in much earlier times by the ancient Romans, and it seems likely that the Arab mathematicians preserved this as they did many other theorems of classical times. Fibonacci brought back quite a lot with him from the Arabic mathematical world.

George E. Duckworth, a professor of classics at Princeton University, in his book <u>Structural Patterns and Proportions in Virgil's Aeneid</u>, says that the Fibonacci Series was intentionally used by Virgil and other Roman poets of his period in their works.[10] If Professor Duckworth is correct, it seems possible that this may have been a part of a larger, bygone mathematical tradition, perhaps quite separate from the temple builders, who are the only ones we know to have been fully aware of the golden ratio. Its possible that there was a much wider world of users of this mathematical form then than is now suspected. Since many ancient writings have complex numerical patterns built into them this kind of numerical coding of patterns in text should not be greatly surprising. There are the Biblical "gemetria," as many other well known complex numerical and positional symmetries, in verse, word and letter sequences of both Old and New Testament, for example. As shown by Davidson and Aldersmith, there are many interesting Egyptian writings, such as that of Manetho's king lists that have carefully coded numerical data. As for the series itself, the Fibonacci Series is so simple, it seems unlikely that it wasn't known in the ancient world, considering the ancients' knowledge of this ratio in other ways.

In the numbers of the Fibonacci Series, *each consecutive pair of numbers* in the series approximates the golden ratio, when divided by the other, with increasing accuracy in higher pairs of numbers. Each successive pair produces a ratio that is alternately larger and smaller than the ideal golden ratio, *converging at infinity* to the true value of the golden ratio.

Using the list of Fibonacci numbers, you can try dividing any consecutive pair by each other and you will find (at least in the higher pairs) a ratio of about 0.618... (or 1.618... if you divide the in opposite direction of larger to smaller numbers.)

These numbers are known by the terms of their series abbreviated as "F Numbers" with a subscript after the F:

$$F_{10} = 55, \quad F_{11} = 89, \quad F_{12} = 144...$$

Also provided here is the series now known as the Lucas Series which would be the next most simple series, known in their series as "L Numbers" with subscripts like the Fibonacci Series. These also eventually converge on the ideal golden ratio proportion much in the same way as the pairs in the Fibonacci Series:

1,3,4,7,11,18,29,47,76,123,322,521,843....

Both of these number series, (as well as many others) arrive at the final ratio which is half of the sum of the square root of five, plus or minus one, divided by two--which is the general equation defining the golden ratio. Using the "plus one" option, this is written in algebraic form:

$$(\sqrt{5}+1)/2 = \varphi \quad (\text{or, phi})$$

8

On your calculator, key in 5, press the square root button, (you should see 2.236...), then add 1, then divide by 2:

(5, square root key)	=	2.236067977...
(add 1, press "=")	=	3.236067977...
(divide by 2, press "=")	=	1.618033989... (φ)

With the golden ratio, one surprise is that the number values of the ratio do not change in reciprocal form, except by adding or subtracting *exactly 1*. "1 : 1.618033989..." is the golden ratio. Also, strangely enough, so is "1 : 0.618033989...", *exactly* the same ratio. This ratio is *exactly the same* in either direction. (The reciprocal is the "1/x" button on your calculator.[11])

(golden ratio)	=	1.618033989... [φ]
(reciprocal or 1/x)	=	0.618033989... [φ^{-1}]

Also very strangely, if you *square* the golden ratio (multiply the number by itself, or "x^2" button), you get the same number with the addition of *another 1:*

(golden ratio)	=	1.618033989... [φ]
(squaring, x^2)	=	2.618033989... [φ^2]

This barely scratches the surface of the odd properties associated with the golden ratio and various golden series of numbers. These are the three fundamental numbers of the golden ratio.

Graphic Construction of the Golden Section in One Dimension, or a Line:

As Euclid used the idea of "the extreme and mean ratio," it can refer to an idea of proportioning of a single line into two segments; such that the whole line relates to the first portion of the line, in the same way as the first portion relates to the second portion:

A---------------------B------------C

Which is to say, if you divided the length from A to C (say 75025) by the length of A to B (say 46368), you would get 1.618... If you then divided the length from A to B (46368) by the length from B to C (say 28657), you will *also* get 1.618..., or the golden ratio (approximately).

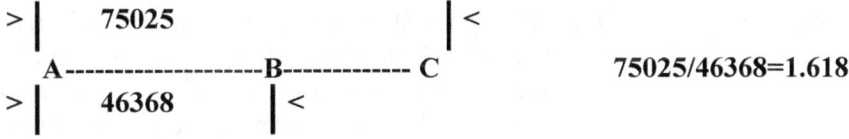

75025/46368=1.618

And:

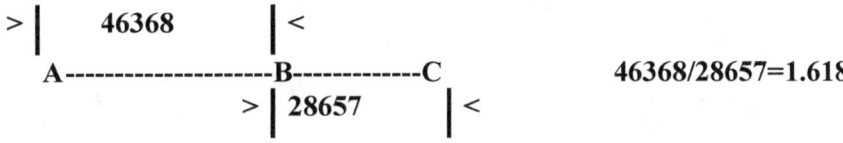

46368/28657=1.618

(28657, 46368 and 75025 are the Fibonacci numbers F_{23}, F_{24} and F_{25} used here for convenience.)

9

So, these three lengths are proportional to each other, which is no big deal mathematically, they *also are with respect to their whole sum,* (and sums of their sums, etc.) which has the effect of turning them into a very special group of numbers. Books have been written about the beautiful and unusual properties of these numbers, and associations have been formed just to study them.

Physical science abounds with Fibonacci relationships, such as the pattern of ray tracing found in the reflections between two surfaces two pieces of glass, to the phenomena of nuclear decay within radioactive elements. Many biological forms follow Fibonacci sequences, such as pine cone spirals, sunflower seed spirals, philotaxis or, leaf placement of various plants, tree limb branching sequences, human body proportions, to mention only a few.

Constructing the Golden Ratio in Two Dimensions as a Rectangle, and in Higher Dimensions:

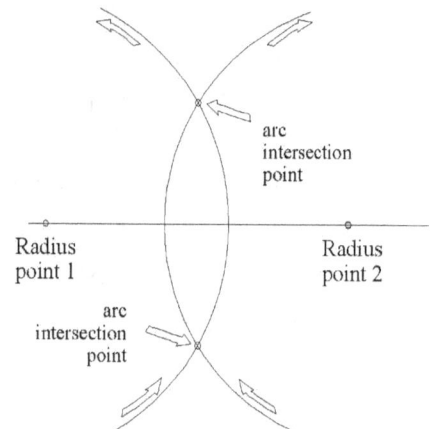

arc
intersection
point

Radius
point 1

Radius
point 2

arc
intersection
point

(III.) Continuing one from one dimensionality, in two dimensions, shapes made of lines proportioned by the golden ratio have some unusual and interesting properties. This is an exercise for those who would like the hands-on experience of developing a golden ratio rectangle. In the following example, the initial suggested dimension of 10 inches (marked vertically, above the line to intersection Point 3 shown here) is given with the idea that it will be big enough to make subsequent details easier to draw for those interested in doing so. But *any* starting size will be all right, and there will be no need to measure--since we may follow the Ancients' restriction of using nothing more than a compass and straight edge. *(We should note here the related Masonic "square and compass" --a symbol for a society deeply interested in geometry).* Draw the horizontal line about 20 inches in length. Make two intersecting arcs whose radius points are on the line, as shown:

Then, draw a line connecting the two intersections, creating a vertical line to begin the corner of a square at Point 3. An intersecting line created in this way--if done correctly--will be much more exactly "square," or at 90° to one another, (at "right angles") to the horizontal line than if done by any other method. Drafting triangles will always tend to be a little warped and slightly sloppy for true geometry. This geometric process will also precisely divide the line between two equal arcs into exactly two parts, a technique which will be used again shortly. (Note here a fundamental *geometric interrelationship* of the "square and compass" idea. The compass *is being used to create* a high precision right angle, or the "square").

Next, spread the compass points to 10 inches if desired, or even some unknown length--many compasses won't open that far. *This beginning length may actually be any length at all and the resulting rectangle will still be the golden ratio if this procedure is followed.* Placing sharp point of the compass at the new intersection of vertical and horizontal lines, mark both lines at your compass-spread length to mark-off the beginnings of a square.

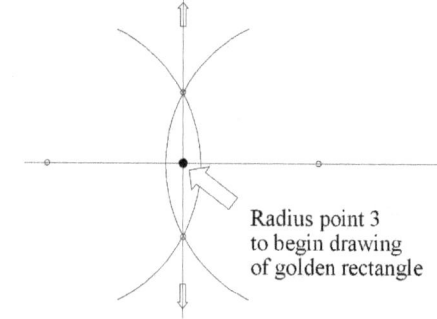

Radius point 3
to begin drawing
of golden rectangle

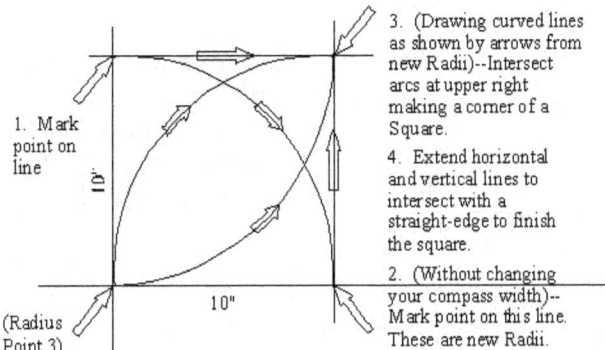

3. (Drawing curved lines as shown by arrows from new Radii)--Intersect arcs at upper right making a corner of a Square.

4. Extend horizontal and vertical lines to intersect with a straight-edge to finish the square.

2. (Without changing your compass width)--Mark point on this line. These are new Radii.

1. Mark point on line

10"

10"

(Radius Point 3)

Whatever your compass spread width is, becomes your fundamental starting length. And, starting from those marks, without changing the compass, intersect again at the upper right:

Now, connect the intersections, creating a practically perfect square. (The need for this fastidiously exact square as a basis will become obvious to those who try this exercise. Unless your drafting is quite good-- you will not be able to correctly complete some of the internal details shown below on the first try).

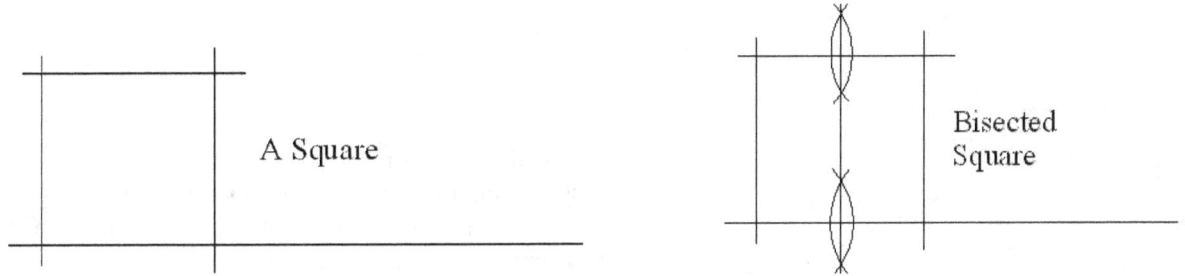

A Square

Bisected Square

Next, using the compass, divide the top and baseline in half. Spread the compass points to somewhat beyond 5 inches, (or beyond half) and make two intersections from the corners (not changing this compass width) and intersect top and base lines as shown above.

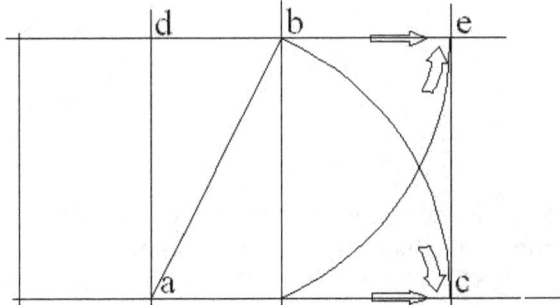

The Golden Rectangle can now be started from edge of this square divided in half, and using a compass with its pin or sharpened point at (a), and its *pencil point diagonally, exactly at corner* (b), draw an arc from (b) to intersect a line which is an extension of the lower edge of the square, at point (c). Then, using this same compass width--starting now from (d), an arc can be made to make point (e). Using a straight edge to intersect, when the top line is extended and likewise intersected at point (e), and points (e) and (c) are connected, and extraneous lines erased, the exterior of a φ-based rectangle is created. (This is the best known method to produce this special rectangle).

The overall rectangle has a ratio of 1:1.618... But the smaller, internal rectangle *also* has the ratio of 1:1.618... to the outer. Take a square off of it's smaller rectangle, and the remaining rectangle is *also* the golden rectangle. The resulting rectangle shows one of many unusual properties of this shape--the smaller vertical rectangle to the lower right of the square has *the same proportions as the overall rectangle.* Since this is true, it also follows that if a *smaller* square, based on the short end dimension of the smaller rectangle is to be removed from this smaller rectangle, *it too* must leave another yet smaller remaining golden rectangle.

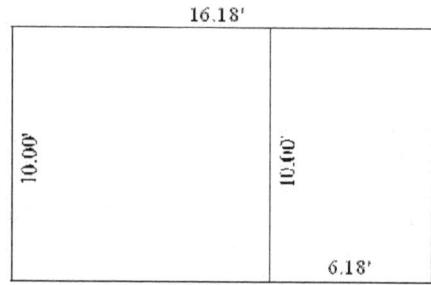

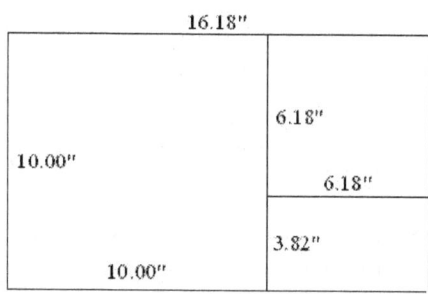

A diagonal drawn from corner to corner of the larger rectangle will then have to intersect *at the corner* of the smaller rectangle, showing the same relationship, a *fundamental angle:*

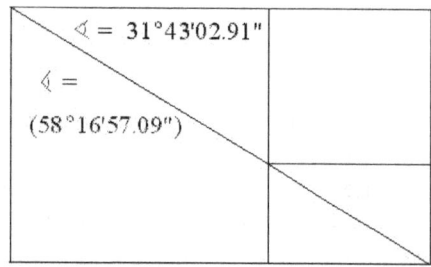

Key in to your calculator: 0.618033989, 2nd Function, tan, (tan[-1] or, "arctan" function) which will produce 31°43'02.91". *(Or, if you are reading 31.71747441° this is in decimal degrees. Press the calculator's "DMS" key and 31°43'02.91" should appear. If you use 1.618033989, you will get the opposite complimentary angle of 58°16'57.09" which is also the golden angle).*

12

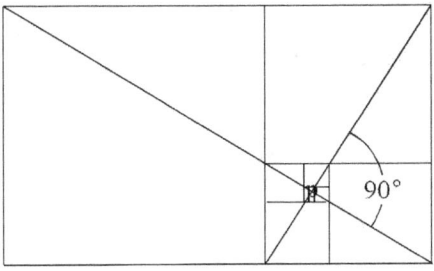

"Whirling Squares"

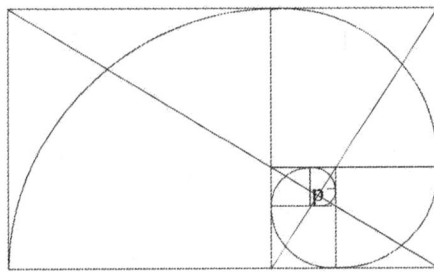

Spira Mirabilis

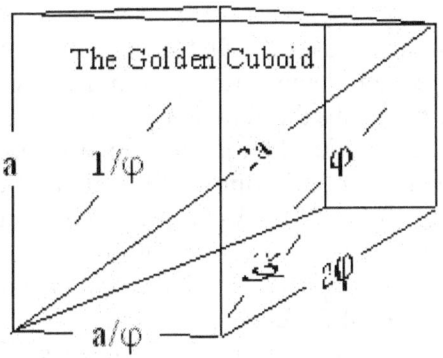

If another diagonal is drawn in the second smaller rectangle to intersect the first diagonal shown, they will intersect at a *right angle or 90 degrees* --a unique property of this shape. This point is called a "pole point" here labeled p.

Ideally, perfect little squares can be continually lopped off of the remaining golden rectangles, converging finally on the pole point at an infinitely small point p. This shape has been called "The Spiral of whirling squares", by J. Hambridge. A beautiful thing can be produced within a golden rectangle called the "Spira Mirabilis" or the marvelous spiral, sometimes called "the Logarithmic Spiral." It looks something like the shell of the Nautilus shell.

Though only approximately displayed here, after creating as many of the increasingly smaller squares as possible within a golden rectangle, one can draw quarter arcs in each square by placing the point of the compass at the interior corner of the squares, starting and ending the arc at the nearest corners. If you can draw more than eight arcs by the use of a compass, your drafting is excellent.

It is possible to make a three dimensional *Golden Cuboid,* or a golden rectangular *box.* Continuing from the above model, its height would be 10"; length 16.18..." and its width would be 6.18...". The ratio of the top (and bottom) rectangle is 1 : 2.618033989..., a number that we saw earlier, which is the square of the golden ratio, or φ^2. A curious thing about this shape, is that *the cubic diagonal,* that is, the interior diagonal line from corner to opposite corner, would be exactly *twice* the dimension of the height. In this case 20.00". This is the three dimensional form of the golden ratio, and doubtless one could continually extend this form into higher dimensions, such as the fourth dimension *(golden tesseracts?)* and so on. The above shape, interestingly, bears an at least superficial similarity in proportions to the Ark of the Covenant chest as described in the Old Testament, as shown by various authors. This shape will be important in later chapters.

Although there are thousands of interesting properties of the Golden ratio, one of the things often overlooked is the fact that this ratio is *fundamental* to mathematics and found deep within all families of math and mathematical structures. For example: *All* of the Platonic Solids can be shown to have three equal, internal, interpenetrating Golden Rectangles, in three perpendicular axi, (from a common origin, at the center of the given Platonic Solid), whose corners will precisely meet at (either) the *corners* or *exact centers* of the planes making up the surfaces of their polyhedra.

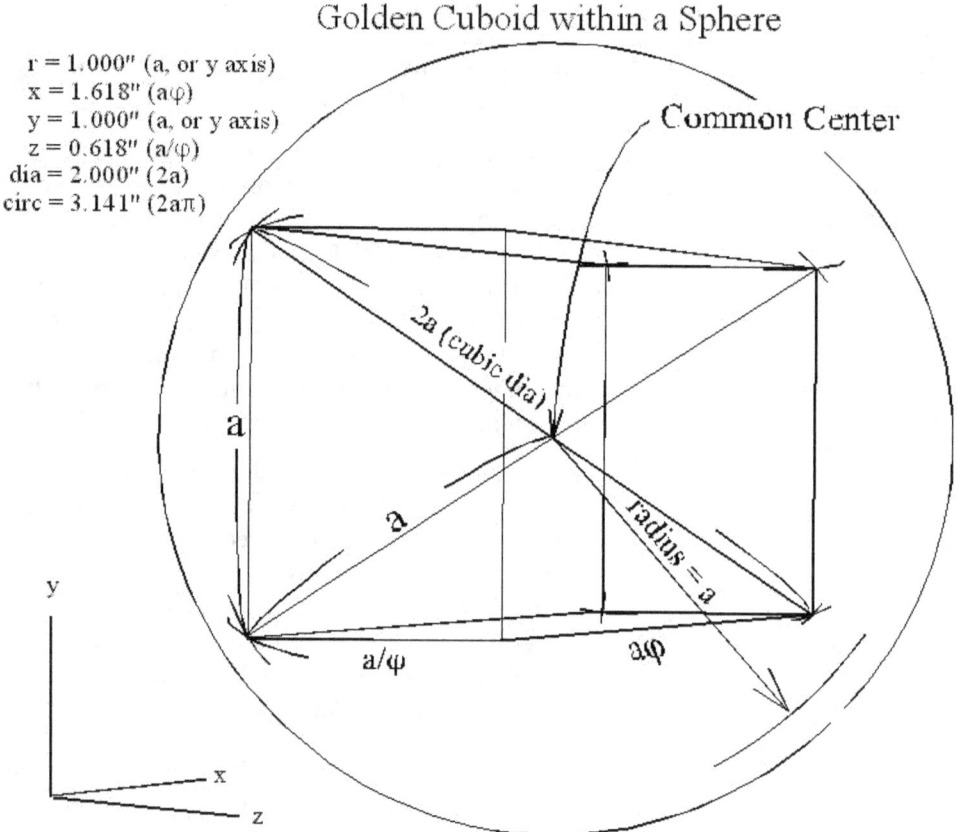

Golden Cuboid within a Sphere

r = 1.000" (a, or y axis)
x = 1.618" (aφ)
y = 1.000" (a, or y axis)
z = 0.618" (a/φ)
dia = 2.000" (2a)
circ = 3.141" (2aπ)

Which brings us to an interesting point made by H.E. Huntley, which uses the first shape before of the Platonic solids, the sphere.[12]   Huntley showed that this shape can demonstrate a simple, and yet fundamental relationship between the golden ratio, φ to π or, pi--being the function pertaining to the circumference of circles and spheres. In this illustration a Golden Cuboid like the one above is created with a side y of the length of one inch (1.0"), a side x having the golden ratio times one inch (1.618...") and a side z of the reciprocal of the golden ratio times one inch, (0.618..."). As above, it has a cubic diagonal of *twice* the length of the height y, or exactly 2 inches. The curious properties of this shape will allow us, mathematically, to place a sphere having a radius of exactly *one inch* centered over this cuboid, where it can be seen that the eight corner points must *exactly intersect* at the edge of the sphere, with the radius exactly at the midpoint of the cubic diagonal. The circumference is therefore 2π, thereby tying the two mathematical concepts together. A thing of elegant but hidden beauty.

Notes:

---

[1] This was probably <u>Look</u> magazine, and it might even have been 1966 though I can not recall for certain. There was a small run of similar articles in a few other magazines at the time. One or another of them highlighted the Danish mathematician-artist Piet Hein, and his special invention of an ellipse-shape, called "the Super Ellipse" which had just become popular among architects in Europe.

[2] Le Corbusier's <u>Le Modulor</u>, ca. 1960. Charles-Edouard Jeanneret, (1887-1965) A Swiss architect, painter and designer.

[3] Nicomachean Ethics, Aristotle, Trans. by Martin Ostwald, 1962.

[4] Ad Quadratum, Lund. Lund finds the first use of the term in De Divina Proportione, by Luca Paccioli in 1509.

[5] See ESOP, Vol. 18, 1989, the Geometry of Stonehenge, by Alban Wall, the Epigraphic Society Occasional Publications, pg 220 (Barry Fell, et al.). This proportion is based on the classic proportion of $\sqrt{2}-1:1$, but there is other evidence for ν and other proportions. The angle of the ancient mid-summer sunrise to true north at Stonehenge is about 51°, calling to mind the function $4/B$ like the Great Pyramid's slope. This may be coincidental, since the Stonehenge builders might have chosen that latitude in ancient Britain so that the lines of solstice and "moon-stice" relationships would occur at 90° to each other--which will only occur at a latitude near Stonehenge, (actually near Avebury) England. This would confine this sunrise angle geographically, as well as through astronomical precession. Is has been said that there is some reason to believe that the Stonehenge builders were aware of the metrology of the Great Pyramid through the Megalithic Yard of 2.72 feet.

[6] Photo from: The Aegean Civilizations, Peter Warren, 2nd. Ed. 1989, pg 87.

[7] Euclid's Elements, A problem in Book II, Proposition 11; a definition in Book VI; and a problem in Book XIII, Proposition 5 .

[8] The Secrets of Ancient Geometry--And Its Use, Vol. II , by Tons Brunes 1967, pg 57. The best source on the subject of $\sqrt{2}$-based proportions or, "Sacred Cut" in ancient construction. See note 5 above.

[9] A Modern Survey of an Ancient Roman villa, by Robert Mangurian and Mary-Ann Ray, in P.O.B.-Point of Beginning magazine, August-September 1993, pg 10. Also see Scientific American, December 1986, A Roman Apartment Complex, by Donald J. Watts and Carol Martin Watts, pg 132.

[10] Mathematical Circus: more games, puzzles, paradoxes and other mathematical entertainments from Scientific American; with thoughts from readers, afterthoughts from the author, and 105 drawings & diagrams, pg 152 et. seq. by Martin Gardiner 1st Ed 1979

[11] What's a reciprocal? It is an important idea in this book and a big part of the design of the dollar. In a calculator, this is the number taken from the other side of a fraction bar (1/x) if it is expressed in tenths, hundredths, thousandths and so on, as a decimal fraction of one, using a decimal point. This will appear as numbers to the right of the decimal place in your calculator. If you want the decimal equivalent of any fraction, say "1/2" or one half, you enter "2" and press "1/x" and you will get "0.5" being five *tenths*, or 5/10. You're supposed to know that when you chose to enter the number 2 on the screen, that this is the "x" in the "1/x" and that it would be converted into a decimal fraction. Some fractions do not divide so easily as a half into ten, such as one third which will yield three tenths, three hundredths, three thousandths, three ten-thousandths..., etc., or, "0.333333333..." If you have divided in the wrong direction in a math problem, the reciprocal is a handy function that will quickly give you the other answer. Even if you don't have a 1/x button on your calculator, you can key in "1 ÷ " a given number then "equals" and this will also produce the reciprocal.

[12] The Divine Proportion, by H.E. Huntley, Dover, 1970 pp 98-100

CHAPTER 2: MEASURING THE DOLLAR

Around autumn of 1973, wandering through bookstores in downtown Chicago, I found a book called the <u>Divine Proportion</u>, by H.E. Huntley. Huntley was a real aficionado of the golden ratio. In his book he enthusiastically discusses multitudes of ideas and principles relating to the golden ratio. For me it was a real eye-opener. Unusual and densely packed, an old fashioned book, it was the classic type of book published by Dover. Huntley was rigorous and mathematical--but not especially difficult; he was lyrical and artistic as well as inspiring. His joy for the subject leaps out at the reader, but I was already there--he was preaching to the choir. Here I began another period of study of this elusive ratio.

Huntley was a first rate mathematician, yet still refreshing to read. For a book of its small size, <u>the Divine Proportion</u> may be the best book written on the golden ratio anywhere. I soon began to see the subtle importance of this ratio just about anywhere I looked, and not just in plane geometry. It was in literature on architecture (as I had seen earlier, from Le Corbusier), design, graphic art, musical chords, advertising the branching sequence of tree limbs, human proportions and elsewhere. I could see it in many advertising logos, the packaging of products, 3" by 5" cards and paperback book proportions. Some of this golden ratio might have been subconscious action by artists rather than by intent. But when found, most of this form seemed to be a clearly intentional effort by artists.

Most constructed shapes in our society are *rectangles*. I was impressed at the time by Huntley's discussion of Fetchner's and Lalo's statistical experiments that tested for people's natural preference for the golden ratio proportion in rectangular shapes. These studies showed that the golden ratio shape (or any similarly proportioned rectangle of about that shape), was the most attractive rectangle to most of the people they surveyed.[1] For a while, perhaps months, I immersed myself in this ratio. This was a period of measuring and computing almost every rectangle in sight, down to doorways and cigarette packs.

I used a six-foot engineer's folding rule in decimal foot units and a small calculator. Being employed by a surveyor these tools were around most of the time. The tools allowed me to measure two sides of a rectangle in decimal feet and divide to find a ratio very rapidly. I looked for golden rectangles everywhere. After a time they seemed to "stand out" to me somehow mentally, and it seemed as though I had become intuitively alert to detecting this ratio when I was looking at artwork. Within a percent or so of the actual ratio of 1:1.618, I began to pick out golden rectangles in many places by eye, such as in advertising or museum art. I had a friend who was also interested in this idea at the time, and after some practice, he could do this trick also. I didn't spend a great deal of time at this, it was mostly an on and off diversion, much as it had been several years before.

****

Starting to Measure the Dollar

I have often wondered how I got started measuring the dollar. It seems like it was one small, odd thing that lead to another, and still another, and so it grew on me. It was re-examinations of small, odd findings--nothing very much in of themselves. When all brought together, the findings seemed to form a large *picture* that demanded some kind of serious thought. This realization did not occur all at once, and I have felt reluctant to take any of this seriously or do anything about it. These findings later avalanched into what became almost a part-time occupation at times, like a peculiar hobby.

Around Christmas of 1973, I think, I became interested in the back side of the dollar bill.[2] I looked long and hard at the dollar--maybe my intuition had picked up something. I started to measure a few places on the bill that looked interesting. It really wasn't the "right" shape over all, being much longer than the sort of rectangles I was mostly interested in.   Yet I had an odd hunch right from the beginning that there was something very unusual or special about the dollar's ornamental design.

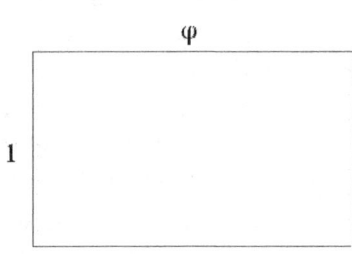

Illus. A

Near the beginning of my investigation, I must have seen a dollar *folded in half* at some point.   My first interest was in the *center area* of the design, and not the edges. Folded in half, it can be seen by eye that part of it--the center part--is close to the golden ratio, if only considering the interior design of lines and filigree. (See Illus. A) I was interested in the golden ratio's use in art and design and I had no expect-ations of finding anything unusual. Having just recently read some books about the fraternity of Freemasonry and early American Revolutionary period, there was another element in my thinking at that time. My first close look at the dollar might have been inspired by a lecture I had heard around that time on the dollar's Masonic symbolism. Having read about Masonic symbolism, I had become aware that the dollar was covered with their mysterious symbols. And of course, on the other side was pictured George Washington, a well known Mason of that period. Masonic lore became a strong undercurrent in my interest. Strange as it may seem, my investigation of the dollar's design eventually converged on far-off events that have their source in the American Revolutionary Period and the Great Pyramid. As we will see later, parts of the dollar's design are tied to that era by evidence of some tell-tale design choices made by the Designer of the dollar. (I was unaware of any of this until the 1980's.)

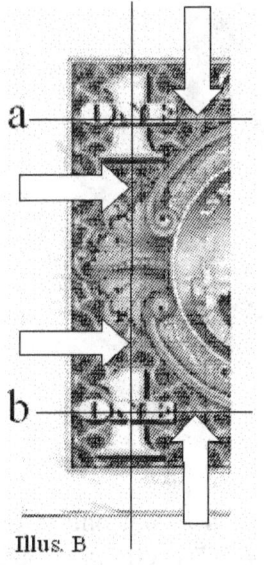

Illus. B

I had a dimestore-type transparent plastic ruler around my apartment to begin with that read in inches and sixteenths along one edge and metric in centimeters and millimeters on the other. The kind that had holes in them for school children's ring binders. I also had some cheap, flat, acetate rulers around the apartment from somewhere. These weren't good rulers. Later, in comparing them one to the other, there were minor differences in what they represented as inches and millimeters. But this didn't bother me at the time. They were cheap transparent rulers, helpful for measuring small things underneath them.   They were not as accurate as drafting scales, but they were regular enough to determine *ratios*, which was all I was interested in at the time. At this point I was only curious, only interested in the proportions and not the exact dimensions involved. I used inches and metric alternately, and might have used printer's points or pica scales if they had been available. After what I had found, it wasn't very long afterwards that I became very interested in what the *exact* measurements were.

The cheap plastic rulers very soon gave way (within days) to a succession of more expensive drafting scales, and somewhat later to machinists' scales and dial calipers. Most of my first measurements, however, were done with drafting scales under strong light with a 5 power magnifying glass.

Strangely enough, the place I started to measure was the Inner Rectangle area on the back face of the dollar, in the filigree. The first rectangle that I was interested in was made up of two *implied* horizontal lines made out of the *focal areas* of the filigree. These have as their corners, what might be called center lines projected at the middle of the four numeral 1s, (not the middle of the text "ONE") intersected by vertical projected center lines from the focal areas of the filigree at the ends of the dollar. (See illus B.)

To my eye, it appeared that there were two golden rectangles end to end within this inner rectangle. And as seen earlier, when folded in half this seemed likely. After some careful measuring and dividing the dimensions, there did indeed appear to be something like a pair of golden rectangles, end to end, on the right and left sides. (see illus B). I now began to wonder about the exact ratios.

In the midst of computing arithmetic ratios from my various measurements, I noticed by accident, *that if this half-rectangle was computed as a golden rectangle,* using a height of 24.33 sixteenths of an inch, which I had measured vertically between the centers of the lines in the filigree, the resulting length would be 39.37 sixteenths. (24.33 1/16ths is the short length of this rectangle between **a** and **b**.) I had become interested in the unit of measure used in the design, and I first looked at sixteenths of an inch as a possible fundamental design unit.

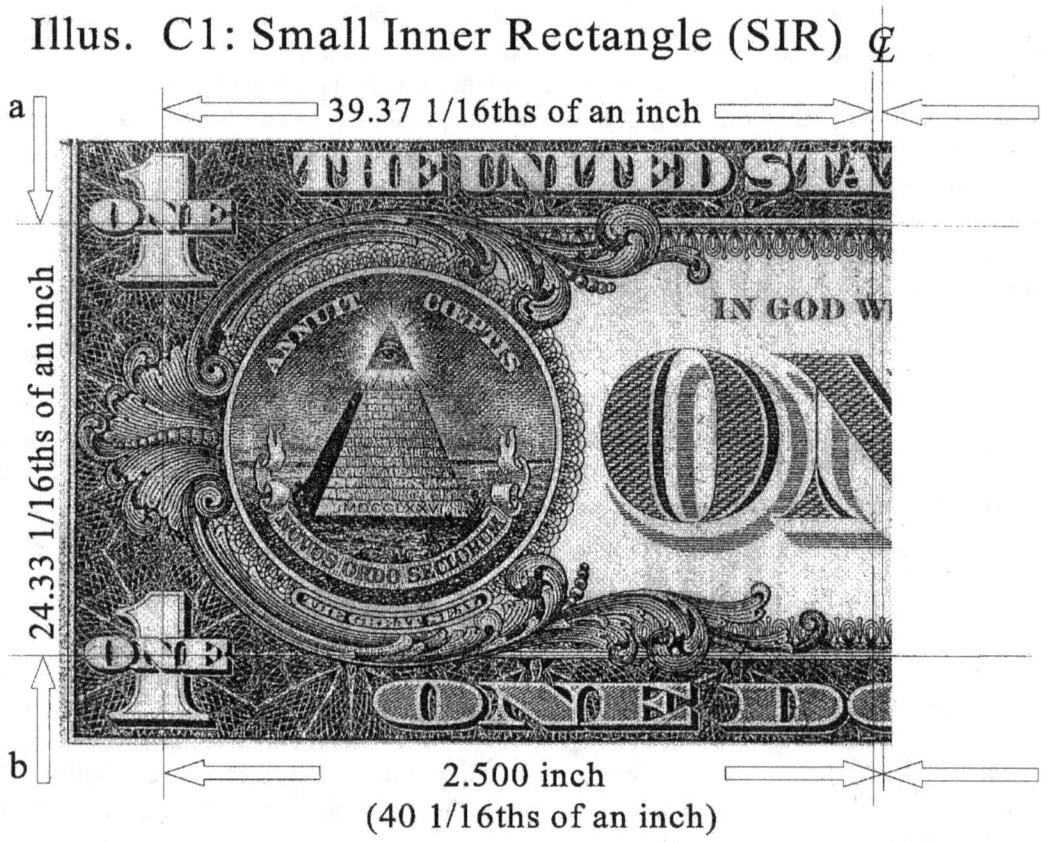

Illus. C1: Small Inner Rectangle (SIR) ₵

a ⟵ 39.37 1/16ths of an inch ⟶

24.33 1/16ths of an inch

b ⟵ 2.500 inch ⟶
(40 1/16ths of an inch)

"39.37 sixteenths of an inch" seemed *really odd*, and stuck out in my mind as not only a peculiar length, why not a more regular 40 1/16ths for example? But the number "39.37" itself stood out, since that would be the *exact number of inches in a meter*, if we were talking about *inches* instead of sixteenths of an inch. This was a peculiar coincidence numerically: only one number out of ten-thousand would be an exact fit for all four places if it was an accident. And trying slightly differing measurement estimations of the height of this filigree area this "39.37 1/16ths" (rounded) turned up repeatedly by multiplying the distance **ab** by the golden ratio. After thinking about it for a while, it occurred to me (after switching around the arithmetic and my point of view), that this length had to *also be a 1/16th of a meter.*[3]

39.37 1/16ths" = 2.460625" *and,* 39.37"/16 = 2.460625" (= 1/16th of a meter: 0.0625m)

Since this depended on the vertical dimension being 24 and a 1/3rd sixteenths of an inch, accurate to about one hundredth of a sixteenth of an inch, (about a half a thousandth or 0.0005") I carefully checked this height with my machinist's scales and a borrowed dial caliper, repeating this result several times. It seemed like a real, but odd coincidence. Were meters being used here in some odd way?

Or, was all of this accidental? Yet, if I accepted this end height of 24.33 1/16ths (**a** to **b**) as measured, and multiplying by the golden ratio, "39.37 1/16ths" was clearly the length that comes out of comp-utation. But notice in Illus. C1, *that this resulting length falls short of the centerline of the design,* (or anything else) if projected from the left end, so this whole idea appeared to be an unimportant and unintentional result. Who would use *a sixteenth* of a meter anyway? Wasn't everything in meters supposed to be *decimal* lengths--wasn't that the idea behind meters? It wouldn't seem as though meters were being used if the "1/16th of a meter" idea didn't make any sense. For the moment, anyway. (It also pointed out another bit of trivia that I hadn't noticed before: a rectangle of *2 feet by 1 meter* is about one percent away from being a golden rectangle.[4]) If I hadn't continued to "sniff around" with a scale in a few other places, I probably would have moved on to something more interesting that weekend than measuring the dollar.

Illus. C2: Small Outer Rectangle (SOR)

Soon, I noticed *three new and very curious things*, and which kept my attention focused on the dollar bill's oddities and the golden ratio.

The first interesting discovery was that upon close inspection, I found *another* golden rectangle, formed by the top and bottom *edges* of the center border area, *centerline* of the dollar and same *left line* as before, centered on the focal points in the filigree. (See Illus. C2) This is symmetrical on the right side of the dollar also. This rectangle however, turned out to be a *very exact* and unambiguous golden rectangle, extending to the centerline. But a slightly larger one than the first, superimposed upon the first rectangle, making a tiny vertical (and apparently equal) margin of space between top and bottom lines of the smaller rectangle. In retrospect, it now seems that this was intended as the first, or most "obvious" beginning clue to encourage investigation, probably the one that was meant to be located first by a future investigator.

So it turned out, there were (at least) *two closely superimposed sets* of inner rectangles of a slightly different height and length. (See illus C2 and C3). On the right and left of the centerline, this second set were rectangles an exact, regular, 2.500 inches in length. These made a lot of sense, lengthwise, from the vertical ends of the filigree, forming an even overall length of five inches--but the vertical distance **c** to **d** did not make a lot of sense for height. (But they can't of course, only one side or the other are going to make sense in regular numbers if this ratio is used.) These lines were composed of the set of plainly visible, straight, printed lines at the top and bottom at the middle area, (**c** to **d**)--projected to the corners at the ONEs, And the vertical line formed in centerline of filigree between the ONEs; and an imaginary line at the exact center line of the bill. This invisible centerline at 2.500" is arrived at by dividing the total span of a measured, exact five inches between the centers of the ONEs by two.

Illus. C3: Small Outer Rectangle, from Centerline and end of Small Inner Rectangle at left; and *outer edge* of ruled border in middle area.

Now, if there was a golden rectangle based on half of that length, or 2.500 inches, *it would have to have* a height dimension of approximately 1.5451 inches. I had originally measured 1 and 35/64ths inches for this outer width of lines, (1.5469"), which (amazingly) appeared to give the golden ratio within about one part in eight hundred, or about a tenth of a percent shy of a true golden rectangle. No doubt about it, *this could not have been accidental,* and was a "real" enough golden rectangle for anyone's standards.

Measurements made somewhat later with a dial caliper showed this height to be right at 1.545 inches. *This clearly established one more or less exact golden rectangle, with great precision.* As time went on, this seemed to me to point a finger at *any discrepancies* with respect to its smaller sister rectangle within it. For instance, having made the one larger rectangle so perfectly, where was the exact end point graphically, of the *first* smaller rectangle? It seemed as though something should have marked this length by a line or some important feature. There didn't seem to be much of anything marking the end of the "39.37 1/16ths" of an inch position.

The second very curious thing revealed itself immediately after this, when considering the relationship of the two rectangles to one another. If you *subtract* the supposed "39.37 sixteenths of an inch" length of the inner rectangle, (2.460625" in decimal form) from two and a half inch length (2.500") of the outer rectangle, *the difference is: 0.039375".* If we ignore the final 5 millionths of an inch, and ignore the decimal place, *the number "3937" has quite separately* popped up out of the relationship between these two rectangles. (This is the small distance e to f in Illus. C3. Notice also, that this small 0.039375" length must then be very nearly *one millimeter* unit in length.) Here again; something that seems pretty much intentional, but somehow quite far-fetched and perhaps absurd. Based only on the two sets of clues this clever (and peculiar) scheme would have to rest entirely on the logic of the information provided by these two arguments:

(1) the designer's elaborately careful *choice of the exact height* of the smaller of the two little rectangles, and

(2) the implied use of the golden ratio as demonstrated by the outer rectangle, *as the proportion to use as a multiplier.*

So if we have one special rectangle, with a height *and* a width with a proportion, but we have only a *height* for the second: another *length* for the second rectangle could be inferred by the special proportions of the first rectangle. When computed, the number "3937" appears again--part of the math *as a pun.*

This self-enclosed and self-defining relationship is fascinating for its logic and serves as a first small look at the Designer's peculiar work. Notice the amount of information conveyed by this seemingly simple relationship, and how subtle the clues are. (you are not convinced? please read on!) Having known something about the golden ratio, I had stumbled onto by accident. Up until this time I had been unaware that anyone made use of the golden ratio in its exact form, other than in academic geometry. Here it had appeared in both geometry and some self-referential oddities implied by arithmetic. But beyond these things, there was *still more:*

The third unusual thing, somewhat after this was the discovery that careful measurement of the *exterior lines* of the dollar design, or the thin outer border lines, revealed a ratio of 1:2.618. (see illus F). I hadn't done much mathematical research into the golden ratio up until that time, not thinking that there was much more to it beyond the oddity of the "golden ratio of 1.618.. and its reciprocal 0.618...". I was unsure of this new ratio's meaning. But I remembered there *was* something about this new number in Huntley's book or somewhere else, and realized I would have to re-read all of them again. I could see that the ending decimal was also ".618," and I was soon to discover (with an accidental punch to the square root button) that this number seemed to be pretty close to the square of 1.618... or *the square of the golden ratio.* (Amazing stuff. It is, of course, *exactly* the square of the golden ratio, as we saw in chapter 1). This was a surprise to me, and I soon backtracked and learned all I could about the math surrounding the golden ratio. Okay, so the design of the dollar definitely involves the golden ratio, and perhaps lots more, I concluded. It seemed that I was on the trail of something, anyway.

Being near Christmas and all, with lots of other things to do, I was near to letting this strange little investigation drop. But I hadn't settled in my mind what--if any--meaning or significance the first

21

theoretical rectangle with a length of "39.37 sixteenths" really had. This had begun to bother me. The Designer's hook had been set. I re-checked the starting dimension. Had I *really* found something? With all the above observations in mind, these related thoughts rattled around in my thinking--though not organized consciously as this list, and here with some added hindsight:

(a)   The more I thought about it there was the 1/16th of a meter coincidence, which with respect to the "39.37 1/16ths of an inch" seeming to be something symbolic and *probably not* a coincidence;

(b)   (b) especially when the subtracted difference in lengths (the math pun) of these two rectangles gave 0.039375". Or was this just an odd, trivial and artefact of the mathematics? Additionally, the *0.039375 inch length* (**e** *to* **f**) seemed to be almost the same as one millimeter, or 0.001 m. [5] If nothing else, an interesting but peculiar joke or number game--But not simple, *not something brought about by chance.* What in the world could this design be trying to say?

(c)   Then, even ignoring the unlikely oddness of those above observations, there were already two indisputably exact, golden rectangle related shapes. At least one golden ratio rectangle and another, longer sort of golden rectangle in a *squared form*, no doubt about it. (The third possible "39.37 1/16ths" shape still seemed tentative or hypothetical at that point, since it was not marked, even though it was becoming increasingly clear that there was some reality to the above implied inch/meter length and pun) and;

(d)   Since the more believable of the two small rectangles by evidence, (the 2.5" x 1.545" or an exact, golden rectangle shape) was just a little larger than the first rectangle, it would make it appear that the Designer was possibly inviting, or forcing a comparison for some reason between these two superimposed rectangles. That is if, the smaller was really a legitimate golden rectangle after all, since its length didn't appear to be marked.[6]

(e)   This clearer 2.5" x 1.545" rectangle also used the same filigree centerline for the end part of its shape as at the end of the smaller inner rectangle--the vertical line between the ONEs, which is to say, perhaps forcing a shared kind of significance.

(f)   Using the above clues, I concluded there could be only two approaches, logically, for examining these sides to see if there were golden rectangles present (restating the earlier first two original arguments):

I. Dividing 2.5 inches (half the total inner rectangular area) by the golden ratio to find a resulting height--for which matching features were found, i.e., the printed lines 1.545" apart with the exact split between ends giving 2.5"--And,

II. Measuring the end height of found features (such as the 24.33 1/16th inch wide focal areas lines as found in the filigree, which apparently represented the width of a smaller implied rectangle) and multiplying that length by the golden ratio, and then looking for whatever features there would be of that length. Which did not, however, appear to match any found length. But did however, lead inescapably by subtraction, to the peculiar length pun of "39.37 1/16ths" or, the odd length difference of 0.039375" of about "one millimeter," or, **e** to **f**, as seen in Illus C3.

It seemed like a lot of thinking for so few clues. And a lot of design work without leading anywhere. But the fact that there was good success by both theory and observation for the first part (f I.), and tantalizing partial evidence of any significance for the second (f II.)--there being only the filigree focal area line at the end and the peculiar length: this seemed to require a really close, second look at the length that should result from the second case. And I wasn't at all sure how to go about this.

I had reached a logical logjam in my thinking. With the above golden relationship being true for the larger rectangle, it seemed as reasonable and equally likely that the two little stretches of line in the filigree, between the ONEs and the design around the seal, seem to bear an equal logical weight as the more well established rectangle. And it might just as well make up the ends of another invisible rectangle. Yet, as shown earlier, when the calculated distance of 39.37 1/16ths was projected from the outer ends, *it irritatingly fell short of the centerline*, causing me to wonder if the whole idea wasn't all ridiculously far fetched. For the smaller rectangle to be *at* the centerline, its origin would have to begin 0.039375" or about "one millimeter" to the right of the end of the inner rectangle's focal area line at the left edge of the filigree, and nothing seemed to be marked over *there*, either.

This itch had to be scratched somehow. Much time elapsed while alternately thinking about or avoiding thinking about the whole idea. But later, after some careful drafting on the surface of the dollar, using a cheap compass, and a sharp pencil (more or less after the fashion of Chapter 1), I managed to delicately plot a golden rectangle geometrically from the left end at the center of the 1s--and while at it, I also laid out the "pole point" by intersecting short and long diagonals. (As shown in chapter 1, of which there would actually be four, not shown, arrayed symmetrically, per rectangle.) And I also laid out some of the "whirling squares" areas, to see if the design incorporated any of these features.

Illus. D: Eye (lower left iris edge)

is tangent to diagonal

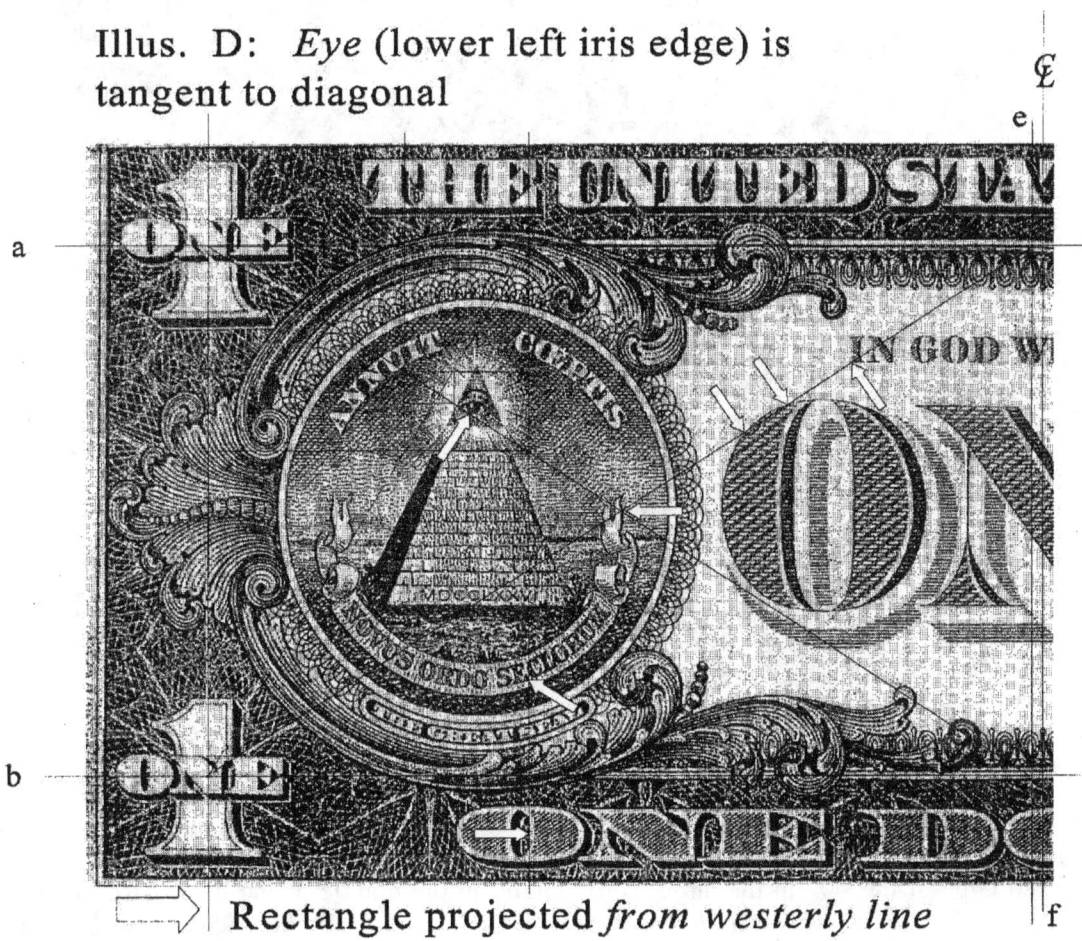

Illus. D: *Eye* (lower left iris edge) is tangent to diagonal

Rectangle projected *from westerly line*

One of the pole points *seemed to fall very near* the "All Seeing Eye". *Now* I was curious, since the only other possible next step as far as I could imagine starting from this **first position** (See Illus. D) was to now re-project this little rectangle *from the centerline* or from the other direction, from the right, instead of from the left end. And then repeat the placement with pole points, diagonals and all--which would doubtless move this pole point *towards the Eye* to a large extent.

When that was accomplished (Illus E.) it appeared that this pole point fell very close to the Eye, though not actually within the Eye itself. With the tiny markings that I had made on the bill, I knew I couldn't be sure, but it seemed that if this rectangle was to be moved to the right *maybe just a tad more*, that the Designer's placement of the Eye might be right on the pole point. Wishful thinking. It really doesn't fall within the Eye, but it appeared so close and my drafting just uncertain and cramped enough, who could tell? Yet, our Designer was a much more subtle artist than that.

Illus. E: *"O"* (lower left edge of ONE)

is tangent to diagonal: Eye *centered*

Notice in Illus. E *that a diagonal does, however, cross the Eye* when the rectangle is moved into this **second position,** as projected from the right. While at the same time, it now *just grazes* or is tangent to the lower left edge of the "O" in the large word "ONE" in the middle of the bill. Also notice, that the "All Seeing Eye" inside the radiant triangle seems now to be *centered* at the middle of what would be the fifth sequentially smallest golden ratio rectangle of the "Whirling Squares," while at the same time, now the *lower right hand corner of the little triangle* appears to be aligned with the long diagonal in this position. The top of the little pyramid (back edge) appears to align to the bottom edge of this rectangle. Also note the *tangent edge* of the little curlicue below the "O" and "N" in ONE at the center of the bill.

This clearly demonstrated something of the truth or correctness of the smaller rectangle theory. Notice that all these relationships are totally hidden, and strictly speaking, non-existent, *only implied by analyzing the line-work of the dollar.* Notice also that there is probably no natural or accidental way to have found this. You would have to discover it through outside effort, and then you would have to have some knowledge of sacred geometry. This is a very subtle design.

At this point we can see that the Designer has really has been trying to say something, and in a very quiet way. This individual was quite conscious of these ancient geometric forms. And he has also revealed a secret code framework built out of the golden ratio shape, and something of a signature method, perhaps to look for elsewhere. Only a small portion of his secret has been uncovered--and this was by means of what I had thought was a very doubtful "39.37 1/16ths" dimension. These alignments-- and many other subtleties--are unlikely to have been noticed without considering some or most of the above assumptions, or if noticed at all, would have little or no context without the golden rectangle. I think this demonstrates the validity of the reasoning process based on the two primary arguments shown above, which looked pretty flimsy at the outset.

My main motive in "moving" this rectangle the 0.039375 inches to the right was really a perplexed, grasping at straws. And it seemed to be offered by the near proximity of the pole point on the long diagonal to the All Seeing Eye as shown in illus. D. I don't think that I would have gone to this much effort if there hadn't been at least one perfect golden rectangle present, along with another rectangle, and *a golden ratio squared,* at the outer border of the bill. That fat "millimeter" length was the key and I was amply rewarded for my curiosity.

****

While we are here, we should take a look at a few things that were heretofore hidden, in light of the revealed geometry. Some of these alignments might be interpretable or guessed at for meaning:

Switching back to **Illus. D**, where the projection of the rectangle originally began from the **left edge**, rather than the centerline: notice that this time we see that upper-left-to-lower-right diagonal does not graze the big "O," *but is seen to graze the lower left outer edge of the smaller circle making up the "iris" of the "All Seeing Eye"*, much after the fashion that it grazed the much larger "O" in Illus. E. Look at *both sets* of diagonals from both projected positions. Note that they meet *together* at the edge of the right "flag" and at one of the points. Alternately one line is tangent and the other is not. (In Illus. D however, note that the other lower-left-to-upper- right diagonal does not appear to be perfectly tangent to the upper left edge of the "O." We will see more of this elsewhere.) It is evident that the Designer left a message for the careful observer. Could we tentatively say that a grazing line *symbolically equates* the All Seeing Eye of the Deity to the idea of "one," or maybe "oneness" from the word that it touches? Biblically, from the Old Testament "...the lord our God is one," is an old theme, and this may be the message here.

Returning to **Illus. E**: At the far lower left corner of the rectangle is centered in the right vertical stroke of the "N" in "ONE," exactly at the middle of a "V" shaped point of the serifed[7] corner of the letter, *together* with it's diagonal from the upper right. The center of the first "X" in MDCCLXXVI lines up on this diagonal. The lower-left-to-upper-right diagonal also grazes the *lower point* of the fork in the flags at the right end of the scroll saying NOVUS ORDO SECLORUM. This point is also important later on in other alignments. Notice how the (vertically extended) left edge of the largest whirling square just grazes the right edge of the heavily serifed "E" in the word "THE" of the "THE UNITED STATES OF AMERICA" above at the top of the bill. Isn't that strange? Notice the line of the right side of the *centered rectangle* when extended up to the "T" in "THE" exactly matches the right edge of the serif--we will see this again elsewhere. This did not happen in Illus. D. It can then be seen that this pair of lines in this centered position exactly *frames* the word "HE" --almost certainly a religious statement about the Deity. When I first identified this, I was astounded at the care and reverence of the Designer.

"O" and "E" symbolism: In Illus. D, notice that the left edge line of the largest whirling square aligns with the lower right corner of an "E" in "SECLORUM" and later exactly tangent with the right edge of the oval empty area in the right side of the "O" in "ONE DOLLAR." This might also be said of the empty oval area in the "C" of "COEPTIS." Compare this with this line shifted in Illus. E—the line runs through the solid area of the "O" on the *right* side of this "O" and above, it crosses through the solid area on the *left* side of the "O" (the "Œ" diphthong) in the word "COEPTIS." (*Cœptis* is the Latin word for "beginnings" or "eager undertakings." The first part *Annuit* means: "nodding ascent;" so the whole is: *Nodding ascent to our undertakings*). Could this alignment be intended to incorporate both "e" and "o" symbolism together? The question is: since the Designer could have placed this word "COEPTIS" almost anywhere, why are these letters aligned exactly like this? Notice in Illus. E, this line is at the *right* edge of the "E" in the word "THE" above, but just at the *left* edge of the "E" in "SEAL" below "SECLORUM." Could the letter E stand for "eye?" Note how that might fit with the Designer's choice of alignment for this very triangular center bar of the capital E. By itself, in the Latin, the "E" as in "E PLURIBUS UNUM" stands for "out of" as in, "out of many, one." Here, the "oneness" idea again, doubtless the text source of the idea. Perhaps all of these E/O things are to carry the EYE/ONE alignment clues to other places in the design. O- and E-alignments occur *all over* the dollar.

"CL" symbolism: Notice in Illus. D that the upper-right-to-lower-left diagonal crosses at the center point of the *baseline* of the Roman Numeral string MDCCLXXVI near the base of the little pyramid, exactly between the C and the L. *Next,* in the shifted geometry of Illus. E, the left edge of the largest whirling square splits a "CL" at the curved *top line* in the word "SECLORUM." The letter symbol "CL" is a common engineering acronym for "Centerline," where both letters are usually combined as a symbol, as I have used in these illustrations. The second "CL" alignment occurs when the rectangle is aligned from the *true centerline* of the dollar. Is something being said about centerlines, or alternate center lines, such as at the left edge of the first square?

"I" alignments: In Illus. D, the left edge line of the rectangle (centered in Illus. E) touches the *top left serif* of the "I" in "ANNUIT" and over to the right, the upper-right-to-lower-left diagonal crosses about the center of the base of the "I" in "IN GOD WE TRUST." *Next,* in Illus. E, the left edge of the centered rectangle touches the *center* of the "I" at its *base*--whereas the diagonal crosses the *bottom right serif* in the "I" in "IN GOD WE TRUST." The left edge of the largest whirling square appears to align with the body of the "I" at the end of the Roman Numerals in Illus. E, or perhaps to its right side. We cannot escape the clear, visual and text pun of "the center and edges of an I," ("the center and edges of an *eye*.") So the Designer makes English puns in addition to math puns. Note the very minimal and precise use of clues for this message. The Designer, as we shall see elsewhere, has a thing about letters and serifs.

Where does the Pole Point, or focus of the whirling squares diagram figure into all of this? From the clear and exact knowledge and use in positioning elements of this design within a structure concordant with the whirling squares construction, the Designer was *undoubtedly aware* of this seemingly mystical point, but it doesn't yet appear to play a part in this design.

The sharp-eyed observer will note that although the fifth sequentially smallest *rectangle* in Illus. E seems to exactly rest on the back of the "floor" at the top of the unfinished area of the little pyramid, and that the Eye is centered within it with respect to left and right sides, the Eye is *not perfectly centered vertically* on the diagonal by some tiny amount. It is slightly *below* the exact center where it crosses the Eye: perhaps the diagonal rests at a point tangent the lower left edge of the *pupil,* as a continuation of the same symbolism of the "O" and iris of the Eye. It might also be said to graze the *elliptical highlight*, or blaze of light at the upper left inside the Iris. This downward displacement of the diagonal is another aspect of the dollar design, *the offset idea,* which will surface elsewhere in many discoveries. In this case, there is evidence of more than one offset. Many will find heaps things not discussed here.

This shows that there really was quite a bit carefully designed into the etching--something that was evidently intended to be puzzled out. After this point I began to be much more careful about taking measurements, using an expensive K+E hundredth of an inch drafting scale a machinist's dial caliper.

****

The Number: 5.655

Moving on to the outer border, the top dimension appeared to be about 5.655 inches in length. The bottom dimension was slightly larger, but the average was still close to this number. My first attempt at measuring this was with the help of a newly acquired and easy to read drafting scale, which measured in hundredths of an inch.

(The final five thousandths of an inch, or 0.005", was at first estimated from the fact that the line appeared to fall right about midway between 5.65 and 5.66 inches. This tiny, ending fraction area of 0.005" in the dollar's length becomes quite a question for me years later. During this period of time, though, the above length was apparently verified on this particular bill with far better tools, many times, and appeared to be reasonably true for two other dollar bills I had been measuring. But it is very important to note here, that it was really by a luck that these bills *were about this size*--they were *slightly small due to age shrinkage*. They had been lying in a drawer of my desk for quite some time; they were uncreased and new looking in an envelope that I had kept there for some reason or other, so I used them for measuring. This is important because the true size of the original printing of a fresh, crisp one dollar bill is actually just a little larger, as we shall see later. Just about the right size larger, as it happens, to where the *length that I had found in inch units, turns out to be about the same in numerical length in Pyramid Inch units in a fresh dollar bill.*

As it happens this threesome of crisp bills had conveniently shrunk to more or less the perfect size in terms of the Designer's ideal arithmetic--if we were to use *inches*. And this is also true of any other paper bills of about the same age in the same environment. This is due to a well known process of shrinking that happens to paper maps and other carefully scaled paper surfaces over time. So, what I am saying is that these bills had shrunk fairly evenly, and by about the same amount as the ratio of the Inch to the Pyramid Inch, or a ratio of about 1:1.0011. And oddly enough, this also meant that for these three bills, no appreciable correction was needed in the initial theoretical mathematics and proportion that I was about to do--and this remained true for quite some years into the future. This difference mostly affects longer dimensions, and is not very noticeable in measurements made in the vertical or height direction.

I was unaware of the Designer's use of "Pyramid Inches," at that time and remained blissfully ignorant of this until I replaced these with several fresh new one dollar bills some years later. This seemingly strange luck of having shrunken bills made a lot of things easier for me to find, things that otherwise I might not have found at all. In reality, it wasn't luck at all. The person who designed the dollar, was a person in the position to *know all about* the shrink characteristics of the paper of our currency and how it would appear in time after a printing--and to later investigators. But--please bear with me--I was unaware of any of this then, so on with the story in inches:)

What was the significance of this prominent length? I wondered. This length of 5.655" was incorporated in the above mentioned golden ratio squared outer rectangle, and just had to be important. (See illus F1).

The ends of the outer rectangle measured about 2.160" in length. Although I could think of a few possible things that were "216" or "2160" with respect to number symbolism, from the even thousandths point of view (e.g., the moon's diameter of 2160 miles, a zodiacal age of supposedly 2160 years, etc.). But I couldn't come up with anything that made sense for this dimension. These outer dimensions didn't seem to be naturally occurring numbers; that is to say, they didn't appear to be regular fractions other than 1131/200ths and 54/25ths which didn't appear to stand for anything so far as I could tell. But it *was* a form of the golden ratio: it was *squared*, being a φ^2 -proportioned rectangle.

27

$$5.655 / 2160 = 2.618055556, \qquad \varphi^2 = 2.618033989...$$

Early on in this little project I had begun to think that the great body of Pyramid lore might be involved in this design, since, if nothing else, there was a pyramid in the dollar's design, and many Masonic authors spoke highly of the Great Pyramid. And several writers (such as Manly Hall) identified this little pyramid in the Great Seal as *specifically* a symbol of the Great Pyramid. There could have been big doubts in the reasoning here, but it was a fruitful starting place. I had earlier begun to learn about the Pyramid of Giza, and the many numerical symbols that were reported to be built into it by means of dimensions, mathematical functions and ratios. I soon concluded there was a real similarity between the dollar's secrets such as I had found and that of the Great Pyramid. This was very much a 'pull myself up by the bootstraps' process, one little piece of information from one place leading to another piece from somewhere else.

Illus. F1: Large Outer Rectangle, (LOR) the outermost rectangle.

Ratio: 5.655/2.160 = 2.618 : 1 or, φ²

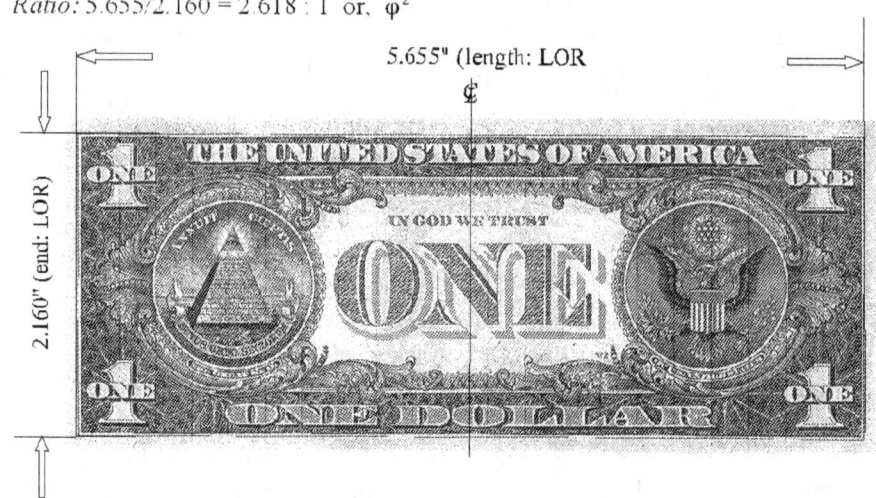

There was a solid tradition in many of the more mystical books on the Great Pyramid, mostly based on the original theories of Piazzi Smyth, a Scottish astronomer of Victorian times. This tradition was later expanded by David Davidson, a theorist and structural engineer, also from Scotland. Smyth was the Astronomer Royal of Scotland from about the time of the 1860's, and he believed that there was a mathematical, prophetic message to be found in the measurements in stonework inside of the hallways and rooms within the Pyramid (See Chap 4).

The Great Pyramid's ancient builders apparently made use of natural functions, such as pi for building proportions, and were evidently aware of decimal fractions [8] as well as squares and square roots of various symbolic numbers. Davidson (as I found out much later), showed that the dimensions of an inner chamber called the Antechamber were intended to show certain of the proportions of the whole Pyramid at one-hundredth scale.[9] Both Smyth and Davidson showed the builders' use of *special offsets* in various places in the Pyramid's walls and floors. And there are other forms of hidden dimensions that one is *left to deduce by calculation* from those dimensions which are measurable. (This last idea is one of the most important of the many shared subtleties between the dollar and the Pyramid, for which we have already seen at least one very clear example in the SIR or, 39.37 1/16ths rectangle).

Looking back on what I had discovered in the dollar, I concluded that I was likely to encounter at least some of this kind of numerical concept built into this design, or at least in regard to whatever the Designer thought was important. Would his message be a short term, temporal sort of thing, or long term and classical? Perhaps religious or political? It wasn't at all obvious, however, what he was trying to get at, with what I had found. But I was fairly sure, intuitively, that I was going to find some "pyramid like" forms of number manipulation. So I set off more or less consciously in that direction.

****

Independence Date Theory (IDT)

After much calculation (on a small, now battered Casio calculator with a green, light-emitting diode display) and much fiddling around with the numbers somewhat willy-nilly, I happened to multiply the measured top dimension 5.655 inches by the constant pi. If it was to be divided by 100--This produced a number that looked a lot like the date of U.S. Independence, perhaps a sort of a decimal "date." And this is completely in accord with the logic used by the Pyramidologists in their use of pyramid dimensions, a very successful first shot in the dark. (See illus. F2)

Pi multiplied by 5.655 gives 17.7657. (Shown here to four places. Thus begins the somewhat silly business of computing dates from the dollar. There are other "dates" like this that we will see later, in different mathematical forms. See Chapter 4.) This number 17.7657 has the remarkable quality of being approximately one-hundredth of the numerical date of U.S. Independence, if it was expressed as a decimal fraction. Not exactly, but what would be about 208 "days" into the year 1776. *(This is where you get out your new calculator.)*

Illus. F2:

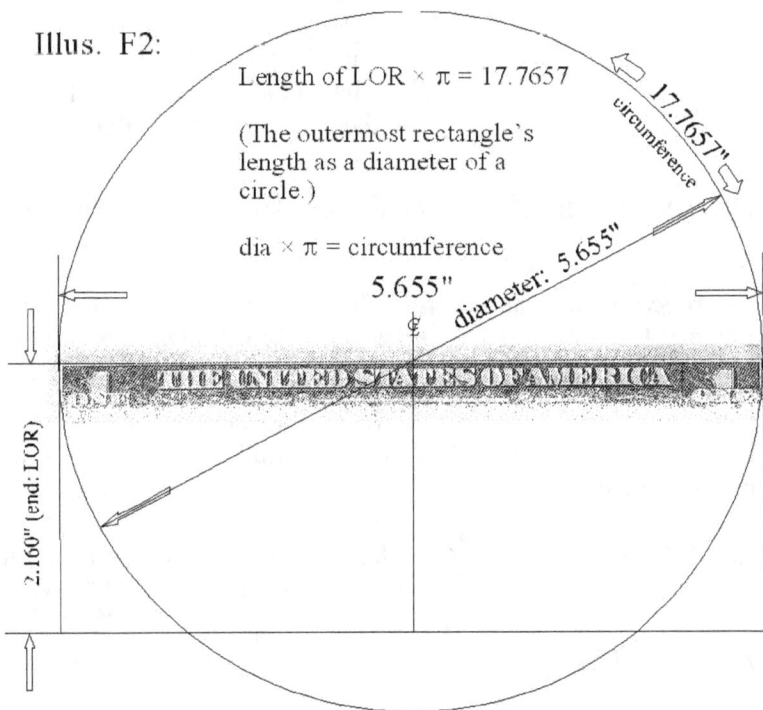

Length of LOR × π = 17.7657

(The outermost rectangle's length as a diameter of a circle.)

dia × π = circumference

5.655"

Working the arithmetic forwards, from the measured dimension:

$$(\pi = 3.141592654...)$$

(m,t) 5.655 x π = 17.765706

(t) 17.765706 x 100 = 1776.5706

then,

The "0.5706" decimal is certainly more than half a year, (i.e., greater than the month "June" at 1776.5). So how much would it actually be if it were computed as calendar days? First we subtract

29

1776.0:

(t)   1776.5706 -1776 = 0.5706

then, multiplying this decimal fraction times the year length--

(t)   0.5706 x 366 = 208.8563

366 being the number of calendar days in the leap year 1776 there being an extra 29th day in February: 365 + 1 = 366 days. The number 0.5706 is the same as 57.06 percent of that year, being about what would be the "two-hundred and eighth day." If this were about the 208th day within the year of 1776, this would correspond to Friday, July 26th 1776. Not right on, but very close to the Declaration of Independence; July 4 would be Day Number 186 of that year. But since we are talking about very small measurements, and since these measurements have been multiplied by more than 300, (i.e., 3.14159 x 100) the difference of 22.8563 "days" if scaled back to the dollar will be *really* small indeed. How much length would this be at the top line of the dollar?

Working the arithmetic backwards from the Declaration of Independence Date:

Suppose instead, *theoretically*, we started from the beginning with "Day Number 186" (or what would be the special Thursday July 4th of 1776) *and worked the whole thing mathematically backward:*

(t)   186 / 366 = 0.5082...

(This creates the decimal fraction for the given date July the 4th Day Number 186 taken to four places)

then, adding this fraction to the year "1776"

(t)   0.5082 + 1776.0 = 1776.5082

(Or, what would be the "correct" whole, decimal date for U.S. Independence: July 4th, 1776)   then--(Dividing by 100) then--

(t)   1776.5082 / 100 = 17.765082

(t)   17.765068 /π = 5.654801...

(And finally dividing by pi. A new "bill length".)

This number is *very* similar to the measured dimension. How much actual difference is there between this new theoretical number and the one as originally measured?

(m)   5.6550   (Apparent measurement, inches)

(t)   5.6548   (Subtracting the calculated "4th of July" or, IDT theory)

         (-)--------------

(t)   0.0002   (Theoretical difference = 2 ten-thousandths of an inch.)

This could not have been a coincidence, I concluded. For the purposes of our ability to measure the lines on the dollar over that distance, *there is no difference*. Obviously, rounding to three decimal places, there could be no real difference at all. If the Designer had intended this second number, I had no way to distinguish it from the first measurement I had read or estimated. I could not, at that time, accurately measure one thousandth of an inch over that distance of paper, let alone two *ten*-thousandths of an inch-- which would be a fifth of that size. To give an idea of just how fine a difference this is--it would be about 1/10th the width of a given printed line's thickness on the dollar bill, with respect to over more than five and a half inches of printed paper, which might very easily expand or contract that much with daily changes in humidity.

I was elated. Who could argue with two ten-thousandths of an inch? If there ever was a definite symbolic significance for a dimension, this sure cried out to be it--or so it seemed at the time. As if to confirm the idea, the base of the little pyramid had the Roman Numeral MDCCLXXVI which is 1776, and it appeared to have a base length which appeared to be about 0.555 inches. If this was so, I thought it might represent the opposite equation:

$$\pi / 5.655 = 0.555542...$$

In other words, *pi divided into 5.655* instead of pi multiplied by 5.655. Yet, on this little pyramid's right hand corner at the base, just where you would want a clear point to measure to, to verify this idea, there is a small bit of "greenery" partially obscuring this corner. A little bush or something. This means that any values taken from there must be approximate, or in some other way calculated from edge lines or other considerations. Here again, another example of our wily Designer's art and cunning artfulness-- which we will revisit later (See Chapter 5).

We should take notice the "reciprocal" nature of the above pair of equations. Even if the above equation does not give a completely correct base dimension, (which we will later see, it does not), the fact that the Designer covered one corner of the little pyramid shows he intends many possibilities to be looked at--one of which may well have been this apparent reciprocal equation.

This *reciprocal idea* will be seen in many and varied forms all through the dollar design. As seen earlier, there is a sort of reciprocal character to the right/left shift of the golden rectangle reference frame, an inverse reference game between two different measurement systems of inch/meters. And there is the tangent line exchange game between the edge of the All Seeing Eye iris/ and the edge of the letter "O" in the word "ONE." And also edges and centers of the letter "I" and other odd places in letters, lines and whatnot shown above.

At this point the Designer's reasoning and choice behind the dollar bill's length appeared "case closed". It seemed that the relationship between measurement and my Independence Date Theory (IDT) were far from coincidental. I was convinced that I knew the one and only answer. As I was to discover much later, this great sense of certainty however, was wrong, or, I should say more accurately, "not entirely true". And in more ways than one.

For quite some time, this wonderful certainty blinded me to any other possible meanings or any deeper exploration of the "5.655" dimension. Since everything was turning out so successfully, this IDT theory blinded me to the need for re-measuring most of the larger, fundamental dimensions of the bill. Having found the one answer, it looked like that was it, and I was stuck with it. It appeared self-evident-- and if nothing else, the date and the base length of the little pyramid seemed to hammer the idea home. And at this point in the gathering of evidence, many may well ask how there could be any other "partly true" answers. Considering the circumstantial evidence, I am sure many of those reading this may have also become completely convinced of this "Independence Date Theory".

But as years went by I finally stumbled across deeper numerical values that were in fact, strangely enough, much superior and far more certain than the Independence Date, as will be seen in the following chapters. However, this was not at all evident to me by means of the measurement process, but only by means of calculation. These later-discovered values differ so very little from the ones above, that one would never guess or measure the difference. Our Designer was into subtleties of math in a big way.

How did the Designer intend the geometry to be discovered from what we know at this point? The key was probably intended to be the outer border. This would yield the curious ratio 1: 2.618, which would then be recognized by some people as part of the golden ratio mathematics. (But this however, would have required a more knowledgeable person than myself. I don't think I would have had any reason to suspect the "1:2.618" ratio was important before my lucky find in the study of the dollar.) Following the recognition of the square of the golden ratio from the dimensions of the outside borderline, the investigator should then be primed to look for the more regular form of the golden ratio, the shorter 1:1.618 rectangle. The next step might be that the investigator notices that the five inch middle area could be neatly divided into two perfect golden rectangles: the 1:1.618 form (SOR). And sometime after this, the puzzling evidence in the centers' areas in the filigree for the smaller rectangle (SIR) would be been found, perhaps then the "meter-inches/16ths of an inch" arithmetic pun, and then the hook would be set. At this point, I think many people would have become interested in the dollar design.

****

As more pressing matters came up, I let my progress in calculations and measuring lapse for quite a while, with the intention of later returning with a fresh view point. I was uncomfortable with all of this and I had the a clear feeling that I wasn't quite clever enough, or mathematically knowledgeable enough to solve this great riddle or even to do justice to it. And I wasn't mathematically knowledgeable at the time, at all. I could see that I was going to have to spend a lot of time analyzing this design before I got to the bottom of it, if ever.

Most parts of the dollar seemed entirely too small for me to measure. With this in mind, I persuaded a friend of mine, who had a darkroom and a photographic enlarger, to photograph the left side of the bill, and enlarge it so that there was exactly 10 centimeters in the enlargement print to one inch on the dollar. This was to simplify the work of measuring, by expanding the reading of 10 centimeters to one inch: where 1 cm would be a tenth of an inch, 1 mm would be one hundredth of an inch, and a tenth of a millimeter would be the equivalent of one thousandth of an inch, etc. Granted, this was a weird scale, (1:3.937 or close to 4x) but it was a handy one in size neatly fitting on to a 8" x 11" photo, and I could easily find good transparent metric scales to use on my photos.

5.655" (length: LOR)

2.160" (end: LOR)

We made several prints, carefully adjusting over many prints by trial and error for the slight shrinkage caused by drying of the paper in the prints. These I found very helpful. After measuring and studying these for a few months, I put it all away--out of sight out of mind.

Notes:

[1.] Divine Proportion H.E. Huntley pg 52, 62-64

[2.] For some reason or other I have never been much interested in the front face of the dollar, or those of any other denomination--or any other currency for that matter. After a few quick looks at other denominations, $5 and $10 notes, my interest in them soon dwindled since they didn't appear in any way out of the ordinary. This may, however, be entirely wrong. But I have never looked any further on other notes or currencies.

[3.] 39.37 is the exact number of inches in the "U.S. Survey Foot" definition as used by the National Geodetic Survey (NGS). Some readers will be familiar with a quite different figure "0.3048 meters equals 1 foot," but this is known as the "International Foot," definition, a slightly smaller standard for a foot, by one part in 500,000. This difference is too small to concern us here--a difference of one foot in 94.6 miles if one were to compare the two types of feet, about 1 hundred-thousandth of an inch in terms of measurement (0.00001") in the total length of the dollar bill. We will, however, look at different kinds of American and British feet in Appendix A.

[4.] What seemed really intriguing, (at least for a while), was the fact I had read somewhere that the Freemasons had a 24 inch ceremonial scale (the "24 Inch Gage"), usually made of metal, typically divided overall into thirds by hinges. I had an idea that this might somehow correspond to the 24.33 sixteenths I had measured along the short end of the SIR inner rectangle. Since Freemasons were involved in France at the time of the Revolution, and this was the period that produced the meter--was there some sort of peculiar connection? I soon abandoned the 24 inch ceremonial scale theory, since all such Masonic scales really were only 24 inches long, as found in Makey's Masonic Encyclopedia, and personal measurement of such scales found in antique shops.

[5.] Or, which is to say, was this is only a trivial result naturally accompanying a rectangle of that length, based on the short dimension of "24.33 1/16ths," when subtracted from the bigger 2.5 length? Or, could it be that the "24.33 1/16ths" height for a rectangle still does represent a very much intentional number game of some sort, such as a mathematical way to add "exclamation points" to this shape? This number trick will not, for example, work for 39.37 1/15ths, or for 39.37 1/17ths, when subtracted from 2.5. If you poke around with these number combinations for a while, it will become clear that it has to do with the properties of the number 16 and 39.37 and apparently, only in this situation. And this is also not an exact relationship, being an excess of 5 in the millionths decimal place. Is this really a "millimeter?" No, but 0.039375 inches is quite close to a single millimeter within a precision of approximately one part in 7,900).

[6.] Now if instead of 1.545" as the height dimension of SOR, *if it was 1.550",* this would actually equal 0.03937 meters! This would have made a stronger and weirder case for the "39.37 1/16ths" idea of SIR at that point in time-- when you consider that this is practically the same, *numerically,* as the pun difference between the lengths of the two rectangles, (0.039375") seen earlier. I was unaware of that possibility at the time. The correct dimension is about 1.545"; --but if measured at the *outer edge* of both of those lines--1.550" *could be* correct, just to give an idea how narrow the lines are (+/-0.002") and how very careful one must be when measuring. Currently I think of this as another possible subtlety by the Designer, such as several other situations we will see later where oddities in math hover in and around line widths and edges.

[7.] "Serifs" are the little stylized edges and hangin-off corners of the capital letter shapes that were handed down to us from ancient Greek and Roman stone carvers. These sculptors cleaned up their lettering errors at the top of their strokes by rounding and polishing where they over-shot the tops of the letters with their chisels, thereby producing the serif--an artful fix of a sloppy error, now a hallowed tradition in the Roman, Greek, Cyrillic and other letter styles.

[8]. <u>The Secrets of the Great Pyramid</u>, by Peter Tompkins, and elsewhere.

[9.] <u>The Great Pyramid and it's Divine Message</u>, by David Davidson, Plate XLIII (Approximately Pg. 208)

CHAPTER 3: PATTERNS AND LIMITS

Graduating to Note Taking

Making Some Rules

Finding Small Differences

Discovering the Great Pyramid's Signature in the Dollar[1]

Progress in investigating the dollar's design has been one of fits and starts, all the way along. For quite some time at the beginning stage of my research, I kept my ideas and discoveries about the dollar almost entirely in my head, working and re-working ideas mostly from memory working with hand calculators. This seemed reasonable at the time, since perhaps I thought I had a good memory for numbers--and, maybe all of this was a passing flight of fancy. I would soon come to my senses. I wanted to solve the puzzle and go do something else. I was digging for treasure and more sparkling baubles kept appearing. There was something to it. But I had no idea of the size of the project I had embarked upon, and it wouldn't quit. Nor, for that matter, did I have any idea what I might do with whatever it was that I was learning about the dollar. I had written down a few numbers and drawings and other things, like questions to myself on scraps of paper, but that was about it. Soon I couldn't keep track of it all. By now there were multitudes of variations of mathematical ideas, and multitudes of measurements. Soon I found myself rediscovering half remembered lines of thought. I knew I should be writing this down, but it was difficult. I didn't like the idea at all, since it seemed like it was going to become a whole lot of work if I was going to track my progress. Working for a surveyor allowed me to see the profound importance of notes and note taking.

Soon I filled a few crabbed notebooks with my intermittent discoveries and measurements. Notes are surprisingly useful for digging through various number patterns. Notes allow you to cast aside unproductive paths. They help uncover superior routes through all the data by allowing you to look at the whole thing. Natural-born researchers probably know this sort of thing right away from birth. But for me, the most important aspect of "notes" was that it simply allowed the evidence to stare you back in the face. Basic stuff for a lot of people, but a real discovery for me.

Once it is there on paper, it won't go away; the evidence dares you to review the yet-hard-to-believe-discovery to see if you think it is still really true. Also, notes allow you to find out what could happen if you were to re-calculate your idea with improved measurements or a different formula. This little paper trail proved to be important for me, since I was initially doubtful about the many things I found. I was enthusiastic in a way, but uncomfortably skeptical. The discoveries seemed like *very* unlikely things to find on a piece of U.S. paper money. How many more new and positively strange things could there be on the dollar anyway? This is to say nothing about the significance of what I was finding. In the beginning I wouldn't allow myself to think about that at all.

Much later, deep into research, I had begun to find my thinking thoroughly fixed on or crystallized around certain theories. I had become unwilling to consider any alternative ideas. Even to the extent of shoveling a small amount of things under the rug, ignoring some of the evidence. I had become unaware of doing this--these ideas had gone to the "file and forget" part of my memory. So it was looking over *the notes* that provided a very important shock on certain crucial subjects--a new look at old thoughts, a revisiting of my original evidence. Not all of my notebooks are worth that much to me now, but they allow me to see where my thinking was earlier. You won't always know the value of what you are looking at right away, *but you might sometime later,* after the passing of months or even years. This happened to me over and over. Without such notes to look through I could have lost a lot of what I originally found and forgot many crucial ideas entirely. This is the "in your face" theory of note keeping.

In good survey note taking, you are supposed to keep a notebook, number its pages, date the observations, draw little sketches, and write down your measurements and arithmetic, showing all your work. You are not supposed to erase your mistakes, instead you are only supposed to draw a line through them. You should also keep track of conditions, or anything that might aid your or someone else's ability to reproduce and verify your findings. It is very important to date your observations and calculations so you can recover what you did, and in the order in which you did them. Most of my note books were the cheap 120 page 3 by 5 spiral bound memo books, convenient kept in a top pocket with a calculator, in case I came up with an idea while doing something else. Since "the light bulb" goes on only occasionally for me, and at very odd times, this was a useful strategy. If you don't write it down, it will go away.

****

At that time, I worked for a surveyor in the Chicago area, who truly enjoyed the odd little complexities of the small lot and block-type surveys. He specialized in the arcane art of interpreting legal descriptions of deeds to parcels of land in city subdivision situations. His intoxicating enthusiasm for these subtle puzzles was infectious. I learned to search for the "Original Intent" from him, and to even-handedly examine the existing traces of evidence found on the ground. There might be about twenty important variables in any land survey, plus loss or ambiguity in old records. It was from this background that I was fortunate enough to start. From hindsight I can see that this had the effect of training me for the search through the dollar's puzzles. I picked up a lot of methodology from him and his equally colorful partners and employees. Much of their style of thinking, ideas, tricks and shortcuts have been very useful to my reasoning process in this investigation.

I had an extra room in an apartment on the north side of Chicago that I had originally set aside for artwork and other hobby projects, and this area eventually became mostly devoted to measuring the dollar. I had evenings and weekends free to spend on this investigation of the dollar if I wanted, but I also had ample distractions with other things going on in my life, so I wasn't always at it and often forgot about it for long periods of time. I suppose it has been like that through all of my research. Since the dollar's riddle had become important to me, at some point or another I was going to try to get serious about the research, and figure this thing out.

****

Questions about the quality of measuring tools:

In the Autumn of 1976, I cleaned off a space on my desk, and spent several months thoroughly re-measuring the original three dollars in my envelope. At first, I mostly used a drafting scale, ruled in 100th of an inch divisions and a small, 5 power stamp collector's pivoting magnifying glass, under bright desk lamp. Soon, I became concerned about the *accuracy* of my drafting scales. How much better were these than the dimestore rulers? I had machine shop experience some years before, so I was already quite sensitive to questions of small scale measurement. Looking over my various scales and rulers, I wondered, "How long is a real *inch* anyway?" I don't think I had ever questioned measurement units before. Where you would locate a true "inch" with any certainty if you really needed to find one? Over time, within a deeper unfolding context, the inch question became quite important.

In the land surveying work with which I had been involved, the only concern had been whether our 100 foot steel tapes matched up with two special, chiseled marks in a concrete walk near the store front of our survey office. The length of a steel tape will change a little bit over time, and small errors will ultimately add up as an accumulative error over any long distance. So you will always want to know what kind of correction to make for measurements, or when to retire an old worn-out tape.

Most steel tapes we had did more or less match this distance when compared to these marks, or at any rate, the differences were quite small. In an older, well used kinky tape this could be as much as plus or minus 0.01 feet, (approximately an 1/8th of an inch), over 100 feet of steel measuring tape. But usually less than half of that. Even when corrected for the temperature at a job, in practice this might amount to a taping error of about one part in twenty thousand, in an older tape. How did we know that these marks in the concrete were any good? These particular marks carved into the sidewalk were said to have been put there as a professional service for our company by a small, grumpy, semi-retired surveyor of some local renown, a Mr. Feeney, with the help of his elderly assistant chain-man, many years earlier. They were said to have used a special nickel-steel alloy (Invar) measuring tape. This unusual (and somewhat fragile) type of tape, if used correctly, can be used to ensure great accuracy. Beyond that, we at the survey company didn't use any other standard check. Many other surveying outfits that I was aware of at the time never bothered to check their tapes at all. Once out of the box, the life of a surveyor's steel tape in Chicago could be short and was soon replaced.

So, the steel surveying tapes were generally good with respect to these marks, on a large scale--but were my drafting scales correct for inches, or did they vary all over the place? The "dimestore rulers"

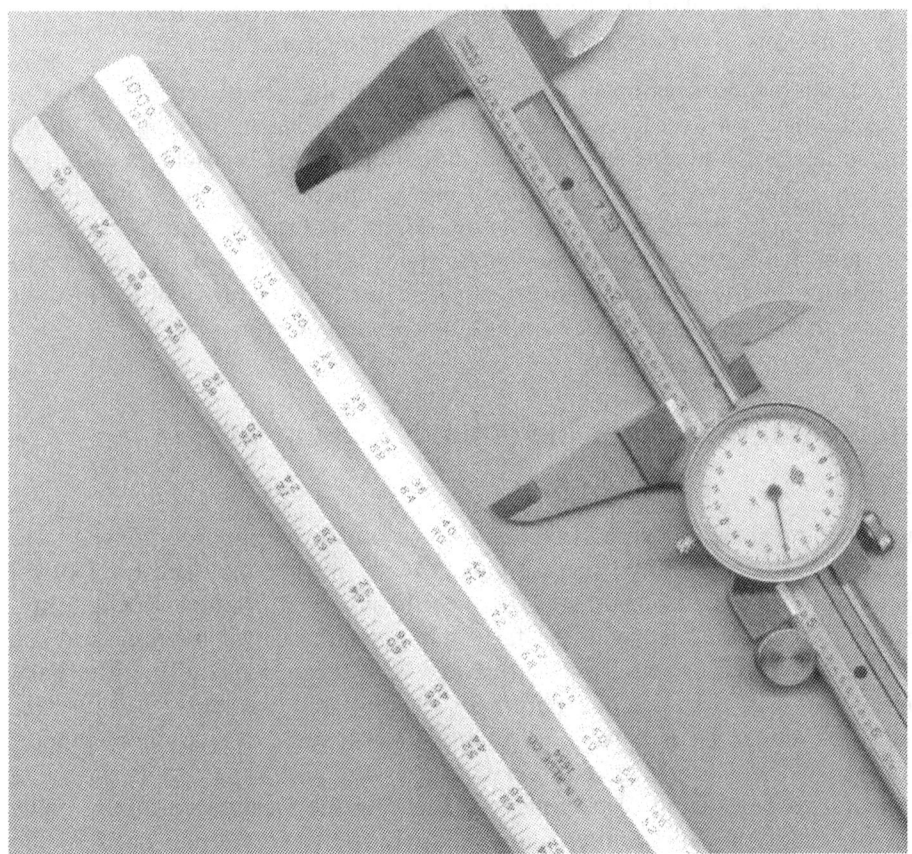

were no good and could not be used. Since I was going to redo most of my previous work on the dollar, I had reached a point where I had to make an effort to know what these values really were. I would be working right at, or at least near, the level of precision where small errors in the tools might be important. Making drafting tools must be a delicate art. If a company that makes drafting scales made a mistake in their length standards for a scale, how would I know? What are their tolerances in the production of drafting scales?

I had a friend named Stash, who was a old-country machinist in the neighborhood who had some gage blocks and several other measuring tools that I could borrow. With the aid of his low-power shop microscope I made many comparisons from the blocks to my scales. "Gage blocks" are especially ground small steel blocks (called Johannsen's Blocks or "Jo Blocks") that came in a red velvet lined box, which were certified for accuracy to plus or minus five, one- hundred thou-sandths of an inch. These blocks (when new) are ground mirror-flat, so much so that if two are touched together they will stick together due to molecular forces or vacuum effects.

All of my expensive scales and dial calipers were surprisingly good with respect to the gage block standard. When compared to the blocks over a distance of one or two inches within the middle of a scale, I couldn't measure any difference at all. With several gage blocks, calipers and other equipment, however, I found some differences in a few cases of a ten thousandth of an inch or less over a length of six inches or so. But even with my good drafting scales, I noted that the *end areas* were occasionally imperfect. Yet most of these small differences were not all that important, since this level of precision was about three times better than that of the surveying steel tape error described above. I was merely hoping to approximate measurements of one thousandth of an inch over features of six inches in length, (one part in six thousand) and I was more than happy with the accuracy of my scales and calipers.

****

Now reasonably certain of my scales, I began to completely re-measure the design, ignoring a lot of the work I had done earlier--basically starting over.

Only a small amount of measurement like this can be performed at one time. As interesting as it was, I found it to be very uncomfortable work. I had to hold completely still, holding my breath. I had to crouch down and line up the dollar, scale and magnifying glass on a flat surface and had to choose some number that I thought was right. Later, I had to compare a multitude of readings, throw some out and make averages--all very tedious. It was a lot like target practice with a rifle--tiny target practice. As I went on, I became uncomfortably conscious of a tricky problem: *the truth* in measurements.

Measuring with scales turned out to be a poor good method of getting measurements. Soon, machinist's dial calipers replaced the hard-to-use scales. I lusted after "measuring microscopes" during that period, sometimes seen in some tool and die shops, the kind with a fixed vertical microscope mounted on a heavy traveling slide. These make high precision measurements over areas of a whole foot. But these were much too expensive for me to buy, or even rent. Hand held magnifying comparitors with built in reticles (cross-hair and measurement markings) also appeared to be attractive for small measurements, but careful examination of these showed surprising inaccuracies and difficulties in their use. With few exceptions, I stuck with dial calipers and occasionally, precision drafting scales.

****

Searching for a fundamental pattern in the dollar's design

General Limitations:

During this period I began to become sharply aware of the *accuracy* problem as opposed to *precision*. We could call these "correctness" and "repeatability" respectively. Many people may see little or no difference between these two ideas, but they are miles apart. There is always a real limit to repeatability. Very fine or precise measurements don't help much with the fat, fuzzy lines that I was beginning to see that the dollar was composed of when viewed at higher magnification. But these lines did seem to have an original intent, or exact basis. In fact, that was the nature of the dollar's problem--to accurately measure and ascertain just what the intended basis was. Past a certain small scale of size, more precise tools *would not* get more accurate answers. I would have to verifiably *predict patterns* in many places in the dollar's hidden design by theory, to be assured of accuracy in guessing the Designer's intent. This problem appeared to be impossibly difficult with smaller features, such as the very small parts of the Great Seal, which mainly required computation and educated guesswork.

I was beginning to realize that my ability to measure in a meaningful sense had limitations. By determining the accuracy of my tools, I was attempting to nail down one side of this problem. However, I began to realize that even if I got a hold of substantially better tools than a dial caliper, which could have been acquired if I had needed them, I would still be stuck with another variable. This problem was the "imperfect" nature of the dollar itself, even if new with sharp printing. Past a certain point in magnification, the dollar's printed lines have a certain thickness, and will appear grainy at the edges. At some certain smallness of size, the meaning of *accuracy* in the measurements would then begin to overtake the meaning of the value of *precision* for measurements. Being able to measure things down to the high precision of fourth decimal place (one ten-thousandth of an inch) is of little use, if our true knowledge of where that location is still unclear at the third place (one thousandth of an inch). So I recognized that ultimately, these lines were *accurate hints*: close approximations made by the Designer of exact mathematical ideas. It seemed that he liked *pure ideas,* and once he gave a hint--like a relationship that appears close to π--it is like poetic licence. He wants you to assume the true, exact number. If you find a graphic relationship, it is as though he had written an equation.

Mathematical Extrapolation Limits:

Next, a fascinating problem of design theory began to unfold at the calculation and analysis steps. Over time, I found several good, competing theoretical mathematical explanations for certain shapes that I had measured. Often these competing imaginary relationships seemed to "cluster" around the edges of the same apparent figure. That is to say, when I could produce a nice theoretical "length" out of mathematics to explain an apparent measured length, I would often find several others just about as good that would still closely fit the figure, for very different but equally good reasons. Sometimes these would fit *near or within the small width of those printed lines.* This result happened often enough that I couldn't brush it off as coincidence. The tiny margins of difference between the various ideas were often smaller, sometimes much smaller, than anyone's ability to actually measure. This was therefore smaller than my ability to verify the Designer's intent in any direct measurement-based way. Now, I had to find good mathematical answers for the things I was measuring. I was forced to figure out *something solid* about the Designer's theory; a bonafide prediction of how he had hooked all of his ideas together.

Later, I became certain that some of this clustering of possible answering formulas was intentional. Which is to say, the Designer appears to have chosen certain specific ranges of values, ratios and other small subtleties of geometry *just for this puzzling effect.* Even to the extent that eventual shrinkage of the paper of the dollar had been accounted for and intentionally used as a device. The Designer seems to have gone to an extraordinary amount of effort to get these effects.

Measurement Limits:

In my situation--far from being privy to the Designer's thoughts--there was the width of the lines, contrast visibility of the caliper jaw edges, proper lighting and the visual angle of observations. These were real challenges to investigation. Over a general range of about two inches or so, a real limit to the fineness of my measurements eventually become clear to me at a point somewhere between the one-thousandth (0.001") and three ten-thousandths of an inch (0.0003"). Much larger or smaller than this length range, the overall accuracy that I was finding fell off somewhat. There was more than enough area within these magnitudes for error or interpretation. But this range suggested that I was probably reliable for one-thousandth of an inch, at any rate, within as far as three inches. Then, beyond all of this, there was another question of *just what it was exactly* that was being measured.

Lines and edges:

When I first started out, it seemed that the only reasonable thing to do was measure from the apparent *middle* of a line on a feature, to the apparent *middle* of a line on the other end of the feature. And think no more about it.

This appeared to be the only legitimate method of measurement that is incontrovertible. As time went on, I found it was very likely that there were certain dimensions the Designer must have intended to be considered from the outside or inside *edges* of certain lines. (Later we will see that the inner border line or LIR, lends itself exclusively to measurement at the *outer edge* of it's printed area, since this is the only clear definition for this rectangle. This edge seems to have been *provided* as a standard reference for measurement in many geometric problems.) These fine points may be hard for some to accept, much as they were for me.

But this peculiar aspect of the design and the above mentioned theoretical "clustering" of possible alternative mathematical ideas will become increasingly evident later on in the story. *Any* pair of clear, discernable edges are measurable, of course, and it appears that the Designer was often willing to use both the center and edges of printed lines. Occasionally he will suppress one edge or end of a feature through obscurity, (such as the inner edge of LIR) or even *both* ends of a figure, to provide a search condition for a puzzle problem.

A Theory of Continuity:

The Designer's work was wonderfully consistent. I began to try to picture a theory for the whole design. What I was searching for was the *intended pattern* underlying the all of the shapes and line work that could predict hidden relationships. What I was looking for was an idea that could be verified as a proof that I was on the right track, more or less like the SIR/SOR sliding alignment arrangement. I became convinced from my beginning experiences that the whole picture must follow a single design concept of some kind. The Designer didn't have to do all of this in a mathematical way, it could all have been "art" and arbitrary design, as I am sure must be true for other bill designs. But it appeared that it was, in fact, mathematical and geometric and it seemed that had to be all of a piece. With the gathering evidence, it seemed as though the whole design was probably solvable from one end to the other as a single unit by a key concept, *if you knew what it was.* I had some ideas, but I lacked proof or a valid means from which to prove them.

****

Figuring out some rules:

At this point, all I had found was in a logical tangle. I sat down and decided to take stock again of what I knew about the dollar's design. I had to prioritize and do a triage of ideas: not all I had found had the same importance. I began to see that this puzzle problem was expanding with complexity with every step I took, and that some general, logical plan was called for to sort all this out. And, I have to say, I have resisted organizing and limiting my thoughts on this project the whole way along. But the acquired clutter of measurements, numbers and tangled reasoning was now unbearable, I couldn't tell what I was looking at any more, and had to do *something.*

Evidence is a lot like loot: its nice and wonderful in a great, gleaming pile. But eventually you will want to count it out into categories: e.g., large bills, bearer-bonds, piles of bullion, heaps of gems and so on. Until you catagorize, who knows what all is present? I gradually began to compose some rules to narrow the field and to tried to focus on the fundamental nature of these puzzles. At first this was an uncon-scious process, but then I began to look over my shoulder self-consciously with respect to what ideas I was accepting and just what I was willing to accept. Some of this feeling was by "osmosis" from working in the land survey world, with it's special rules of evidence--and I was fully aware that I could fool myself. I didn't know of any standard guidelines for this sort of thing, but I had to do something rational. And I had to *justify* to myself how I was wasting my own time. The organization I finally settled on helped quite a lot. It now appears it was crucial to the process of thinking these puzzles out. Re-organizing my thinking and evidence, I found that many theoretical ideas I had begun with still seemed justified, and even now appear to have been realistic ideas. Many others, however--now mercifully forgotten--were completely unjustified and went to the trash.

Justifications and Rules:

As a sort of cannon of fundamental examples I had already seen up to this point, these are some types of "justification by example" (J) that I began to use to lead and filter the search. These were avenues of research gradually arrived at by experience that I decided were reasonable to look into. Most of the following we have seen in one form or another already. These provided general tests for the direction of dubious ideas. These tests were excuses or justification for following certain broad lines of investigation or accepting some kinds of mathematical extrapolation. The tests became general definitions supported by early observations and surprises. Some of these categories overlap a bit, and this list was then later added to, subtracted from, and occasionally modified as time went on. Apart from many hidden assumptions, this is what I ended up with around the time of the writing of this book:

J1: The "**shown vs. hidden**" example: phi (φ) ratio and the Small Inner Rectangle/Small Outer Rectangle arrangement and math pun. Apparently this is a fundamental *puzzle principle* of the Designer: First there is a *basis (a)*; a clearly measurable graphic with a proportion or symmetry. Second, a following *hint (b)*; which is from another associated graphic feature with an only partly corresponding clue, such as a contrasting length. Another form is found through reversing symmetries (see J3), where the regularity and symmetry is the (a) part or *base* of the puzzle, and the reversal or irregularity is the (b) part holding a *hint* that must be studied. This is then should be followed by discoveries that demonstrate a validation of the puzzle by geometric construction, alignments and measured or computed lengths. (See the evidence and procedure in the first two arguments of Chapter 2 that cover the SIR/SOR rectangles.)

J2: The example of "**motion**"; geometric translation (virtual motion or "sliding") of shapes. This is demonstrated a by clear, *alternate alignment* of a mathematical framework (or angular orientation) to otherwise seemingly meaningless design features, as seen in the alternate alignment positions of the inner rectangle. (See Illus. D and E of Chapter 2, another aspect of SIR/SOR.) In time, the discovery of the concept of "initial conditions as opposed to shrinkage" appeared to fit in here fairly well.

J3: The various examples of **reciprocal** mathematical equations i.e, $1/x$, $1/x \cdot 10$ and $1/x \div 10$, $(1/x \cdot 10)^2$ and $\sqrt{(1/x \cdot 10)}$, which we will see more of later. A big theme in the dollar. Also fitting in here are: backwards equations; inverse alignments; or even symbolically inverse relationships such as the All Seeing Eye's Iris edge/edge of the "O" in the central word ONE; opposite symmetries, etc. (See the edge alignments in Illus. D and E of Chapter 2); This is perhaps *the most important design idea* in the Designer's scheme--possibly the root idea behind J1, J2 and J3.

J4: The distinct appearance of an element of **arcane mathematical lore** or very unusual mathematical knowledge. A good example is the phi squared (φ^2) ratio found in the Large Outer Rectangle (LOR). This is a ratio one wouldn't expect to find just anywhere--an obscure ratio based on a function the classical ratio squared. There are clear, but very unlikely mathematical relationships, as opposed to superficial or random relationships, such as the unique proportions of the Great Pyramid (π) and so on. Also, there are such special ratios such as φ, $\sqrt{2}$, $\sqrt{2}-1$, $\sqrt{5}$, the log e and log 10 to be found in various places. We will see special ratios of scale, such as 2 fold, 10 fold and n^2. Davidson's Displacement Factor (286.1) and the odd scale of Pyramid Inch values like 1:1.0011 will eventually fit in here in time.

J5: The **abstract math projection** example: "pi (π) and Independence Date" relationship-- *purely mathematical entities* being based only on (I) a single measured graphic dimension, together with (II) a mathematical function and (III) an easily identifiable resulting mathematical identity of some kind. There is, for example, no seventeen inch circumference circle *actually shown* in the dollar design. *But* a perfectly credible arithmetic result that looks like a numerical calendar date that appears when divided by 100 like "July 4th, 1776," having it's origins in the use of π, and a measured graphic length used as a diameter. This may be considered as a J5 verification. (Other date-based numbers surface in the dollar's design later. Also, a use of clear proportions, ratios, powers, strings of powers, roots, logarithms, etc., that will produce recognizable ending number related to dollar lore. Many startling examples can be found in Chapter 4 onward, and in the Appendix A and even Appendix B.)

J6: alignments/offsets--(a) The repeated examples of "**alignments** at the edges," small, telling alignments to O's, E's, letter serifs, vertical or horizontal, tangents and diagonal lines, as well as (b) small, peculiar **offsets** at regular distances found in the design, that the Designer used to facilitate some message. Perhaps these two are not really the same thing. But they appeared to be a functionally unified theme in the Designer's work, and I have generally thought of this as one idea, since these involve a kind of (a: basis) touching or (b: hint) *an implied touching* by the addition, subtraction (or even angular rotation) of a special line at crucial points in the dollar design.

J7: Messages, themes or motifs found, to be looked for again. The pun: "the center and edges of an I" (center and edges of an *eye*) is this sort of message. (See Illus. D and E in Chapter 2.) Some of the repeated alignments suggest a sort of "symbol language" that may more fully reveal itself in time. (See the discussion for the "O's," "E's" and "I's," etc., in Chapter 2.) These, coupled with multiple references to certain numbers in dimensions, angles, ratios, trigonometry functions and other odd, but consistent facts point to a language or at least a motif containing a message. We should be alert for religious and philosophical symbols or sectarian connections: i.e., Hebrew, Christian, Masonic, Deistic, Pyramid lore based, or whatever other school of thought might happen to appear in the design of the dollar. For example, certain geometric figures and dimensions were found to apparently make reference to specific Biblical passages through the means of Pyramid based lore. (See Chapter 3 pg 74.)

I will sometimes note these search justifications beside my investigations and discoveries or in discussion of them. Later in my investigation, I found myself vaguely scoring discoveries in my mind by the number of J relationships that appeared to be met. I have never been completely sure that this scoring made sense, but I have always followed evidence that appeared to be compounded forms of these coincidences. The more the coincidence, or the more a finding seemed to stand out, has generally lead me to conclude that I was closer to being on the trail of something. But these above thought pictures were not enough. The above "J" argument values I began to reduce to acceptable "rules of engagement," hopefully a tighter yes/no test of acceptability to narrow the unruly clutter of ideas and apparent findings that I had.

By the 1980's I had finally settled on some rules distilled from above justifications and some surveying ideas. None of this seems important until desperately needed. These started out as vague notions more or less jumbled together as one idea, later resolving into rules as different problems came up:

Rule 1. Expect and accept only exactitude for problem conditions.

Observation: It seemed evident that the Designer was solidly committed to true graphic precision in whatever puzzles or message he left us. All of the graphics appeared to be constructed within the precision of a machinist's or diemaker's ability to measure. From this observation, (R1) I decided to *dismiss out of hand* any route of inquiry on puzzles not based on an exact relationship. For an acceptable problem condition for a graphic geometric relationship, I concluded that a minimum a *measurable* relationship should be as clear for precision and accuracy at plus or minus one thousandth of an inch or so, (+/- 0.001"). Occasionally, as an exception, if an obscure problem condition appeared to lend itself to exact inferences, with eventual graphic proof, Rule 1 could be ignored for an immediate reliance on Rule 2 for a starting place.   But I wasn't going to let inferences be too far fetched. I would try to follow Occam's Razor [1] to a certain extent.

Rule 2. Expect and accept only mathematical rigor.

Observation: Exact natural functions, such as squaring, roots, arithmetic reciprocals, pi and phi and so on were evidently being used correctly and with great care. In places that there were precise graphic geometry elements on the dollar, there was a parallel mathematical relationship. The *exact mathematical statement* seemed to be the crucial idea, with the graphics given as clues. Therefore, (R2) it appeared safe to say that this design invites almost any sort of exact and reasonable mathematical inference from known functions or constants that would have been technically available at the time of the dollar's design composition in the 1930's. And this is if, and only if, these are based on good measurement supporting reasonable inferences made from the design or lore related to it. It appeared reasonable to place an arbitrary lower limit on acceptable *mathematical* accuracy having a ratio of about one part in a thousand (1:1000) for scoring a hit in locating a mathematical identity, such as a function or a recognizable "date" as we have already seen, or some other important recognizable number.

Rule 3. Expect and allow only definite graphic proofs for confirmation of ideas.

Observation: With the detailed and multiple graphic alignments evident in the sideways shift of the Small Inner Rectangle, the Designer showed that he is *not at all* making vague hints--so, *when you do* figure out what the Designer had in mind, he rewards the investigator with *unmistakable, and exact confirmations of his intent by clear graphics*, as seen in Illus. E. Therefore: (R3) I would accept as valid *only definite, graphically based demonstrations of intent*, measurable and/or mathematical. This is the Designer's apparent modus operandi. No general approximations and no excuses for why an idea didn't work. (1) Any acceptable problem conditions would have to meet the three digit rule above, per R1; and (2) an acceptable error ratio for a possible proof of some kind, that is as close as either of these above two limits in R2 and R3, and (3) clear graphic evidence within the above limits to support the conclusion.

Error Equation use:

I often use simple "error equations" to see what I am doing. Many are given in this book to make logical points. (In statistics there are a lot of possible subtleties of distinction and many fancy equations that the purist might like, but we will not get in to those here. An error equation is just a simple mathematical "magnifying glass" to track precision and likelihood.) The above rules involve error measurement in several ways. Rule 1 could be called the simple *measured standard* for problem condition limits; Rule 2, a *calculated difference* to the theory limit; and Rule 3, a limit of physical demonstration using *error ratio* to theory. I look at *error of closing ratios* as a kind of guide, there being little else one can use to measure how "close" your theory is to reality of measurement. The *closing error* is the tiny measured difference to theory: *the error ratio* is it's division into the whole. The first R1 standard can be found by subtraction, and R2 and R3 are found by division to produce ratios. Here below, for example--the use of a ratio under R3, in a statement like:

"my theory is as close as one part in 2500."

This statement would mean that as *a test of the idea*, you could take the theory number "t" divided by the measurement "m" as found in reality, which would give a fraction of 1/2500. Lets say I measured 5.002 inches on the face of the dollar for some feature (like the *measured distance* between two lines), *but* that I had calculated for some reason that it "should" be 5.000 inches. (Or, *I thought or theorized* the Designer was trying to say "five inches"in his graphics.) How close was my guess?

$(m / t - 1)/x$ m = 5.002 and t = 5.000   (This is the Simple Error Equation and data.)
     let's say, 5.002"/5.000" = 1.0004,   (5.002" divide 5.000 equals 1.0004)   then,
     1.0004 - 1 = 0.0004   (1.0004 minus 1 equals 0.0004)   then,
     0.0004/x = 2500   (1 divide 0.0004 equals 2500)
which is 1/2500th, a one in two-thousand-five-hundredth part difference or, a ratio of "1:2500".

(Here of course, I am assuming that the Designer made no technical mistakes, and means what he says with his line-work and our measurements are good.) So, I would have to say that by R3 that this was an OK guess--this is better than twice as good as one part in a thousand. But suppose instead as a second case, "m" was instead measured to be 5.008". In a comparison to theory like the one above, this would give only 1:650, and one part in 650 *would not be close enough* to demonstrate anything important by the three adopted rules. I would want to be about twice as close as this to theory, wanting a number at least like 5.005 as "t" for Rule 3. (Which is to say, accepting these rules in this second case of 5.008", my theory must be *wrong* by my own standards.) The standard of 1:1000 was really too loose. In time I came to expect far higher standards in results for a real "hit" from what I actually found on the dollar, but this was a *minimal* guide for acceptance.

This is a useful tool in trying to figure out whether something really matches a theory. It gives you a representative fraction of the error and this fraction can be compared to other error fractions. This method can also be looked at as a way to guess the likelihood of whether one would happen to land on the number in question by accident. Higher error fractions, say, "one in ten-thousand" or "one in five-hundred thousand" are *a lot less likely* to be happenstance, maybe showing very real intent by the engraver.

The above rules were arbitrary limits, but I wanted some kind of "go or no go" way to test the validity of what I was finding. Otherwise I would be accepting too much. In reality, I mostly found myself wanting to accept much less than what I was finding. These rules forced me to a higher standard of research. As time went on, most of the really interesting things that I have found greatly exceeded the above standards. (Over time, and with experience, I came to believe another side of this: That is, I thought it was safe to say, from what I had come to know of the length-language of the Designer, *the minimal numerical precision in length that he was working to* was probably one thousandth of an inch (+/-0.001"), like a "standard unit." It is even possible that he might have been working closer than one-thousandth of an inch in certain cases. Most, if not all, of his puzzles appear to work out well using even thousandths of an inch.) Not shown here, were the many early, dearly cherished ideas that were then nixed by using these rules--but nothing that couldn't be done without.

But lots of bigger questions forced their way to the surface in my mind.

(1) *Who* designed the dollar?

(2) What on earth prompted the Designer to create such a complicated, secret pattern?

(3) What was ultimately being *said* with this design, or what is its intended purpose?

(4) Also, who else knows about this dollar design mystery, or how widely is it known?

(5) Does the U.S. Treasury Department know of any of the hidden parts of the design?

It seemed that most of these questions were unanswerable, except the first. But so far as I could tell at that time, the various books I had found on U.S. currency--while informative about the various original *engravers* for the dollar--but none gave any clue to the identity of the actual Designer. At this point I think that for questions (2) and (3) it might be said that perhaps the Designer was making something of mathematical beauty like the stained glass of French cathedrals--a thing of religious, patriotic or philosophical devotion.

Finding little of use in any of the information I had looked up, and not having much patience for research, I followed up other clues that related to the dollar. At the time--strangely enough--I didn't really care to know who the Designer was, since I didn't want to loose sight of what I was doing or divide my investigation of the design. What I was finding was obviously not something that our Designer was likely to have admitted to in any case.

I tried many mathematical schemes and multitudes of variations of those schemes trying to find a clue or a proper connection to any tradition that might help explain the Designer's intent. During that time I became more familiar with pyramid literature, the golden ratio and many other related arcana. In time I assembled a list of incontrovertible discoveries and I began to try to build on them to find a pattern.

****

Small Differences--Beginning to Analyze Interior Patterns:

The Large Outer Rectangle (LOR), Length Diagonal,

and the Small Inner Rectangle (SIR): Length

This was the first evidence I had of a pattern, or a string of ideas. Some patterns on the dollar were fairly easy to see. The Large Outer Rectangle (the outer border of the dollar) produced **a measured and calculated diagonal which was the *square* of the *length* of the Small Inner Rectangle,** or very nearly so. (See Illus G1)

(measured, m)	$a =$	$\sqrt{2.160^2 + 5.655^2} =$	6.054" and,
(theoretical, t)	$b =$	$(39.37\ 1/16\text{ths})^2 =$	
(t)		$2.460625^2 =$	6.055 (to three places.)

This left a small error of about a thousandth being one part in five-thousand. (Right at R1, but well within R2 and R3.) What was intended here? Was this small discrepancy important? If one "officially" used the "39.37 1/16ths" number as a basis, (making another theoretical diagonal of about 6.053470404") then what Outer dimension length would it make from a similar rectangle? Using the same proportions and working backward from the above figures to make proportion c:

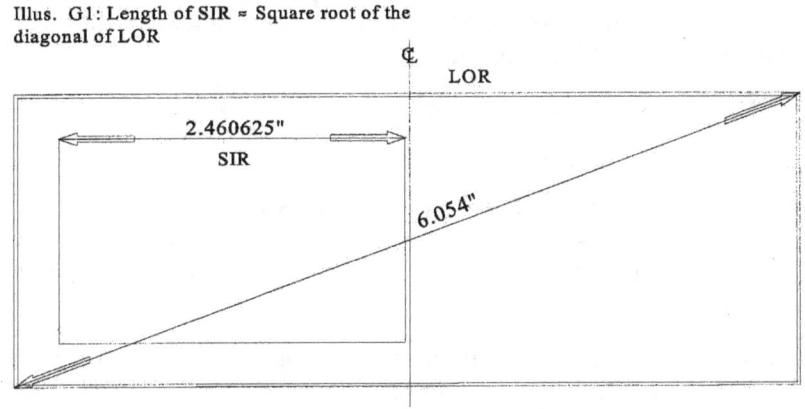

Illus. G1: Length of SIR = Square root of the diagonal of LOR

¢

LOR

2.460625"

SIR

6.054"

(t)	$a = 6.053480404"$ and $b = 6.054675391"$:
(t)	$b\ /\ a = 1.000197405 = c$

Then, c, a scale ratio, multiplied by the apparent LOR length: (or, what would become of the length of the same kind of rectangle having this slightly larger diagonal. This process is called **"scaling"** and is like using the ratio to "inflate" or "deflate" a shape--while keeping its proportions--to see if that results in a better explanation for the measured evidence.):

(t)	$c \cdot 5.655 = 5.656116325"\ (=d)$

This new number meant absolutely nothing to me, at all. (But it *might* have provided a clue for some other things that I found later, if I had played around with it a little. See Chapter 6, "Logarithms.") It didn't verify or improve the Independence Date theory--since to be closer to that figure, a slightly *smaller number* would be required, rather than this slightly larger figure.

This was an unresolvable quandary: The two rectangles' lengths seemed to imply each other from the larger rectangle's diagonal--since the Small Inner Rectangle's length was near the square root of this diagonal. But on the other hand, if you used the measured rectangle's computed diagonal (or a very similar diagonal using φ^2 as a proportion and the Independence Date length), you wouldn't get "39.37 1/16ths" as a length. (I had become somewhat fond of this "39.37 1/16ths" number.)

If the Independence Date Theory length is subtracted from c above:

(t)	(d =)	5.656116325"
(t)	(IDT =)	5.654801219"
	(-) -----------------	
(t)	0.0013"	

Here we get a very small theoretical dimension of about thirteen ten-thousandths, (or, about one and a third thousandths) leaving us with one of the many tiny remainders found in small differences between similar features. One could say that this relationship is quite good from research Rules 1 and 3--it is based on two (now) seemingly well known and well defined discoveries of clearly intentional numerical and geometric relationships. But they really don't agree. After quite a bit of calculation, I found few possible meanings beyond a superficial stringing together of these shapes. Here and there "almost" proportions stuck out, begging for investigation.

Around this time I found another of what was to be many, good alternative sources of the length of the Large Outer Rectangle--which I must say, in this case, I have completely ignored: The **diagonal of a square four inches by four inches** is (See Illus G2):

(t)	$\sqrt{(4^2 + 4^2)} =$	5.656854248" = a		= 5.657" (three places)

(t)	(a =)	5.656854248"	
(t)	(IDT)	5.654801219"	(from above)
	(-)----------------		
	0.002053030"		or, about 0.002"

So, this new dimension "a" is about two thousandths longer than the Independence Day Theory length, and is otherwise unremarkable. I felt certain, intuitively, that I could safely discard this as being any part of the source of the dollar's length. (Here we see some of my prejudice and bias.) Yet I can not really say that any measurements were good enough to have ruled this out as an important design idea source. Okay, it was pretty close to my original measurement, and yes, perhaps should have counted as a possible origin in my reasoning, what with it being a *fairly simple explanation* for the Designer's choice of the length of the dollar bill. But it seems like a really *dull idea*. The stuffy old Bishop of Occam might well have to stop right here.

Illus. G2: Length of LOR ≈Diagonal of 4" × 4"

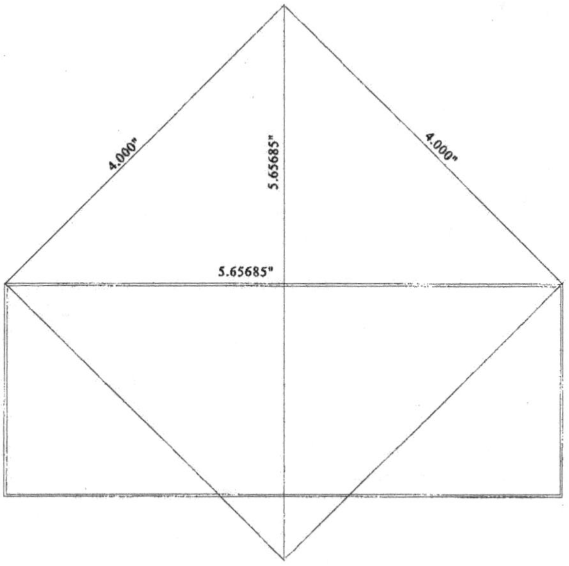

But this idea just didn't interest me in a theoretical sense when I found it, and frankly, still doesn't. Looks like a coincidence. But who am I to say? It fulfills many of the justifications, excuses and rules listed above. And Occam's Razor. Here it is anyway, and it may well be important somehow.

****

The Large Inner Rectangle (LIR) and Its Proportion:

In many ways, this inner rectangle turned out to be much more important than the outer LOR. As we will see later, the Designer used this rectangle almost exclusively as the basis from which to launch the extraordinary multitude of geometric riddles within the interior area of the dollar design. My initial thoughts about the Inner Border (LIR) were that it must be a proportional, concentric form of the first, or Large Outer Rectangle's border. Since the Outer Border now had a known proportion being $1:\varphi^2$, one would need only determine its smaller size and know its proportions, and that would be all there was to it. Yet this is not at all what the Designer had in mind: Its proportion *was not at all the square of the golden ratio*, and it appeared to have a simple, *even integer* diagonal of exactly 6 inches. Yet it is every bit as interesting as LOR. (These dimensions were originally measured at the *outer edge*, as the mid-point of printed line width was difficult to reliably measure to. Later investigation shows that this outer edge was, in fact, the intended reference line for several puzzles.)

Its dimensions appeared to be about 5.612" by 2.121" and had a diagonal of exactly 6.000" or, an intended exact six inches. This rectangle's ratio was apparently the *square root of seven*:

Illus. G3: Large Inner Rectangle (LIR)
Diagonal = 6.000" x = 5.612" y = 2.121"
Ratio: √7 : 1   (or, 2.645751311 : 1)

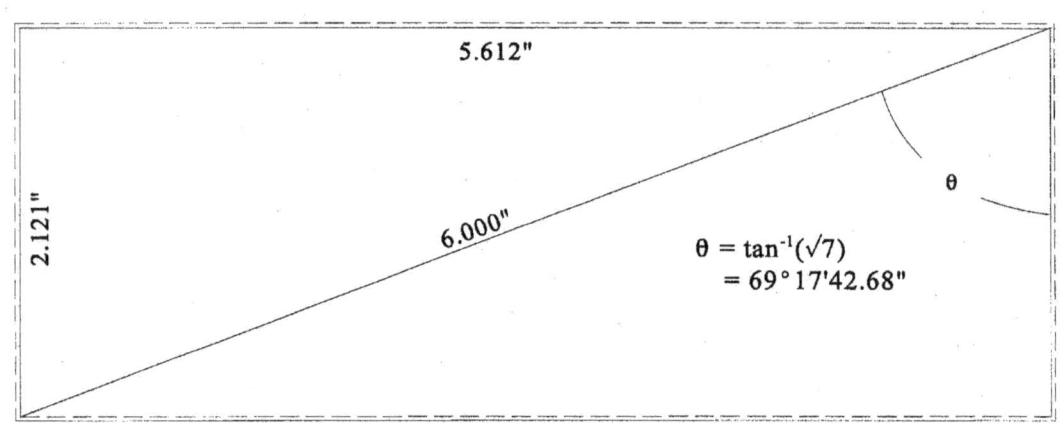

(m)   5.612 / 2.121 =   2.6459217 =   a

a^2 =   7.00090   …or about 7.

(Which is to say, about 7, within an error ratio of about one part in 1:7800--"a" then, looks a *lot* like an intended square root of seven.)

If this rectangle had a diagonal of 6" as measured, it followed that if these two dimensions were squared and added together--no matter what they were--they would have to equal 36, since 36 is the square of 6.

We know this is from **Pythagoras' wondrous theorem**: Any true rectangle will have a diagonal, and this diagonal if squared will be *the same number* as the result of adding together the squares of the lengths of the *two sides* of the rectangle--whatever they are. The square of 2.121 was about an *even* 4.5. Subtracting 4.5 from 36 would leave the long dimension of this rectangle the square root of the remaining 31.5. The roots of these two numbers must be the true lengths of this rectangle. (Get out your calculator campers:)

(m)   Given measurements: x = 5.612" and y = 2.121"

*(See Illus. G3: "x" is always length East going left to right
and "y" is always North or up the page as height,)*

(m, t) 2.121^2 = 4.498641,   or, very close to 4.5;

If the measured diagonal *is actually* 6.000", and if this is *really a rectangle,* then the sum of the squares of the two sides *has to be* 36. It would also appear that these squares must be *even fractions.* Therefore:

(t)   36 - 4.5 = 31.5,   we can subtract *the square of the x side* from the sum of
                                            the squares of the measured dimensions which will leave
                                            the square of the long side or, the square of the side y.
                                            Then, the square root of both sides of the squares will
                                            give us:

48

(t)	$\sqrt{31.5} = 5.612486080...$"	for the x dimension, (length) and
(t)	$\sqrt{4.5} = 2.121320344...$"	for the y dimension, (height);
		Dividing the two proposed sides x/y:
(t)	$\sqrt{31.5} / \sqrt{4.5} =$	or, the ratio of the two measured sides must have been
	2.645751311...	intended as an *exact square root of seven* = $\sqrt{7}$:

The diagonal from the above x and y: $\sqrt{(x^2+y^2)} = 6.000000000$", apparently the same as the exact 6.000" diagonal as measured. Other similar rectangles could give a diagonal of 6, but this is the best explanation that fit the facts. This seemed a reasonable solution for the intent of the Designer here, since this was the simplest explanation for this rectangle's apparent dimensions and closely matched the measured height and width. (See Illus. G3.) Since the two sides together have a clear intended ratio of $\sqrt{7}$ the internal angle may be easily and exactly computed as $\tan^{-1}(\sqrt{7})$ or an exact, theoretical angle of 69°17'42.68" (see below).

This is not really a weird rectangle, just an very unlikely one, though I suppose it isn't any more unlikely than any of these other rectangles. But it *does* have a sort of "classical" character of an even integer diagonal and a ratio taken from the root of a regular number. This rectangle has an exact integer diagonal of 6, which is a little odd, when you realize that most any diagonal at random will probably have a *fractional length* regardless of its exterior lengths. And it has a simple, exact ratio derived from the root of a round, exact integer of 7, which is also a little odd, considering that almost any of a multitude of simple fractional ratios is more likely. One could compose many different rectangles like this with different integer numbers--but this is the one found here.

From my point of view, its strangeness is from the fact that LIR's true nature was *totally unexpected*, and it required *calculation and careful measurement* to distinguish it's character from that of LOR just outside of it. Here, I think, at a beginning point of investigation, you might say the Designer put us on notice: Things may not be as they seem. At all outward appearances by eye, one is given no obvious reason to suspect otherwise. This seems to be the general character of most of the Designer's work. Hidden, though not too deeply hidden, but with really solid contrasts and often found out from subtle differences. (J1)

The Designer makes you *work* to discover the very different idea going on here from LOR. Perhaps something of importance is being "said" in a specific way (J7) about the numbers "6" and "7." It should be noted that the well known "thirteens" in various forms, are found all over the dollar, and of course these two elements: 6 plus 7 are 13. Could we take this further? In Biblically-based number lore, "6" is often said to be "the number of man." This is widely extrapolated by many writers from the account of man being created on the "sixth day" in Genesis, where later on the "seventh day" God then rested. "Seven" then, is also universally associated with the Deity. So are we looking at a symbolic geometric statement? Using these two symbols, the *length* within this *form,* (diagonal and angle function) the Designer might be saying something like: "man, within a sacred framework." This sort of idea is really not all that far out symbolically--but we really don't know what he is trying to say. (I should say, I am not convinced by this first numerical-symbol theory, and I don't think the idea can be taken very far. As we will see later, there is probably a lot more depth and specifics to this choice of these ratios, and maybe a far better reason for this specific choice of relations. See Chapter 4. "Gift Wrapping.")

The fact that *it was made to seem* that it might just as well have been the square of the golden ratio proportion by appearances, like the Outer Border LOR, raised some other interesting alternative ideas (J1): *What would happen, for instance,* if we take the clear diagonal of "6 inches" as the controlling factor, *but instead* form a theoretical rectangle around it using the *square of the golden ratio*? A rectangle of two quite similar but slightly differing dimensions can be found this way:

(t)	$(\sqrt{5}+1)/2^2$	= 2.618033989	= φ^2

Ken McGrath

(press 5 then √ button, then + 1 =, then ÷ 2 = this creates φ, then press the x^2 button creating φ^2. Then:

(t) $\tan^{-1}(\varphi^2)$ = 69°05'41.43"   (= θ)

(with φ^2 still on the calculator screen: Press the "tan⁻¹" button--usually the "second function" button followed by the "tan" button on most calculators; then the D°M'S" button to see the minutes and seconds. Be careful not to use this D°M'S" form of this number for any later calculations--you have to go back into "decimal degrees" since calculators *only use decimal degrees.* The symbol "θ" or theta is just a commonly used Greek letter to stand for a *given angle* so that you don't have to write out the whole thing out in an equation, it is just an abbreviation or marker for the specific angle being talked about. The above angle is the fundamental angle of the square of the golden ratio rectangle, which is also happens to be the angle of the Long Diagonals based on the LOR border lines.) Then:

(t)   sinθ · 6   = 5.605034154   = x'
(t)   cosθ · 6   = 2.140932538   = y'

(With the *decimal form of the φ^2 angle* on the calculator screen (69.09484255°), store in memory, and press the "sin" button, then press "x" 6, then the "=" button. Then, recall the angle from memory and repeat this process using the "cos" button. The apostrophe (') shown here next to the x and y only says that these are a new variation of the originals.)

The two above dimensions x' and y' could become important somewhere since this would *be an alternate Large Inner Rectangle* (LIR') that might have been hinted at to provide the length of some sort of small shift or angular clue. This could be a right/left shift just as well as may be up/down, or both, being simple subtractions from these numbers. But we should also note the other fascinating surprise-- and seemingly weird fact about this new, alternative rectangle LIR' is that its *area is precisely an even 12 square inches!* Try it yourself: take x' and y' from above and multiply:

x' · y' = 12.0000000"

Isn't that strange? This is one of those odd things that has to do with the properties of the golden ratio.

(In the calculations above, the **sine and cosine**-- "sin and cos" on your calculator--are the two aspects of one route to *finding the sides* of a rectangle when only the angle and the diagonal length are known. They will only work with ratios smaller than one, or a fractional number appearing on the right side of the decimal place. The **tangent** function ("tan") on the other hand, is the other route, which will give the *angle of a diagonal* with respect to the sides if only the fraction based on the division between the known sides is given. In one case above, the *square root of seven* was the fraction used to get the angle of the Large Inner Rectangle *based on its sides*, and in this second hypothetical case, there was a given fraction that was the *square of the golden ratio,* which was used to get *the length and height of these sides* from a known *diagonal length* of 6. The little " ⁻¹ " or *minus one* superscript after "tan" means the "reciprocal function," which when seen together with "tan" is called the "arc tangent," the function that *extracts the characteristic angle* from a fraction. That's not all there is to it, but the above methods can be used for all *right triangles* found in rectangles, and in reality these three things are almost all one needs to know to get started experimenting in trigonometry. Check your calculator: almost all scientific calculators now have "polar and rectangular conversion," which is a more automatized form of this found somewhere in key functions, which is a *much* quicker form of the above process, saving a lot of time.)

****

A first look at The Great Seal and a possible King's Chamber Related Ratio:

At first glance, in rough numbers only, the length between the apparent centerlines of the sides of the Great Seal, and the height of the Small Inner Rectangle seemed to make a new rectangle having about a √5 proportion. Here again is a single integer root similar to the √7 shown above, but in this case is an elemental part of the mathematics of the golden ratio. This would be interesting if true, since it is also a fundamental proportion within the height and width of the *East and West walls* of the King's Chamber of the Pyramid of Giza. (This is *not*, however, the ratio of the King's Chamber *room*: the East and West walls have a related (√5)/2 ratio, or about 1:1.11803...) These and related proportions also appear in other Egyptian ruins, I believe, (such as the square root of 5 minus 1). But the √5 is best known from the center room of the Pyramid--and here it seemed that there might be a symbolic connection to the fundamental center location between the faces of the Great Seal.

Illus. G4: BSC1

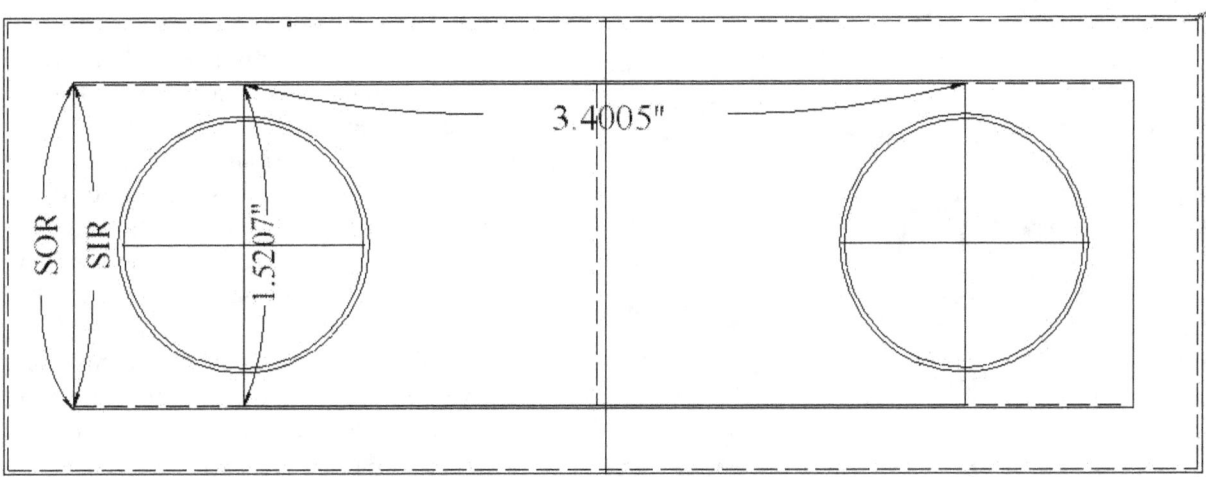

My first few attempts to measure the distance Between the Seals centers, (BSC) came up with about 3.400", but I was not at all happy in the beginning with the uniformity of my measurements. At first I measured between both right edges and both left edges of the Seals, and after a time concluded that this distance was 3.4" or maybe some tiny fraction less--on the order of half a hundredth of an inch, or about 0.005" less. *If*, however, this distance actually turned out to be about 3.4005", (say, it was later found to be "shifted") by what would amount to about a whole hundredth of an inch *longer*, the Seal Faces would be at a point where an invisible or virtual rectangle having a 1:√5 proportion could exist between the extended horizontal lines of the Small Inner Rectangle (SIR) and the centers of the Great Seal faces:

Given: √5 = 2.236067977;

(t)   1.520749883 (= the theoretical form of the height of SIR or, φ · 39.37 1/16ths of an inch at filigree focal points, or, about 24.33 1/16ths of an inch.):

(t)   1.520749883 · √5 = 3.400500116 = BSC1
              (=BSC1: Between Seal Centers #1 See Illus. G4)

This was an attractive idea, but that is--*if*; if there were some solid clues to "slide" the Seal faces apart a tiny bit. But "One hundredth of an inch" is *quite wide of the requirements made in Rule 1*, to call it any kind of discovery. Since it mostly lies within the other parameters of J1 through J5, it is a clue close enough to start hunting with, if nothing else--with a lot of wishful thinking. And, this was where another series of tricky problems begin to reveal themselves in the Great Seal and the Long Diagonals.

****

The Great Seal with respect to the Long Diagonals:

In re-measuring the dollar under better conditions and with a more methodical approach to reasoning, my conceptual targets were beginning to change, and ideas about *what to measure* had changed quite a bit. What I wanted now was "cut and dried solutions" to an endless collection of possible variations to the dollar's geometry. Soon I was to find that you can make no assumptions about the nature of this design; you must measure it thoroughly and check very thoroughly. I had begun to uncover a bewildering assortment of hidden geometric shapes and numerical ideas and was beginning to develop a catalog of them. Here progress happened mostly by the "flying by the seat of the pants" intuition within calculation rather than by a truly orderly search. Although mostly baffled, I was gradually coming to know some of the Designer's tricks, or at least thought so. Returning time after time to the Seal faces and Long Diagonals over several years of examination, I was beginning to learn something of the possible extent of subtlety that the Designer was capable of.

The Designer's Use of Symmetry:

Apparent design symmetry, which we take for granted in other designs, is subtly tweaked *all over the dollar's design*. This is because the Designer obviously liked to use one side of design data to give a bit of information. Foe instance, a line or point that one has measured to, with respect to the centerline of the design, is often followed by a slightly non-matching design feature as measured on the opposite side of the picture. (This is J1 "doctrine.") You might in ordinary circumstances-- anywhere else--assume this opposite length to be the *same*, as being "just the opposite from the other side of the picture." And, any tiny discrepancy when found might look like a small mistake on the part of the Designer. But later, measurement, calculation and reflection on possible configurations usually show that small differences are *puzzle elements* that have been concealed in these relationships. Eventually, the nature of the contrivance surfaces--and will often appear elsewhere in the dollar design in a similar form.

This is an *offset asymmetry.* It is often discovered in the design and is intended as a *forced* subtraction as a measurement check, or to otherwise force the investigator *to have to compute* certain measurements to get a single dimension or an overall dimension. This could also happen by forcing one to measure a thing from both sides, two different ways, from one end and the other from the middle, or in some other kind of inverted way. Sometimes it is only *the math* (J3) that is inverted or made tricky in some

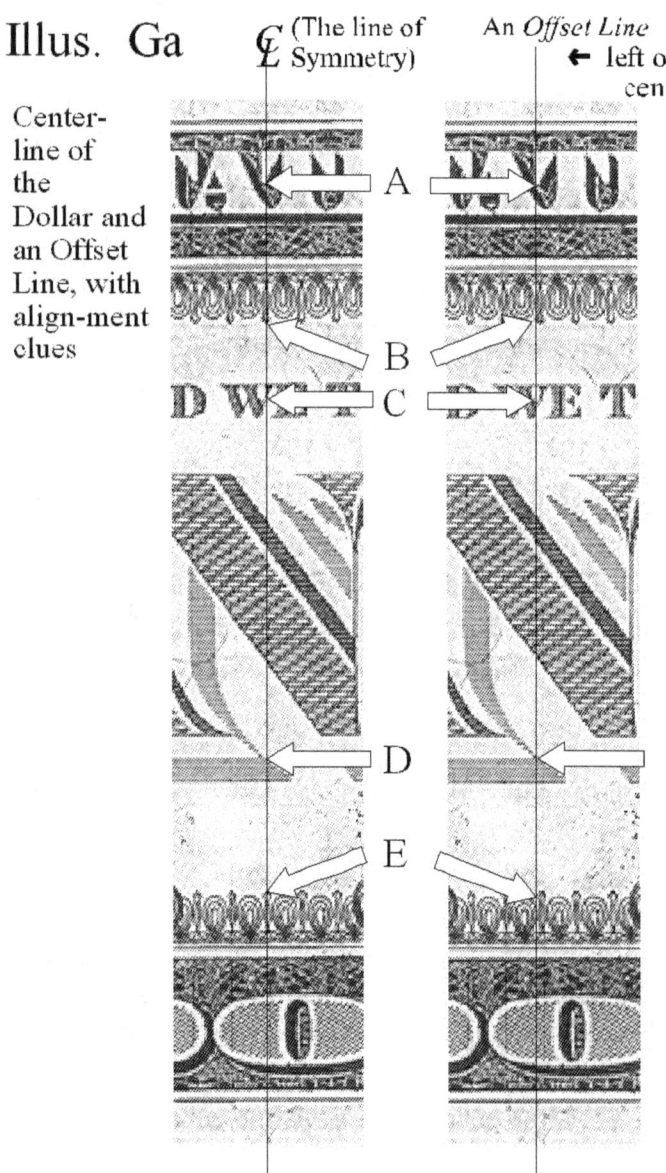

Illus. Ga

₵ (The line of
Ł Symmetry)

An *Offset Line*
← left of
center

Center-
line of
the
Dollar and
an Offset
Line, with
align-ment
clues

A

B

C

D

E

way or other, rather than measurements. But it is the multitude of offset differences that are the biggest difficulty and a key for understanding this design.

If for example, there are two ways to measure and determine the centerline of any feature on the dollar, I have (repeatedly) discovered that you must measure them *both*, even if the other measurement appears to be redundant. If there are two elements so similar in appearance that they must be the same, *measure them both, anyway*. The Designer has typically provided two slightly differing sets of length *on either side of a given line of symmetry or regularity* which are not accidental differences. One side of these usually represents half of a true, or imaginary symmetry, the "whole" of which might be called a basis, and the other side providing a clue through the discovery of a small difference. Here the "a" element of J1 is applied to the concept of symmetry and the "b" element to the discovered asymmetrical aspect.

The centerline of the dollar, (CL) or true line of symmetry, is marked by two opposite tips of the ornamentation in the millwork (B-left and E-left) of Illus Ga. This centerline is paralleled by a small offset alignment of points. This begins at the left edge of the left serif of the second letter "T" in *STATES* in THE UNITED STATES OF AMERICA at the top of the bill (A-right). Note the right side of the "W"of WE in GOD WE TRUST at the *point of the inner white area* of it's downward pointing v-shape (C-right). *V-shaped themes* appear elsewhere on the dollar. Note the *inner white area* again--another repeating theme--also exactly at the inner area of shadowed serif of "N" (D-right). A few features on the dollar appear to have to have been made normally, or made completely symmetrical by the Designer so that some puzzles can be figured out, such as the two central, opposite tips of the ornamentation in the millwork (B-left and E-left). These function as the simple and direct "basis."

For a good, clear example of his J1-type *modus operandi*, that is easily visible to the naked eye, look at the "horizon" behind the little Pyramid on the left side of the Seal. (Use a magnifying glass.) See Illus. Gb below. Notice on the right side of the Pyramid, that this apparent horizon lines up about at *the middle* of the fifth step. But on the *left* side this lines up (following shadowed side's perspective) at what might be the middle of the *fourth step*, or alternately, if projected across directly to the front face, two dimensionally, centering about at the *beginning* of the fifth step and top of the fourth. We might speculate about what this all means symbolically with respect to the steps. But if we thought that these lines of the horizon marked something to measure to, it is clear, at any rate, from a measurement point of view, that there are at the very least *three* possible beginning points. I have no doubt, though, that one of these is intended as a "base", and that some symmetry- or proportion-based reason will eventually emerge from calculation and measurement. (It is interesting to note that unless we used *the average* of these somewhat sloppy lines, this "horizon" will not be useable evidence following my rules, due to the requirements of Rule 3. Perhaps it was intended as a more vague message such as "a new Heaven and a new Earth" of changed horizons or something similar in an apocalyptic sense. What could differing horizons be? But I rest assured that it will be accounted for.)

In a far less visible but also quite interesting place, note the filigree at the upper right side of the bill. There is a *small jog* in the vertical line at the upper right hand corner--and *also another* small, yet more visible jog on the horizontal line made out of the filigree near the upper right corner (See Illus Gc). This is in the lines that correspond to the top and end of SIR on the *right side* of the dollar. In the second case however, it is a real, or precise hint of some sort, given for an alternative rectangle's height or some other special aspect in that region. I do not know what was intended with the variable "horizons" inside of the Seal or these jogs, but the "horizons" are definitely not as easily measurable potentially, as the sharp, tiny jogs in the filigree.

Illus. Gb

Not so easy to see, but quite important later, is the difference between the top and bottom dimensions of the dollar, or, the top and bottom lengths of LOR (not shown). At some point, it became evident that the bottom dimension was definitely in the neighborhood of two thousandths of an inch (+/-0.002") longer than the top dimension. When I first discovered and confirmed this, it seemed really annoying--I had what I thought was the solution already worked out with the Independence Date theory (IDT). But as irritating as it was, I was beginning to see that this kind of thing was part of the program. The Designer had apparently indicated something by the top dimension, and was evidently signaling some other idea the wanted to show with the bottom length. It was a long time before I figured out what this was. It was a "J1" situation, clear and simple. But until that time, I mostly ignored this otherwise glaring difference. I winked at it and shoveled it under the rug. (See Chapter 5: A Digression: 1936 and the Bottom Line Length of LOR)

Illus. Gc

The Character and Position of the Seal Faces:

Let's look at the ideas behind "circles." If you have a circle--as the Seal faces at first appeared to be--it would seem that they would be universally symmetrical, and that the centerlines of their location in both axi should be easy to locate. About all you need know about *any* circle is where its edges are, and with the edges, the diameter can be found at its widest measurable width. Later you can divide the width in two to determine its radius, which is its centerline or half width. This *would all be true, if these were circles.* Now, let's take a look at the two sides of the dollar's Great Seal design: first appearances only. Here we will initially look at the logic of the problem, and later examine particulars of measured dimensions.

Illus. G5: Elliptical Seal Faces

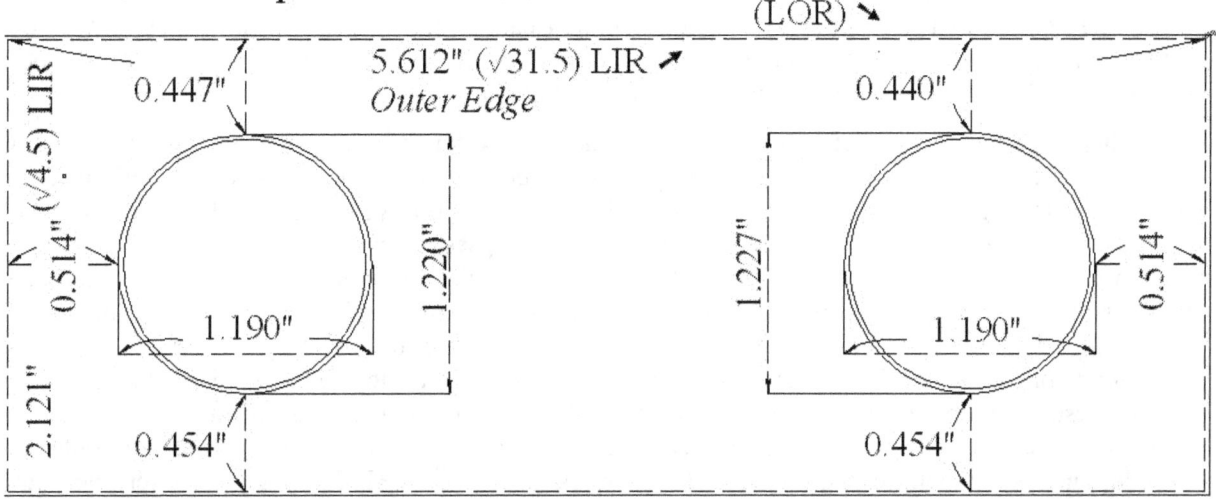

Apart from their contents of pyramid, eagle and so on, these circular features first appeared by eye to be duplicate images of each other: two regular and perfectly normal pairs of concentric circles. From the point of view of the poor engraver, these would be simple enough to construct with fairly regular drafting techniques--perhaps with a compass or a vertical milling machine. But that's not what's on the dollar--something *much* more difficult was created here. Following careful measurements, I made the observations shown in Illus G5. Within the rectangle of LIR all dimensions were made to the center of printed lines. Measurements to LIR were made to the outer edge of the printed lines.

Observations:

1. Both faces of the Seal *are actually elliptical*, (that is, although very round, regular and circular in appearance, they are *both slightly taller* than they are wide.)

2. These Seal faces *are not the same size*, Obverse and Reverse. (The right side or Obverse is slightly *taller* or 1.227" than 1.220" on the Reverse or left face--a difference of 0.007". This difference is repeated elsewhere and becomes very is important later.)

3. Yet both Seal faces apparently *have the same horizontal width*. (1.190")

4. This is complicated by observations that seemed to show that *they are closely tangent to, or grazing, the lines of the Long Diagonals*. These are the invisible diagonal lines that one could draw from the far corners of the dollar to the other. (See Illus. H1 and H2)

5. Both *lower edges* of outer ellipses appear to be tangent to the diagonals of LOR rectangle

6. (+/- 0.0005"); and both ellipses check as being symmetrical from the center, right and left; and both have the same width; yet their *upper edge intersection with respect to these diagonals* are dramatically different on the right and left due to their differing heights: See (2) above: the ellipses' *centers are vertically offset* by a small amount: 0.007" divided by 2 or, 0.0035".

7. I think it is fair to assume that both the diagonal lines from the outer rectangle corners are intended to be thought of as (a) *symmetrical in angles*; and (b) oriented at a *common center* of the dollar, vertically and horizontally; and (c) that the vertical centerlines of the Seal faces are truly straight-up-and-down *vertical* and (d) *symmetrical* right and left from the center; as well as (e) for the ellipses being mathematically lawful, symmetrical and otherwise graphically correct in shape. These underlying assumptions are supported by measurement.

From the above observations, two general considerations and constraints must be true:

(I)   If the centers of the ellipses are located symmetrically right to left, but have different vertical widths, *and are still tangent on at least two lines and on two sides*, then their centers must be at different *vertical* heights if Observations (3) and (4) above are true.

(II)   When it comes to computing exactly where they might touch a tangent line, there is a peculiar character of ellipses *which are not at all simple like circles.* The center of any circle will always be 90° or square to a tangent line no matter what point on the circle you choose. That's the way it is with circles, but this is *not at all* the case for ellipses. Ellipses have a *bias*. For an ellipse, its center can only be square to a tangent diagonal at four points on the major axis and minor axis, at the widest and narrowest points. That is to say, at the exactly horizontal and vertical points, only. For the Seal faces, any diagonal to ellipse tangent point or intercept will follow a *much more complicated mathematical rule* than that of circles, if not exactly flat vertical or horizontal. I didn't like this at all. *All* of these intercepts to the diagonals occurred in the non-regular or problem areas on an ellipse. Yet all of the elliptical variables in this case are completely measurable as simple horizontal and vertical dimensions with respect to the rectangular framework of the LIR rectangle border lines, since their axi are perpendicular to these lines by Observation (6). They may be measured from the horizontal and vertical edges as latitudes and departures in x and y, and are then calculable and interpretable through various well known formulas. (See Illus. G5)

What the Designer is clearly asking us to do, is to use symmetry and other clues to reconstruct his hidden rectangular form (or forms) to reveal a mathematical relationship. With all these given clues, there may easily be several such relationships. When the primary form (the "base"theory of J1) is precisely subtracted from other discovered tweaked forms, some significant difference should appear. This could allow some discovery like the SIR slipping rectangle seen earlier, or something else. But these new clues were to be quite small in size, (0.007" vertically, between faces.)   How the Designer might set the stage to graphically demonstrate any problem solutions did not at all seem clear at this point.

Obviously the best strategy would be to chart the problem variables first, and then measure all the evidence. When only considering *regular* diagonals taken from the separate rectangles--at the very least-- there are sixteen implied solutions needed of elliptical intercepts, and some study for each. (That is: *Two* sets of diagonals, being tangent (or not) in *two ways* to *four* separate ellipses and in at least one unique way each.) There are also *irregular* diagonals, having their origins on different rectangles at opposing LOR/LIR corners; but that is a more complex question. As above, since this will say nothing about the Seal Face symmetry with respect to the bill itself, one must measure from the borders to give the ellipses' centers a position by subtraction. Once all of this evidence is secured, these problems are solvable and significant forms can be analyzed.   (Due to space limitations and the complexity of the subject, we will only look at a small part of ellipse problems in this book. Much of this discussion is out of order chronologically by several years, though a an early point in the 80's I had at least uncovered the basic line symmetries that we will see shortly. Important Caveat: In the illustrations H4 to H7 some theoretical dimensions are supplied based on computer aided drafting (CAD), which is a close approximation of calculated dimensions--about as close as perhaps 0.00005". The data for the CAD is supplied in Illus. G5, the "LOR" is the DFT theory of Chapter 4. *These graphics are not intended as a rigorous treatment however.* Some people may wish to complete these mathematically. For those into the math of ellipses, there is much work to be done. My suggestion is to create a coordinate list from these dimensions, experiment with elliptical intercept formulas and insert these into a spread sheet program. Before attempting to solve these problems, please read on to the later theory developments shown in Chapter 4 with respect to the LOR rectangle's dimensions per the DFT theory.)

Tangent Areas:

In **Illus. H1**, the lower edge of both Outer Ellipses appear to be exactly tangent or just touching the diagonal of LOR lines. Yet at the top right and left points on these ellipses where one would think they might also touch, the lines actually pass inside the curve to some extent. (Note: the incursion on the Left is difficult to see at this scale.) Since one set of ellipses on one side is larger, these intercepts cannot be exactly the same for both sides. On the Right Seal Face the diagonal *falls about half way between the Inner and Outer Ellipses*. (A J1 clue.) We have seen something like this in another form before with the diagonals at the "39.37/16ths" or, SIR golden rectangle. (see the top of the "O" of the word ONE in

Illus D. Perhaps the same kind of inference can be made with some certainty here. This relationship could be a clue indicating a possible horizontal or vertical shift of some sort for the "O.")

Illus. H1: Diagonals of LOR, tangent/intersecting Seal Faces   5.655"

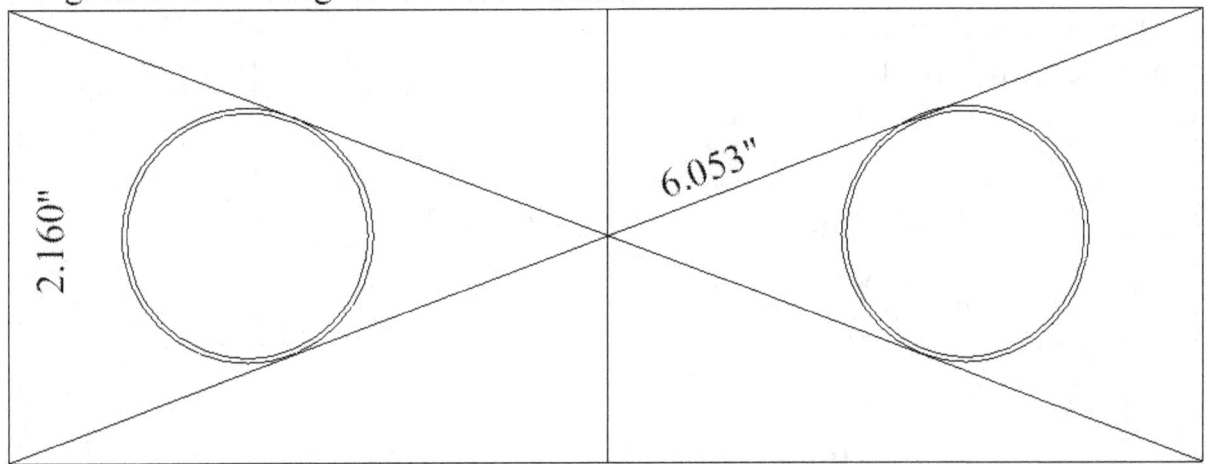

Try it yourself: press a straight edge--a clear plastic drafting triangle would be best--diagonally from exact opposite corner to exact opposite corner of the Large Outer Rectangular (LOR) borderline on a fresh new dollar bill. You will find that it just grazes the edges along the lower rim of either Outer Ellipse. The *first impression* is that both diagonals are tangent above the Seal at left and both tangent below, but intersecting at the top on the right. However, a closer look will reveal that at the upper right, the line seems to divide the space between the Inner and Outer Ellipses into about *half* at the Right Face and a smaller fractional distance of that at the Left (at the apparent tangent point.) Lines and at this scale are hard to see on the left, but this will be clearer in Illus. H4. (See H4 and H5).

Illus. H2: Diagonals of LIR, tangent to Seal Faces

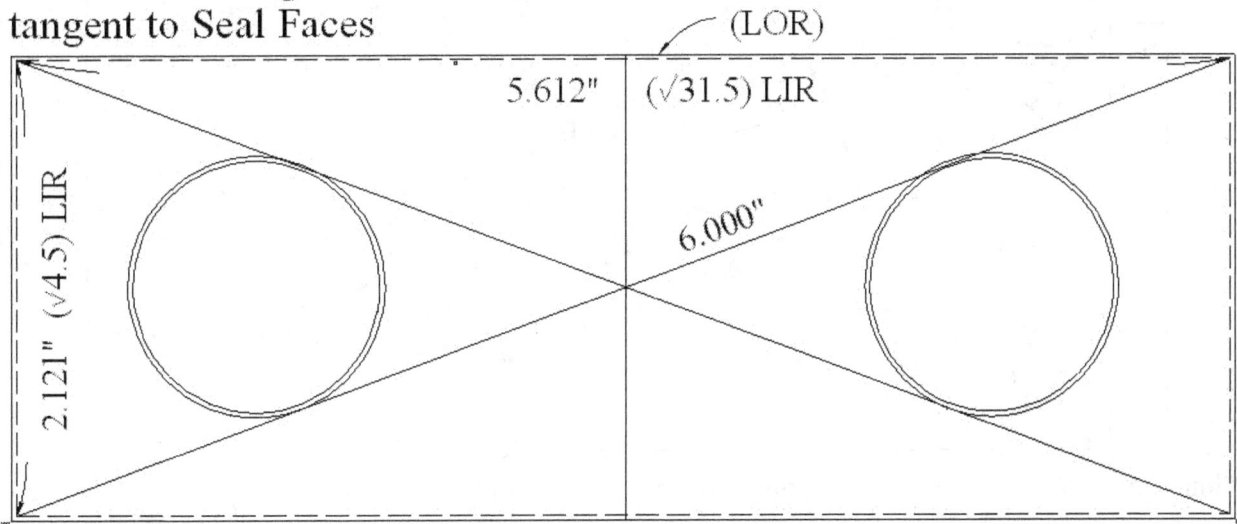

Now look at **Illus H2**. *This second set of diagonals begin at the opposite corners of the Inner Border of LIR.* (Notice in Illus H1, the other diagonals originate at the *far outer corners* of the Large Outer Rectangle, LOR. We will call the first Long Diagonals, Outer or LDO, and the second pair based on the Inner Rectangle, Long Diagonals, Inner, or LDI.) Even though there are only slight differences in their origin and having only tiny differences in their angles, this second set of long diagonals show *a very different resulting intersection* where they touch. Here again, appearances are deceptive. At the Right Face, the LDI diagonal appears tangent to the inner ellipse. But it is not: it is a small fraction of this width away, somewhat shy of actually touching as will be seen below. As noted above, the vertical symmetry of the Seal Faces is offset. What do these look like in detail? (Illus. H3 shows the four general areas of interest.)

Illus. H3: Intersection areas: diagonals on Great Seal faces

(All dimensions in H4 to H7 are theoretical, derived from CAD projection of measured data.)

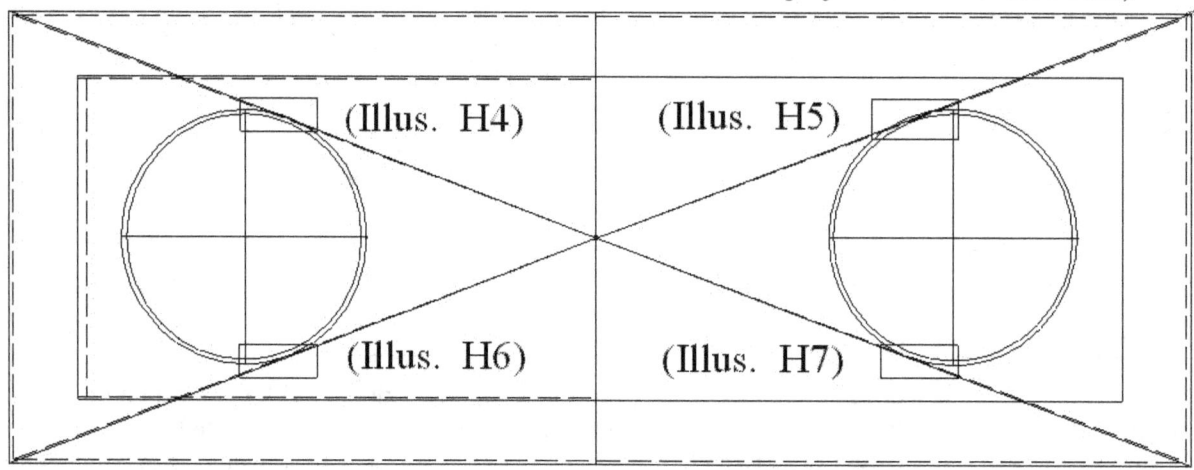

Illus. H4: Upper Intersection Area of **Left Seal** Face

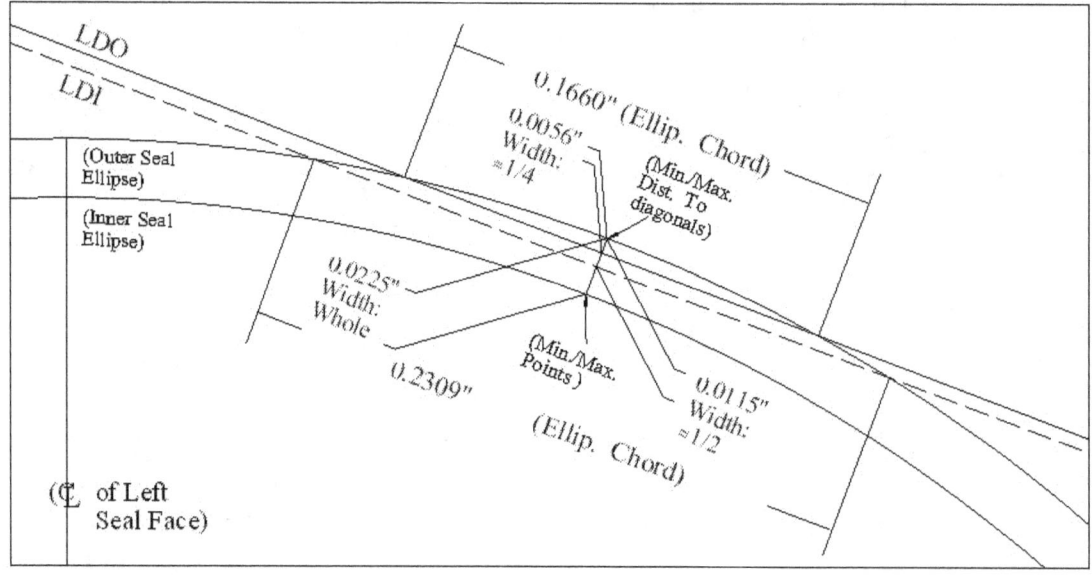

Widths being Whole; ½ and ¼ at midpoint—symmetry *above LDI line*

In Illus. H4, we see that one of the elliptical chord lengths, LDI, **(dashed line)** seems to *exactly split* the width at the minimum/maximum width point. (Strictly speaking, this is not one width but an average of two sets of widths *very* near to one another for both LDO and LDI , being the *points of least difference in width with respect to both ellipses and diagonals,* or what I have called the "Max./Min. Points." Another way to think of these is that their origin begins at the *nearest point* on the inner ellipse to the diagonals. These tiny differences are *very* small, and for our purposes here, we will have to ignore this distinction. A "chord" is the name for a line length inside of a curve--in this case an ellipse.) Next we see that the LDO diagonal **(solid line)** appears to split this width in *half once again--a quarter of the total width.*

Also note that what could be called a line of symmetry--the half width point--is on the *LDI diagonal,* **(dashed line)** and that the other splitting diagonal lies *above* this line. Amazingly, this line order and symmetry is completely and precisely *reversed on the other side,* at the Right Seal face.

This arrangement could not have been accidental, it is an exact relationship of lines to these two pairs of ellipses forming the faces of the Great Seal (J1 and J3). In the second case on the right, the *other diagonal, or, LDO* **(solid line)** has the honor of being at the half width point of symmetry. Whereas inversely, the LDI diagonal **(dashed line)** is made to split the minimum/maximum width--*again into quarters, but this time below the line of symmetry.* Clearly, *something* is being indicated. Here is a remarkable example of something like an inverse "sliding" as discussed in J2. Ultimately, we will be able to say that this geometric setting of diagonals to Seal borders demonstrates all of J1 through J7 arguments, especially that of J6. Those who skipped ahead to Chapter 4 will note that one elliptical chord--the *longest* of the upper four chord lengths--will be familiar. (See Illus H5)

Illus. H5: Upper Intersection Area of **Right Seal** Face

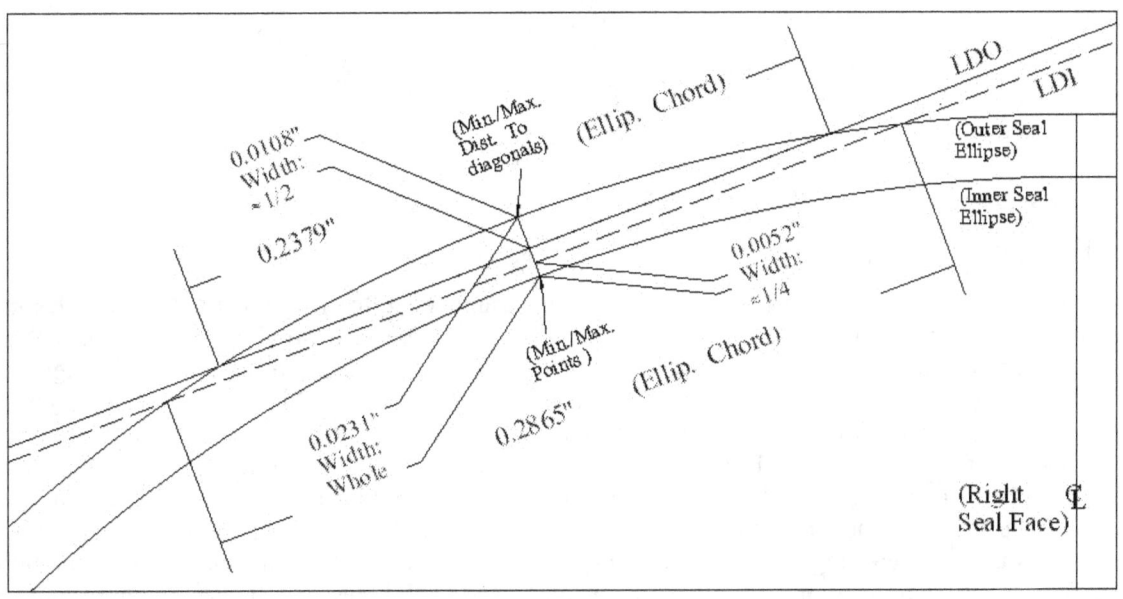

Widths being Whole; ½ and ¼ at mid point—symmetry *opposite below LDO line*

Now we will look at the details of the other two areas: H6 and H7. Logically, if the outer ellipses are found to be about tangent to the LDO diagonal, then the LDI diagonals must intercept *somewhat inside* the outer ellipses, since they form a narrower angle than the LDO diagonals. On the dollar this is not easily visible--both lines are close to tangency in appearance. With a transparent straight-edge, you will just be able to detect this difference to the LDI diagonal on a crisp, new dollar. And although this is not completely visible on Illus. H2, there is, in fact, a small intercept chord. Since these are illustrations of mathematical lines, or as centers of the rough, grainy-lined, printed ellipses and invisible diagonals, we can see more details here than we would on a true magnified face of the dollar. To actually measure this on the dollar would be very difficult. These details are impossibly small and the printed lines are relatively large, about 0.012". I have made no attempt to show this relationship, but there may well be further subtleties of alignment, such as the inside and outside edges of the ellipses, since the printed lines are substantial fractions of the splitting proportions at the lower Min./Max. Points. (This could be best approached through photographic techniques, or a high resolution scan, imported and overlaid in accurate CAD drawings. See Illus H6 and H7.) As we can see from these bottom two areas H6 and H7, both LDO diagonals appear to have been made to be tangent or *very nearly tangent* to both outer ellipses. But they are *not* tangent--there still appears to be a tiny gap for some reason or other, and therein may lie a puzzle or puzzle element of some kind.

Illus. H6: Lower Intersection Area
of **Left Seal Face**

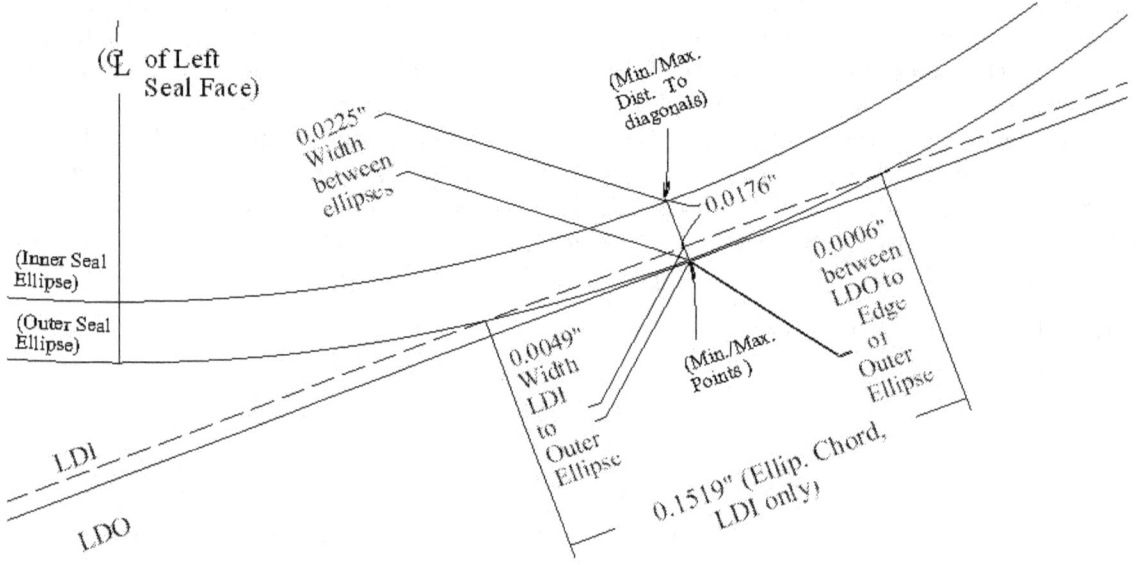

LDO Tangent =/- 0.0001" to Outer Ellipse

LDI intersects Outer Ellipse

Imagine the incredible care that must have gone into this arrangement, to ensure that the symmetries in those areas were created, and--as we will see below--*special lengths* were also provided for. Also, the *spacing* between the Seal's centers turn out to have been carefully chosen to provide another special length, adding another sort of complexity. This means that the *size* and *character* of these ellipses were carefully chosen for this task. I think its fair to say that there are few people who are equal to the task of creating an arrangement of requirements like this even today. But there is a limit. Not *all* of these dimensions and angles can be significant. These interconnecting forms can be manipulated to produce exactitude for certain symbolic numbers, but various parts must be unimportant and derivative or at best lucky approximations of significant numbers. It would be interesting to know how many intentional relationships and subtleties can be built into such a geometric problem. In light of the idea of interconnected subtleties, note these six chord lengths as a sum of their total length:

(t)	Upper Left LDO chord:	0.1660"
(t)	Upper Left LDI chord:	0.2309"
(t)	Upper Right LDO chord:	0.2379"
(t)	Upper Right LDI chord:	0.2865"
(t)	Lower Left LDI chord:	0.1519"
(t)	Lower Right LDI chord:	0.1485"

$$(+)----------$$

(t)   sum: 1.2217"

1.2217" is numerically very *similar* to the vertical semi-diameter of the Left Seal Face outer ellipse. But at one only part in seven hundred to the measured dimension, the relationship is only be close, perhaps coincidental: $(1.2217"_t / 1.220"_m - 1) / x = 1:718$. When I first discovered these relationships I didn't see their importance, and had reason to wonder if I had measured or arranged the problem wrong. Surprisingly, I had not. Later on in Chapter 4 we will see that this number *appears elsewhere* by independent means--and that, amazingly, this sum must been the result of careful arrangement of this geometry and was a precisely made choice on the part of the Designer, as will be seen in Chapter 4 as "$z^2/10$" (J6).

Illus. H7: Lower intersection area of Right Seal Face

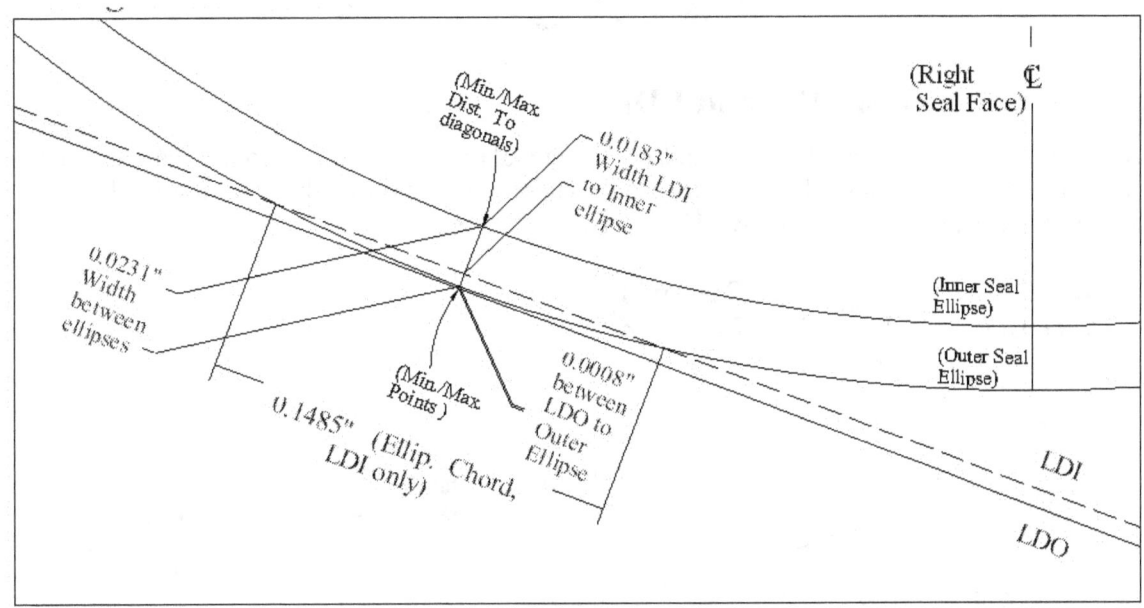

LDO Tangent +/- 0.001" to Outer Ellipse

LDI intersects Outer Ellipse

Looking back to the central Large "O" in "ONE" in the last chapter (pgs 23 and 24) which also aligned at a tangent to a diagonal, this too, *is also not a circle but also an ellipse*. And it's upper edge-- just like the Seal sides--lies within the upper length of the SIR diagonal. Yet *unlike* both ellipses of the Great Seal, its major and minor axes are *opposite* in orientation (apparently, another kind of J3.). Which is to say, both Inner and Outer Ellipses of the Great Seal are a little taller than they are wide--whereas the Large "O" is slightly *wider* right to left than it is tall. This "O" has a horizontal Major Axis that appears to be *half* the height of the vertical Major Axis of the left hand Outer Ellipse. (Here again is some strange little clue that apparently encourages the researcher to make some other kind of manipulation of the evidence. Is there a hint of a 90˚ angular rotation, or a J1-exchange? This is a really fascinating problem, one which I have not solved.)

Now compare the diagonal found in the small golden rectangle SIR, of the last chapter, which also grazes the upper left edge of the "O" that begins the word "ONE" in the middle of the dollar. Like the other diagonals it is slightly within the body of the "O," but the other of these two diagonals does not and is tangent as we saw before. Also notice the alignment of the lower left serif in the letter "I" beginning "IN GOD WE TRUST" to this line (See illus E and D). Next, we see these diagonals doing a similar alignment switch. Obviously, some sort of comparable thing is going on in these larger diagonals and rectangles with respect to the Great Seal faces.

Can we assume that whenever something lines up with the edge of an "O" (or any o-like thing), it might be important? This pattern has become so abundantly clear that one might just as well independently study the possibilities presented by the various "Os" (or "Es," for that matter) found all over the bill.

With respect to the diagonals, the way the Designer might have arranged which diagonal would touch the "O" in "ONE" might have been accomplished by simple vertical or horizontal displacement from the line of symmetry. But later though, we will see that *other constraints also* circumscribe the position of this "O," so it is not really an altogether simple displacement to position this "O." Looking at the previous relationship, it seems that this is another clear indication of an offset message--whatever it may be--and that it may be connected with some larger mystery involving the Great Seal's values, or absolute heights and widths of the ellipses.

* * * * * * * * * * *

The *Angle* between LDO and LDI

The internal angle between the LDO and LDI diagonals is computable through the tangent function since we know the dimensions of their sides, and then by subtraction of the resulting angles. The very minute difference is just a little more than 12 minutes of a degree:

Illus. I1: Angle Between Diagonal
Lines LDO and LDI *(from LOR and
LIR corner points. Not to scale.)*

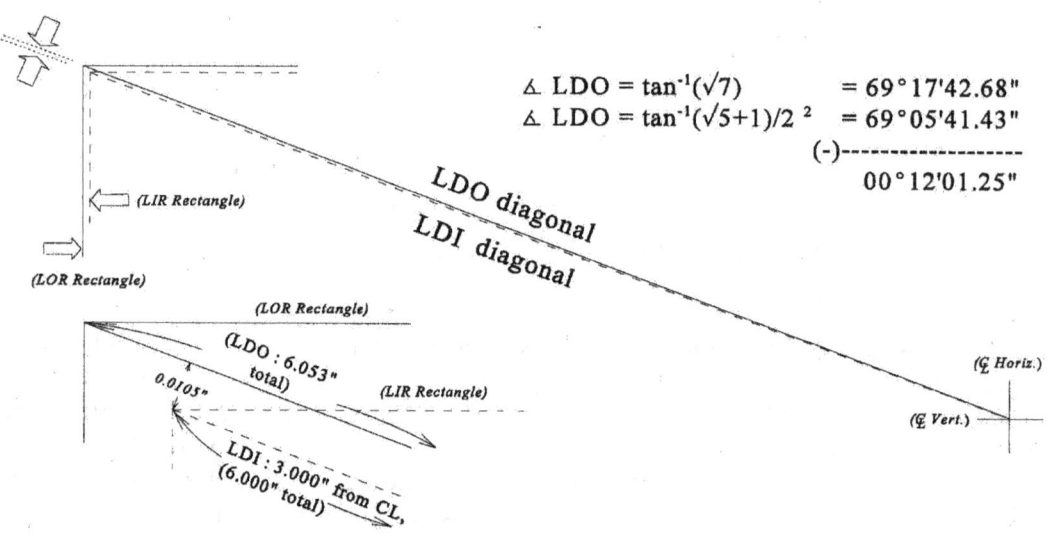

$\angle$ LDO = $\tan^{-1}(\sqrt{7})$ = 69°17'42.68"
$\angle$ LDO = $\tan^{-1}(\sqrt{5}+1)/2\ ^2$ = 69°05'41.43"
(-)--------------------
00°12'01.25"

(t)	the $\sqrt{7}$ angle:	69°17'42.68"	(LIR rectangle, the angle for the LDI diagonal.)
(t)	the φ^2 angle:	69°05'41.43"	(LOR rectangle, the angle for the LDO diagonal.)
		(-)------------------	
		00°12'01.25"	Twelve minutes, one and a quarter seconds.

In other words, we have computed the angles by the arc tangent of the rectangles' fractions from measured dimensions and subtracted to find the angle of this tiny, invisible sliver.

Over a diagonal of six inches of length, the little 12 minute angle would make a difference perpendicular to this line of about two hundredths of an inch:

(t) $\tan(0°12'01.25") \cdot 6"$ =
          $0.003496732 \cdot 6"$ $= 0.02098"$

As we will see in later developments, this little angular difference seems small, but jammed with information,. Here again, a reader who had skipped ahead to Chapter 4 would immediately recognize something special in the little tangent value "0.00349," which I did not see at the time. We will eventually see that even the little perpendicular difference of 0.02098" may be meaningful as part of a general mathematical motif.

****

The Pyramidal Form:

During the late 70's to early 80's I succeeded to a small extent in finding various patterns that were built into the little Pyramid's courses.

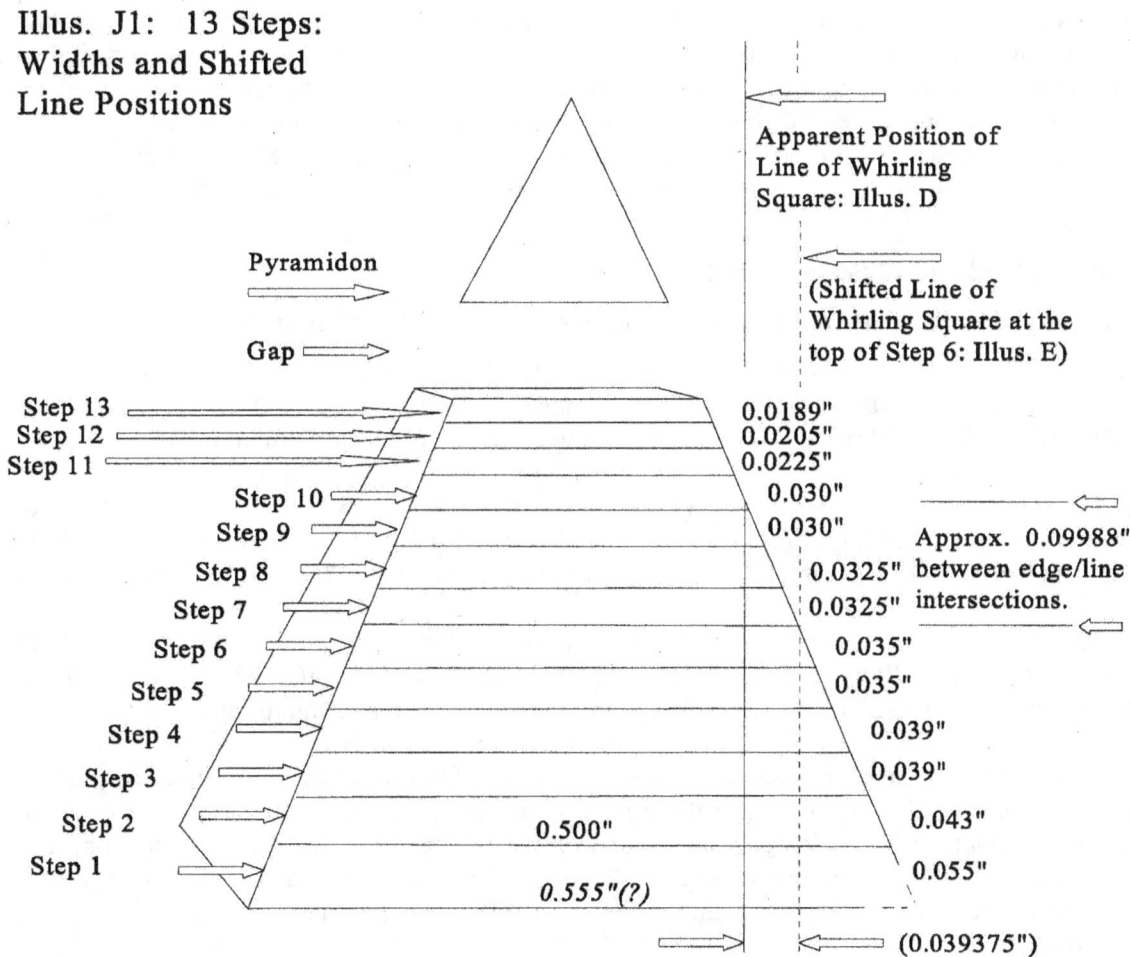

**Illus. J1: 13 Steps:
Widths and Shifted
Line Positions**

Pyramidon

Gap

Apparent Position of
Line of Whirling
Square: Illus. D

(Shifted Line of
Whirling Square at the
top of Step 6: Illus. E)

Step 13 — 0.0189"
Step 12 — 0.0205"
Step 11 — 0.0225"
Step 10 — 0.030"
Step 9 — 0.030"
Step 8 — 0.0325"
Step 7 — 0.0325"
Step 6 — 0.035"
Step 5 — 0.035"
Step 4 — 0.039"
Step 3 — 0.039"
Step 2 — 0.043"
Step 1 — 0.055"

Approx. 0.09988"
between edge/line
intersections.

0.500"

0.555"(?)

(0.039375")

Looking at the *top three courses* of the little pyramid, I noticed that they were radically different in character to the rest of the courses, very much out of proportion over all. And these last three seem to have been set up to announce something of interest. The imaginary left line of the edge of the *first whirling square* shown in Illus. D was quite near this change in architecture at the dividing line of "stone courses" at the top of Step 10 or bottom of Step 11 (later I found to be closer actually to the bottom of Step 10, about 0.00487" above Step 9.). It was this coincidence that first drew my attention to this area, yet the difference in courses was not immediately obvious or important until the position of this line raised the question and made them stand out. Looking at the steps on the pyramid as printed on the dollar there isn't much to see. They are only slightly different in appearance to the eye. But through measurement, big differences begin to surface in all of the steps. (See Illus. J1 and K1 below.)

The 13 Steps and Step Fraction Theory:

These last three courses, the 11th, 12th and 13th, are odd with respect to the other steps. They are clearly a non-linear pattern atop Step 10: all three being quite a bit too small for the whole. Of course these should decrease in size going up, but they are out of place even for this top area. This shows up very clearly when their widths are graphed, as well as other oddities in other steps and groups of steps. (See Table 1)

These steps turned out (surprisingly) to be the 110th, the 120th and 130th *mathematical divisions of the length of the Small Inner Rectangle*, respectively (SIR, see pgs 41 and 18.) Notice the decimal, or ten-fold shift of values with respect to the course number. Note the switch: it is a *vertical* reference (the height of steps) to a *horizontal* length (the width of SIR rectangle.) Also note the reference being made to an otherwise very obscure rectangle, the SIR, a proportion to something, strictly speaking, is invisible and almost entirely unmarked. If you think about it, the SIR discovery must be solved *first* before the mathematical significance of these three little steps will stand out. Although the lines dividing the steps are a little rough and wide, the proportions stand out quite clearly. Note how closely these little step heights match these ideal proportions with respect to the length of the SIR rectangle: (J1, J5 and J6) These measurements are (of necessity) estimates made to the apparent center of the course dividing lines. (See Illus. J1.)

Steps 11, 12 and 13 (as measured, m)

Divided into the (theoretical, t) SIR Length of 39.37 1/16ths of an inch:

Step 11   (m/t)   0.0225" / 2.460625"   = 1/109.36
If the intent was a 1/110th of SIR: (t)   0.022369318"   (0.0224")   = (1/110)
a difference of: 0.00013" being 0.0001" in rounding at the ten-thousandth of an inch.

Step 12   (m/t)   0.0205" / 2.460625"   = 1/120.03
If the intent was a 1/120th of SIR: (t)   0.020505208"   (0.0205")   = (1/120)
a difference of: 0.000005" being 0.000 *or no difference* in rounding at the ten-thousandth of an inch.

Step 13   (m/t)   0.019" / 2.460625"   = 1/129.50
If the intent was a 1/130th of SIR: (t)   0.018927884"   (0.0190")   = (1/130)
a difference of: 0.000072" being 0.0001" in rounding at the ten-thousandth of an inch.

These are *really* small differences with respect to this fraction theory, which seems quite successful. The closing error value of the above three measurements are 1:170, 1:4000 and 1:259--the average of three being about 1:1500 which could be called acceptable per Rule 3. (As a comparison, if instead we try the same error procedure using the 2.500" length of SOR, we will net a total average error of only 1:120. This clearly reinforces the impression that SIR was the intended length for these implied proportions.)

The 2.460625" used above is the same "39.37 1/16ths" length of Chap 2. *If instead* we used the theoretical IDT derived form (which is to say, the square root of the IDT rectangle diagonal), 2.46034...", these fractional values will give practically the same results; there being only about 3 ten-thousandths difference between the two over all hypothetical lengths of SIR. (This is not a defining difference--much too small--so it doesn't seem that the proportions can be used in backward calculation to see which of the two slightly different SIR-type lengths is preferable.)

What could all this mean? At first it appeared that these measured steps sum to about 0.062" or about a 1/40th part of the Small Inner Rectangle's length. And this was suspiciously close a 1/39.37th part of SIR, which would be 0.0625" or a 1/16th of an inch division as we saw earlier (See pg 18 et seq.) This would be only about a half of a thousandth (0.0005") difference, so this could easily be another interesting subtlety to work with. (i.e., first we find a slightly "fat" millimeter, and now maybe, a "short" 1/16th? -- J3). But I don't take this idea seriously at this point in time--more compelling ideas surfaced later.

Why did the Designer make this fraction arrangement with the steps? Since this little unfinished Pyramid's progressive courses are widely thought to symbolize the progress of American Civilization, (Manly P. Hall, et al), was this some sort of a statement about America's history? Up to when? To some point in the future? Were there specific historical meanings for these fractional parts? Just how or why would the Small Inner Rectangle's length figure into the little Pyramid's height?

One of the problems with this idea was that the United States could not really be said to be "finished" up to any given point or level, like the little pyramid's flat top. That is, unless the top tier is intended to indicate some specific point of time past or future. I don't believe the steps ever had any traditional symbolic meaning in terms of time, they were just the thirteen steps for the thirteen colonial States, and they weren't specifically named. None the less, it surely seemed likely that a political or historic statement might be hidden here. The Pyramidologist school of thought claims that there is a prophetic time line in the halls and gallery in the Pyramid of Giza. This is said to stretch from the very distant past to the present day. Was the Designer trying to make a similar statement with this little pyramid? "1776" appeared to be the obvious starting point, what with it being written just above the line of the base of the little Pyramid. As far as I could tell in 1980, there wasn't any pattern of important or historic events to be found on the little pyramid. We will return to this later more successfully from another kind of approach in Chapter 5. (For the past, at least, there is symbolic meaning: see pgs 98, 100 and 130 et seq.)

I saw what seemed to be a possible hint of prophetic character that might have been implied from the three numbers found in the ratios, arising from the curious fact that $110 + 120 + 130 = 360$. *"360 days"* is widely thought by students of Biblical prophetic lore to be the length of the *"solar prophetic year"* as computed as regular days. These are supposed to be summed through many regular biblical or historical years. This is said to been the form used by the prophets of the Bible such as Daniel of the Old Testament. But in addition to this, is the interesting fact that the *sum of the squares* of the numbers of our steps are the number of whole days in a *regular year*:

$$11 \cdot 10 + 12 \cdot 10 + 13 \cdot 10 \qquad =$$
$$110 + 120 + 130 \qquad = 360$$
But also as squares:

$$11^2 + 12^2 + 13^2 \qquad =$$
$$121 + 144 + 169 \qquad = 365$$

So, this could be coincidental, I thought. But this may have been intended as a sort of literary device for a prophetic year, using number symbolism. [2]

From Table 1, we can see that there are *three types* of "steps" with at least three sloping rates of change in this Pyramid's increments. (These "rates of change" are merely my convention for distinguishing these different kinds of courses on an exaggerated graphing scale, created by graphing the sizes and connecting the middles of the columns with straight lines.)

Steps 1 and 2:

The first and second steps show one "rate of change" between them, from big to middle range in size, and these seem to be somehow independent from the rest of the pattern. Can two steps really be a *rate of change?* The first step contains the "1776" notation in Roman numerals, and as noted earlier, its base appears to measure approximately "0.555" in length, but with the tiny bush or shrub somewhat obscuring the right end we cannot entirely be sure. (See dashed line area at lower right corner of the little pyramid. There is much more to this question, as we shall see in Chapter 4.) The step length is about ten-fold of the height of this step, or, 0.055". The second step is much smaller in size, 0.043" and *may be intended to be a tenth of the total unfinished height* of the Pyramid, being measured to lie between 0.4320" to 0.4325". The base of this second step is apparently intended to be 0.500" *or exactly one half inch.* (Perhaps a tenth of the length of the right and left SOR rectangles totaling 5 inches?)

We saw above that the heights of the final three steps 0.0225" + 0.0205" + 0.0190" sum to about 0.062". Dividing into the unfinished Pyramid's height of 0.4325"--gives a fraction of about 1/7th. Were it *exactly* "1/7th" of this height, it would equal about 0.06178" which is suspiciously similar to 1/10th of the reciprocal of the golden ratio or, 0.0618..." (J3, J4) Now, *if* this is the case, then the unfinished height might be calculated to be 0.0618..." multiplied by 7, or 0.43262...":

(t) $(1/10 \cdot \varphi) \cdot 7 = 0.432623786...$"

which would be quite believable--being only be about a ten-thousandth of an inch longer than the largest estimate that varies between 0.4320" and 0.4325". But this is by no means necessarily correct--we will look at better theory on this unfinished height later. Why *these three lengths* for the top steps? This question (a real albatross) remained with me until late in 1990, when I stumbled across a solution (See Chapter 4, pgs 169 and 170.)

Steps 3 to 10:

The next eight steps down--a very separate category--are in progressively smaller *pairs of heights,* (also giving a non-linear slope with respect to the graphing of the respective heights. See Table 1 and Illus. J1.) Except that they progress evenly as a non-linear slope and grouped *in pairs* of the same size, nothing else seems to stand out in this series of steps. But what is interesting though, is that the *opposite shifted position* of the Small Inner Rectangle shown in Illus E, (See also Illus. J1 above) we see that the line of the first "whirling square" *cuts neatly between the pairs of steps,* at the edge line, at the top of the 6th course and bottom of the 7th. The line cuts at the dividing point between the *upper two pairs* and the *lower two pairs* of steps, thereby appearing to point to some sort of definite significance at that place (J6).

Does this middle point *only* mark the line position of the whirling square? It is as though the bottom two steps and top three steps were perhaps important to our Designer as a place to bear their own messages. But these *inner eight* may have served only to *mark the opposite position* of the crossing of the line of the square. Is it possible, after the Designer laid out these special sets of upper and lower groups of steps, he then divided the remaining area into eight steps--exactly four above and four below--using some unknown proportional scheme, so as to mark this line? But these steps are found in *Pairs of pairs,* and they don't appear to follow a linear pattern. Notice that even the four *differences* between pairs (beginning at Step 2) produce regular *pairs of differences as well:*

Step Widths: Difference Pair I	(2-3)	0.043"	minus	0.039"	=	0.004"
	(4-5)	0.039"	minus	0.035" and:	=	0.004"
	(6-7)	0.035"	minus	0.0325"	=	0.0025"
Difference Pair II						

$$(8-9) \quad 0.0325" \quad minus \quad 0.030" \quad = \quad 0.0025"$$

(A *Pair* of Pairs)

Were they only made regular so that the other steps or lines would stand out? Why in *pairs of heights* and *pairs of differences?* Something like this symmetry appears above in the diagonal intercepts to ellipses shown in Illus. H4 and H5. If the intercepts are any guide, the "pointer" could be the longest of these course lengths, or 0.500" (the beginning difference) at the top of Step 1, bottom of Step 2. But that doesn't appear to "point" to much yet. The whole of the Steps' mathematical series do not appear to have any seamless explanation, but there may yet be an easy one that I can't see. More on this later, (pg 169 and 170.) In Chapter 4 we will look at another interesting idea touching on another possible symbolism of increasing mathematical powers that these steps may indicate symbolically. (See pgs 142 and 143.)

****

Ken McGrath

Other Groupings of Steps:

But there are many other things of interest here. From the top of the first step at the base course (having the even 0.500" or, half inch length) to the line cutting at the sixth course lie *five* steps. If the oddly large, bottom step is momentarily ignored, we are left with twelve steps above, being divided by the cut at Step 6 by the shifted whirling squares line, and the next at Step 10 (marking the odd set of three above) by first whirling squares line position (See Table 1):

<u>Group I</u> From the top, one might notice that there is a sequential and familiar pattern of numbers for a right triangle of *three, four and five:*

3, (top of 13 to top of 10); 4, (top of 10 to top of 6); and 5, (top of 6 to top of 1).

And:

<u>Group II</u> The steps can *also* be said to make another more arcane and ceremonial division of the numbers of *three, five and seven steps:*

3, (top of 13 to top of 10); 5, (top of 6 to top of 1); and 7, (top of 13 to top of 6).

First, the *three, four and five* rectangle is sometimes called the Pythagorean Triangle. It among the ancient working tools of the Builders to produce a square corner, since these three lengths end to end will create a perfect right angle in one corner. This is of interest to anyone concerned with ancient geometry and this pattern seems numerically straight-forward and un-ambiguous as a symbol. The three particular edge *lengths* will not work as a right triangle. (However these three lengths as clues to other triangles is intriguing: 0.062", 0.125" and 0.191".) Secondly, there is a curious possibility in the Group II pattern: a possible arcane Mystery symbol. Students of Masonic lore will promptly recognize *"the three, five and seven steps"* (and even nominally as vertical steps). We can either look at all of these groupings as a

J6-relationship (the mathematical idea) or alternately as a J7-thing (a symbol having Masonic character). (There are some of us, who might object to this arbitrary segregation of Step 1, and immediately point to the *another* possible sequence of 3,5 and 8, being the fourth, fifth and sixth Fibonacci Numbers that relate to φ-based *number series*. We might then propose a possible Group III from all thirteen steps.) Once again, a casual first glance at the little Pyramid reveals no pattern at all, just decreasingly smaller steps. Just what did the Designer intend here? Does this guy give us our money's worth or what?

****

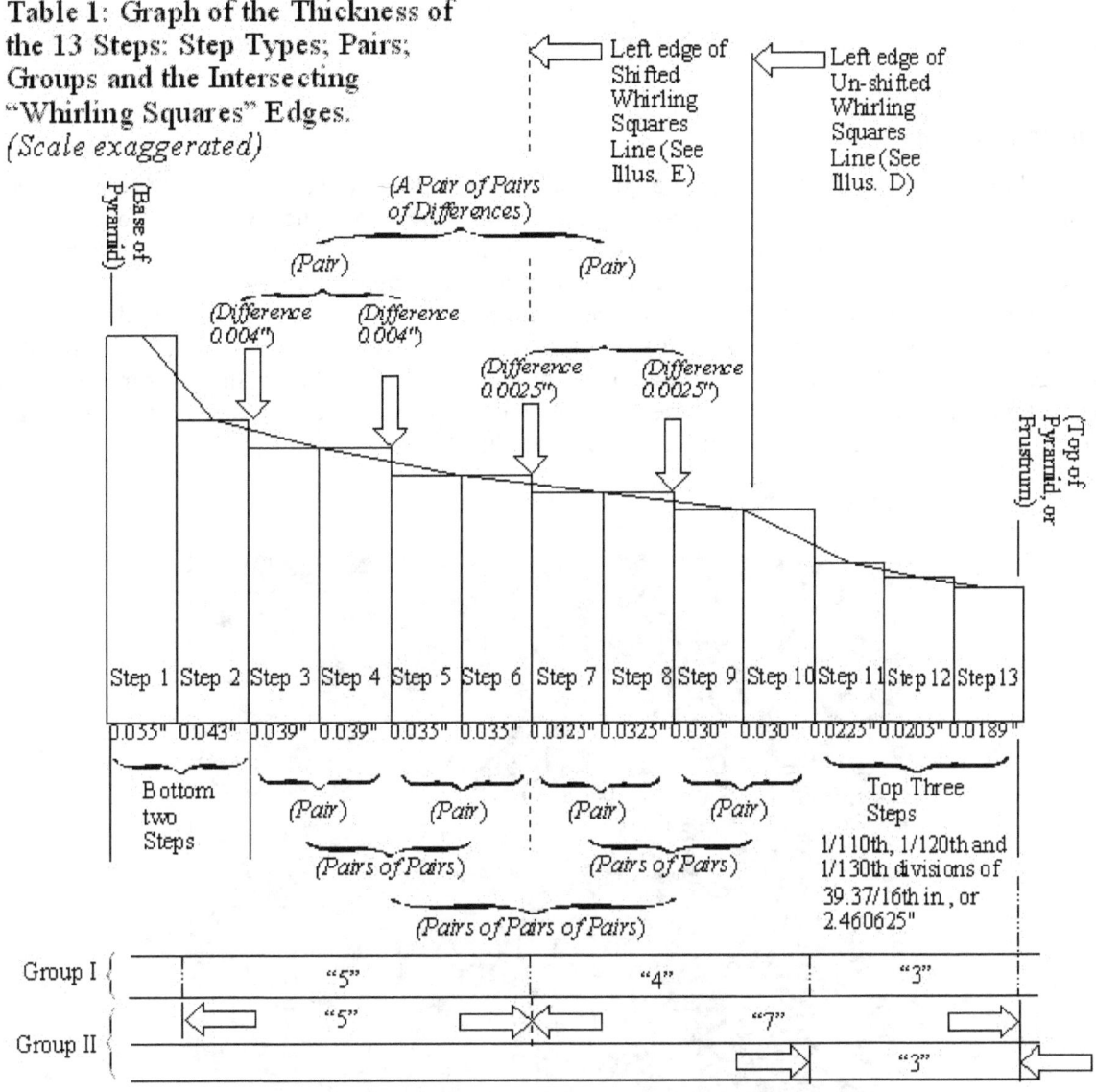

Table 1: Graph of the Thickness of the 13 Steps: Step Types; Pairs; Groups and the Intersecting "Whirling Squares" Edges. *(Scale exaggerated)*

I needed more insightful calculations rather than more and better measurements. The cheap four function Casio calculator gave way to a TI30 (a then popular Texas Instruments calculator with a red light emitting diode display that ate lots of batteries), and then to a succession of more sophisticated calculators--not all very useful in their arrangement of functions. At least ten places, several memories, a square, square root and easily used trigonometry functions were a must.

After work at the office, (in the late 1970's) I had the occasional use of a programmable Wang computer to solve triangles and other shapes in survey work, but this was very slow and it was a take-a-number situation with other users. This was a wonderfully noisy machine, a continuous din of sharp clacking and rattling of printing pins. This dinosaur was of the age that required you to feed it a group of cardboard punch cards to load the program *each time* you did a kind of calculation. And, these cards had become dog-eared, dirty, grubby and hard to use. Having to ferry the data back and forth to this machine, I had a hard time keeping my train of thought. This was a period at the edge of the *calculation table book* and *computers*. There was no such thing as a "PC computer" then. The term *"IBM computer"* as understood then, meant a giant machine that filled rooms and *floors* at the university with its own air conditioning plant and a team of very important looking guys with white lab coats. *Really.* And it wasn't all that long ago. Soon hand calculators improved and got cheaper. The best results came from doing calculations on a hand calculator following intuition, from instant to instant as ideas took shape.

****

The Pyramidon:

Careful measurement and drawings of angles forced me to discover some further oddities on the little Pyramid design. The Pyramidon, or Eye-and-Triangle structure above the Pyramid, is one of the chief features of the dollar's design that draws people's attention. Although I have learned much about the Pyramidon, I have never felt that I have a complete understanding of it mathematically. And it seems reasonable that if the steps of the little Pyramid require a prior understanding of other hidden things (such as the Small Inner Rectangle or SIR), the Pyramidon may require the same or other kinds of background knowledge. This turned out to be true. None of the measurement clues did anything for me at that time, I just wasn't ready. But I was ready for some of the geometry. When certain telltale angles and alignments didn't appear to fit, another collection discoveries was at hand.

Illus. K1

The first geometry clues appeared sometime in 1977, when using a straight-edge to project the right hand edge line of the little pyramid's "rock face" to look for alignments outside the seal.

To my complete astonishment, I saw that the lower-right corner of the Pyramidon protruded *slightly past* my metal straight-edge--about to the halfway point on it's tiny slope edge. Like many past surprises, and many to come, my first reaction was small, sharp annoyance, immediately followed by a smile of awe and recognition.

70

Although this difference is invisible at first without the aid of a straight edge, it stands out a little after you become aware of it--almost like the peculiar top three steps. (Does this edge line up with anything? Yes, it lines up with the left edge of a "flower" near the base of the little pyramid. See Illus. K2)

After checking thoroughly, and marking the apparent position of this angle, I repeated this projection, now with the right edge of the Pyramidon and saw that this was a seriously different angle, amounting to quite a bit more than a degree. (Compare Illus K1 and K2.)

Looking at the Pyramidon, *it was an entirely different kind of design* than the seemingly regular Pyramid below it. Although it looks as though it might be a continuation of the lower "stone" structure with a gap above the thirteenth step, it is not even similar. If you look at it closely, it is two dimensional in character, quite asymmetrical--a scalene triangle, indeterminately suspended in space, with no depth clues at all. The Eye, of course, does have artistic depth, but it is still "on" or "in" this triangle.

Illus. K2: Angular Difference: Edge Lines of Pyramid and Pyramidon

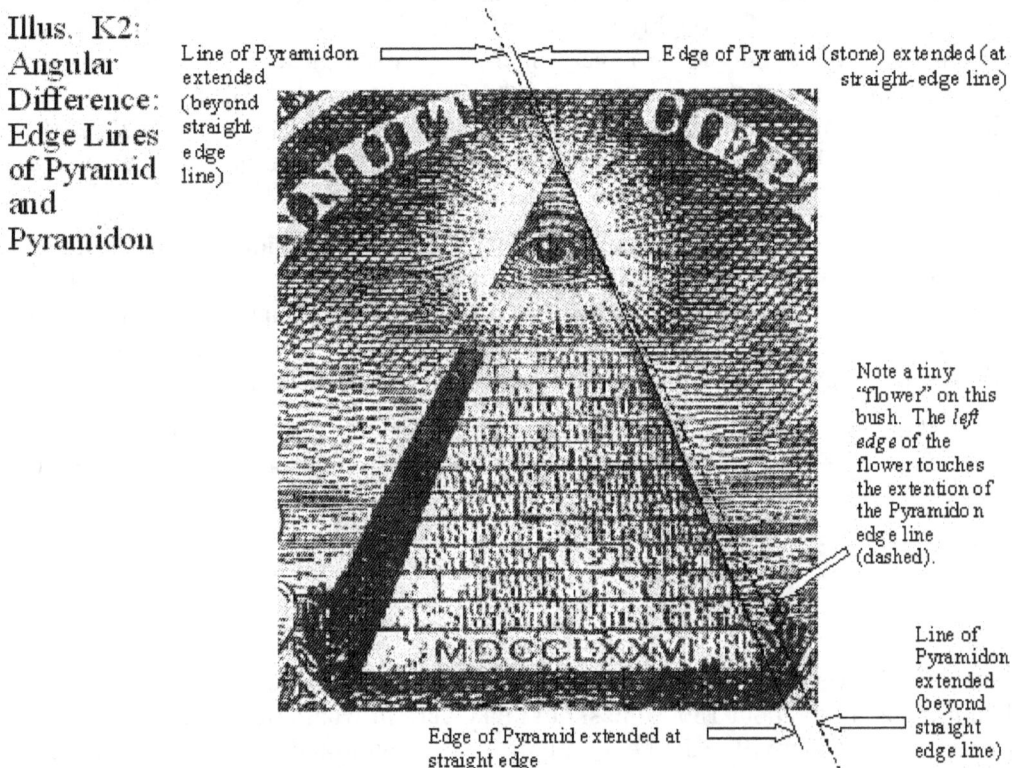

Line of Pyramidon extended (beyond straight edge line)

Edge of Pyramid (stone) extended (at straight-edge line)

Note a tiny "flower" on this bush. The *left edge* of the flower touches the extention of the Pyramidon edge line (dashed).

Line of Pyramidon extended (beyond straight edge line)

Edge of Pyramid extended at straight edge

At some point around this time, my Uncle showed me a small optical device mounted on a small stand. It was a piece of used technical junk that he had acquired somewhere. Apparently it was originally intended to measure the angles of tiny crystals of geological samples, such as sand grains. This microscope-like tool had about 20 power optical magnification and was a high quality instrument. It had centered cross-hairs and a vernier to measure horizontal angles. It was very much like a surveyor's transit, but oriented up and down, vertically. It could measure angles circularly about a point from above, accurately to about a minute of arc, and he lent me the use if it for a while.

At first, this instrument seemed ideal for measuring small angles of the size of the tiny Pyramidon, but it turned out to be hard to use and a little deceptive. What I did find, proved more odd the more I used this tool. Note in Illus. K2, that the extension of the right edge line of the Pyramidon is substantially wider than the edge lines of the "stone" structure. Yet a look back at Illus K1, gives little or no hint of this fact at first glance. Again, an example of the wily Designer's artwork.

Note the little "flower" in Illus K2--also unnoticed at first look. Its only a white area, but it is at the end of a plant, and at the edge of a line projection. Is it a *rose*? e.g., the Rose of the Rosicrucians? It could be, but it's too small to tell. (J7, or a possible message.) This might be a geometric metaphor for the Latin term *sub rosa* ("under the rose"), a phrase denoting a thing holding a secret, such as it came to mean in the 1600's in the time of the appearance of the German Rosicrucians. This was from the time when a public place like a tavern with a picture of a *rose over the door* was supposed to have indicated that one could speak freely without being reported to the authorities. One would then be: *under the Rose.*) This "flower" atop the sprig of greenery has several other alignments not shown in this work.

The slope angle measured from the *base* of the little Pyramid, appeared to be about 68°30', at first appearing symmetrical on both sides--and internally about 43° at what would be the projected apex at a point. This projected point falls *within* the triangle of the Pyramidon. Which is to say, the "virtual" apex of the sides is *not the same* as the Pyramidon's apex, but somewhat below and within it. The Pyramidon though, *is not* a symmetrical figure, and (after much measurement and calculation) appeared, at first, to have an internal angle of about 51°55 +/- 03' at its apex. Its right slope angle was about 65°30' +/-03' and its left slope angle came out to be roughly 62°30' +/-03' or thereabouts--this wasn't very accurate, but it did show asymmetry. This Pyramidon was very interesting, since this angle appeared to be within 0°04' (or four minutes of a degree) of the slope angle of the Great Pyramid of Giza. Further examinations of the Pyramidon's placement showed that it *wasn't centered on the same centerline as the little Pyramid.* It was a tiny amount to the left of the little Pyramid's centerline, from the projected extended lines of the unfinished Pyramid.

Around this time I had become aware that the Great Pyramid had a similar, small offset marked at the middle of it's base, and this opened my research into what this sort of thing might mean in the Pyramid itself. I thought I would try to figure out what this was first, before approaching this tiny offset on the dollar. (This was the source of curiosity for the research behind the section called Entasis. I have not, however, solved the tiny pyramid/pyramidon offset question on the dollar.)

Much later, after more careful measurement and multitudes of trial calculations, my best estimate of internal angles and lengths were as follows (Also, see Chapter 5, ASE1 or, All Seeing Eye #1):

(t)   51°50'23" for the top;
(t)   62°34'36" lower left: and
(t)   65°34'00" at lower right.

These numbers, however, are one of several *theoretical solution* of this triangle. Although these angles will add up neatly to about 180°, one should take care in what one is prepared to believe of these numbers. *One minute of arc* will make about a half of a *ten*-thousandth of an inch on the dollar, which cannot be reliably measured. The printed lines that this figure are made up of are two-hundred times fatter than this, and the grainy edge varies between ten and twenty times this wide.

This theory seemed to match the facts as I had found them for the little triangle, and this theory remained my belief for quite some time. But these three lengths and angles are only a proposed *mathematical proportion.* I used a large photo enlargement for the measurement of these proportions-- and so the *dimensions here are approximate and I now think they are slightly in error.* But the *proportions and angles* here are fairly good--I originally thought the minute of arc was believable in this photo scaling. This solution ignores and bypasses several of my early trial ideas, and we will look at this triangle again, later. Note that the *apex angle* is very similar to the *slope angle* of the Great Pyramid of

Giza. Whether or not this is the exact solution to this printed triangle, it is enough to know that this apex is substantially similar to the angle found at the slope from the base of the Great Pyramid. (See Illus. K3 and K4.)

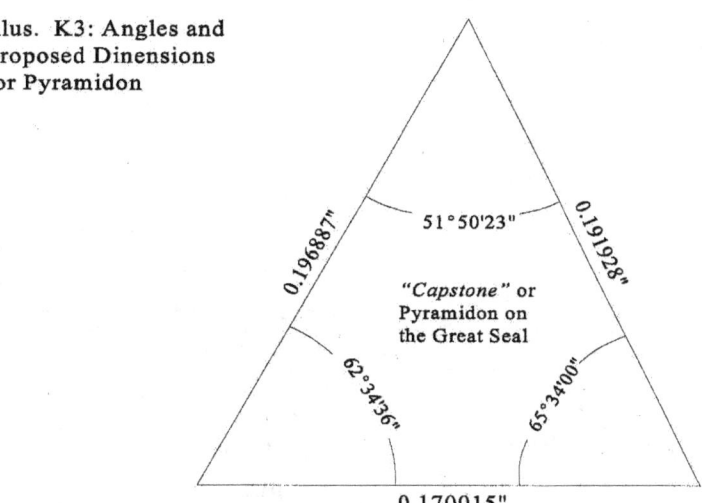

Illus. K3: Angles and
Proposed Dinensions
for Pyramidon

51°50'23"

0.196887"

0.191928"

"Capstone" or
Pyramidon on
the Great Seal

62°3436"

65°3400"

0.170015"

This shape was faintly irritating. The Designer was obviously trying to make a hint but it was not at all clear what could be accomplished with this shape. The guy just couldn't keep from tweaking shapes-- it was asymmetrical, didn't match the pyramid below it, the centers didn't line-up, nor were their origins at the same point. Maybe this was to allude to the difference between the worldly life (the lower stone pyramid) and the sacred (the Pyramidon above).

I knew that the apex angle 51°50" was the side-wise slope angle of the Great Pyramid, *so this was out of place* (J1, J2 and angular J3). I had seen this kind of thing before, elsewhere. I don't know why, but took a long time to imagine the transposition of these angles and consider what they might mean. But what with switching these angle ideas around in my mind with a knowledge of the architecture of the Great Pyramid, an English language construct came to mind as the answer to the puzzle.

* * * * * * * * * * * *

Pyramid Geometry and I Peter 2:6 Theory

In the midst of this complex inter-relationship of triangles, centerlines and various slopes there appears the first solid relationship connecting this design to the lore of the Pyramid of Giza, by the angle of about 51°50", but *also to Biblical* as well as *Masonic* symbology, simultaneously (J6, J7). The Great Pyramid of Giza has been measured many times and written about by many authors, most all of whom agree on a slope angle of about 51°51" to 51°50" (See pgs 77, 89 and 200). And with this evidence, a classic message can be puzzled out here on the dollar. Whereas the slope angle of 51°50' on the Great Pyramid of Giza *rises from a horizontal base--like a cornerstone*, the same angle is *formed at Pyramidon's apex--as a capstone*. (Note J1, J3. See Illus K4.) Following this identification of an angular message, there is an easy temptation here to read from a passage from New Testament at I Peter 2:6 to 2:7 which seems to precisely parallel the geometric argument.

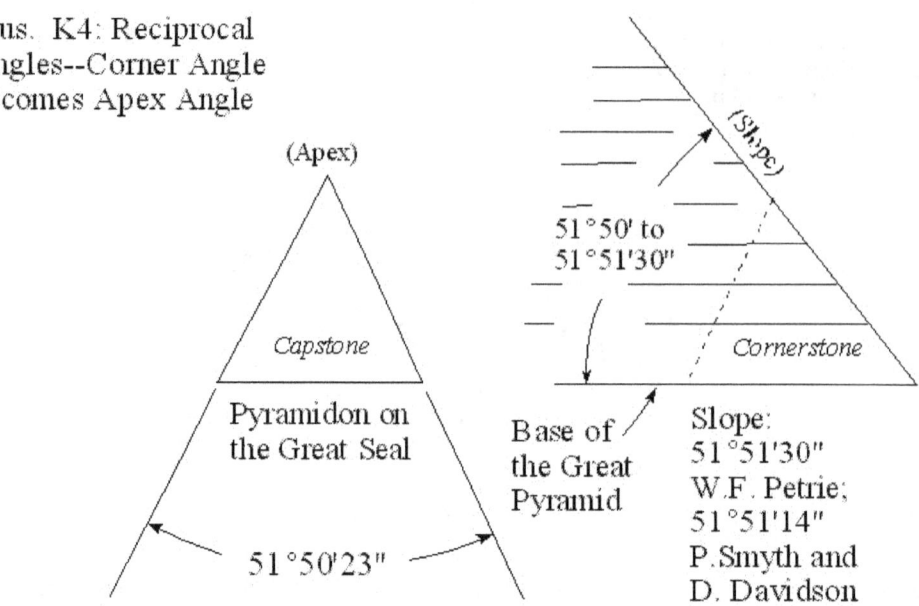

Illus. K4: Reciprocal Angles--Corner Angle becomes Apex Angle

(Apex)

Capstone

Pyramidon on the Great Seal

51°50'23"

51°50' to 51°51'30"

(Slope)

Cornerstone

Base of the Great Pyramid

Slope: 51°51'30" W.F. Petrie; 51°51'14" P. Smyth and D. Davidson

Peter opens the verse by writing of a "living stone" and shortly thereafter continues on with a related recitation taken from Isaiah 28:16:

"See, I lay a stone in Zion,

a chosen and precious *cornerstone*..."

and, in the very next verse, he then references Psalms 118:22:

"The stone that the builders rejected

has become the *capstone*,"

Which seems to me to be much more than coincidental symbolism, considering the likely motives of the Designer. What is really fascinating is, that *without a word or letter of text*, at least three symbolic traditions were accurately called upon *and* linked together by the use of geometry alone--but very precise geometry and really quite minimal graphic clues. Once again, a reciprocal idea of the J2-type, though it was not our Designer's idea for once.

This must all be seen with *respect to us, the belated observer*, who must have been intended as a most unusual and specialized target audience. That is, those who can (1) observe the measurements, (2) interpret or understand these sources, and (3) translate them for symbolic meaning. So here, some of the Designer's message begins to take shape in a definite form. And yet no specific sectarian device surfaced here, apart from narrowing it down to a likelihood of generally Christian symbolism, together with Masonic lore and Pyramid lore. Strictly speaking, Occam's Razor probably does not require *Christian* symbolism as an explanation, since both connecting passages appear originally out of the *Old Testament* from the Hebrews. But what little that can be seen speaks volumes, and Peter is almost certainly the source of inspiration. (Also see Matt. 21:42, Rom. 9:33, Acts 4:11, Eph. 2:20, 1 Peter 2:3-8.)

There is even a another possible connection to the Old Testament verse in Psalms that we will look at in Chapter 5. Later we will briefly consider the verse *Psalms 118:22* in another possible connection to the Pyramidon, considered as a measurable number (See pg 166 illus P14).

****

What do we know about the Designer and his theme at this point? He must have had a planned agenda of revelations, and the arrangement is suggestive of substantial logistical organization and hours of thought and calculation. We knew that he was aware of mathematical functions, and used them to quietly draw attention to parts of the dollar. The J1-puzzle form is, evidently, not limited to rectilinear forms but can be angular in character, as seen just above.

The Designer's ideas appear in unfolding stages, after the simpler problems are solved. We saw this with the outside LOR-rectangle's diagonal, having a square root being nearly the length of the SIR-rectangle. And then, the little pyramid's top three steps as exact divisions of the SIR-rectangle length. But we don't know yet why he likes this length, nor the LOR-rectangle length at this point. But there is a connecting trail in these relationships.

We can safely infer that the Designer was familiar with the Bible. The connecting reference of both the cornerstone and capstone are clearly New Testament. Yet this has to be considered together with his method of demonstration through unusual symbolic artwork. No verses were cited and there are no connecting arrows or legends to do this trick. It was done through the creation of precisely measurable symbolic shapes with angular identities that, once recognized, are unmistakable.

Unmistakable? Well, no. That is an exaggeration--it took me a long time to grasp the idea. The evidence is there, if one is aware of the significance of certain architectural details of the Great Pyramid of Giza, namely, the slope angle. Only *this angle* (51°50" to 51°51") can identify what would be a cornerstone at the base of the Great Pyramid, and that this particular angle would be out of place if found anywhere else. If it is then found at the apex of a pyramidon, which is also nominally a "capstone," then the cornerstone/capstone exchange from I Peter 2:6 to 2:7 becomes comes to light. This relationship is only unmistakable if *you* know of this part of the Bible, *as well as* technical lore of the Great Pyramid.

Another thing to take into account, is the technical aspect of "precisely measurable symbolic shapes" discussed above. It would be one thing to fully know both of the Pyramid and Biblical references, and quite another to detect them. It seems likely to me that the Designer had intended these angles to be measured by someone with a protractor on a photographic enlargement. At that point, the well prepared observer might be ready to put the message together.

In terms of the Designer's allegiances, it is quite likely that he is a believer in God, and specifically a Christian. On page 24, Illus. E, the centered, fifth rectangle exactly aligns with the letters to spell "HE" at the top of the dollar, which appears to be a religious tribute to the Deity. In the cornerstone/capstone reference to Peter, we can narrow this down to Christian. Yet not just Christian. He appears to favor a connecting theme to the Great Pyramid lore that comes down to us from Piazzi Smyth. (Beyond this, there is an interesting Masonic connection. There is evidence of his using the Masonic symbol "G" in a φ-based diagonal alignment beginning at the lower lefthand corner of the LOR rectangle to the *inside serif* of the letter G in "IN GOD WE TRUST". Therefore, the golden ratio angle: 31°43'02.91", the first letter of the word GOD, or : G, the fundamental symbol of the Supreme Being and geometry of Freemasonry, are all connected together in one relationship. But for lack of space, this alignment is not shown in this book.) So we can see that in his point of view, the Deity is adorned by certain geometric concepts (φ and π), a very Masonic-type of idea. And these are found together with a Christian interest or viewpoint. Perhaps with this individual we might say all of these concepts are "factored in" together.

The Great Pyramid's Signature:

The mathematical "footprint" of the Great Pyramid appears elsewhere on the dollar in quite a few places. Although, again, somewhat out of order here chronologically, research into the oddly placed Seal faces finally produced a useful puzzle-piece for interpreting the middle area of the design: The distance between Seal faces. The clue lay in the implied distance between the apparent centerlines of the Seal faces, as measured from between the left and right edges of the Large Inner Rectangle's borders to the outer edges of the outer ellipses; and the half-widths of the outer ellipses; when then subtracted from the length of the Large Inner Rectangle. When this distance is compared with *the height* of Large Outer Rectangle (LOR) the Great Pyramid's geometric profile appears:

Between Seal Centers #2 (BSC2):

(m)	0.514"	=	LIR to the outer edge of the Lt. Outer Ellipse (See Illus. G5)
(m)	0.514"	=	LIR to the outer edge of the Rt. Outer Ellipse (See Illus. G5)
(m)	0.595"	=	1/2 of Lt. Outer Ellipse (centerline: 1/2 of 1.19" Illus. G5)
(m)	0.595"	=	1/2 of Rt. Outer Ellipse (centerline: 1/2 of 1.19" Illus. G5)
	(+)----------		
	2.218"		

Then--

(t)	5.61248608..."= SOR (or, $\sqrt{31.5}$ as per theory) (m = 5.612")	
(m)	2.218"	
	(+)----------	
(t)	3.39448608..." = BSC2	

When *physically measured*, (from the right-hand outer ellipse to right-hand outer ellipse of these Seal faces, and from left-hand to left-hand outer ellipses to the outer edges at the ends of LIR) I got about 3.395". This is about right for matching this theoretical length BSC2. (But this does not approach the earlier proposed BSC1 which would be a whole hundredth of an inch wider than this.)

Notice that I am using the *mathematical* LIR length as a basis, where it is used in conjunction with measured dimensions simply taken to the *round thousandth of an inch*--as though *these were the intended numbers to be used*. Although I won't dispute my idea of the LIR fractional length of $\sqrt{31.5}$" it *still remains theory*, as well as any figures derived from it. This is one of the few dimensions on the dollar that I never felt any reason to doubt that I knew the Designer's intent. If the Designer intended this $\sqrt{31.5}$" length, he couldn't have identifiably printed this ideal ratio length without a formula. But he did leave convincing formula clues, such as the regular proportion $\sqrt{7}$ and regular diagonal 6", showing that he did intend this figure, which really can be taken as a formula with an *exact solution*. One might say that this clear mathematical regularity forms a "base" in the J1 theory, allowing us some small amount of certainty from which to measure. I am using the J4 implied mathematical argument to say the Designer is promoting a precise idea in decimal places beyond his ability to print. BSC2, based on the above side tie assumptions and the ideal LIR length, turned out to be a reliable working dimension, which lead to many interesting places.

Working with this number, I first noticed it was about six tenths (0.6) of the length of the Large Outer Rectangle, which at the time I believed was the IDT length (5.654801219"--See Chapter 3). I happened to multiply 0.6 times the IDT length and got a similar number to my estimate BSC2:

(t)	(IDT)	5.654801219 · 0.6 = 3.392880731 = BSC3	Now turning this around:
(t)	(BSC2)	3.39448608... / 5.654801219 = 0.600283891	which is almost "0.6"--

which is pretty similar. How similar?

BSC2 - BSC3 = 0.001605348"   or, a little more that one and a half thousandths
                               of an inch, at about one part in two-thousand.

I wondered if this second number was what I was really measuring, and if this new number 3.392880731 (BSC3) was the actual intended length between centers.

At some point around then, I had tried using the height of the Large Outer Rectangle (Independence Date Theory) as the height of a small π-proportioned pyramid to find the length of a side--and found *another*, very familiar length:

This is the height of the LOR rectangle (IDT) seen earlier:

(t)   (IDT)   5.6548..." / φ^2 =

(t)   5.6548..." / 2.618033989...   =   2.159941866" = h

Then, using the Squaring the Circle formula, the "Pi-derived" pyramid formula:

(t) $(2h \cdot \pi)/4 = 3.392828749"$ =   BSC4

or, amazingly close to the length of "BSC3." Just how close? This is the obviously same number to about the fourth decimal place:

(t)   (BSC3; or, IDT · 0.6)   3.392880731"

(t)   (BSC4; from $(2h \cdot \pi)/4$)   3.392828749"

(-) -------------------

(t)   0.000052   or what would be

about *half* a ten-thousandth of an inch, or, one part in 65,000. That's a pretty small difference, indeed. And this leads us to the (now well founded) suspicion that the centerlines of the Seals may be the edges of the footings of a precise, but completely invisible pyramid--a profile of the Great Pyramid.

For reasons based on measurement, the actual measured centers of the Seal faces are *still somewhat wide of this by about two thousandths of an inch.* (i.e., a thousandth of an inch at either side.) And all reasonable theories should be derived from, and depend upon measurement. Just what kind of invisible slope angle would appear, then, if we used the measured dimension--the slightly wider *BSC2 length*? (J3, See Illus L1)

To find out, we need to divide BSC2 in half (to make a right triangle) and then divide into the height h, to find their fractional value, and then find the Arc Tangent of this fraction to find the slope angle:

(t)   (BSC2)   3.39448608..." / 2 = 1.69724304"   = b   and:

(t)   (h; being the LOR (IDT) length/φ^2)   2.159941866"   = h   then:

$\tan^{-1}(h/b)$ $= \theta$

$\tan^{-1}(1.272617896)$ $= \theta$

(decimal degrees)   = 51.84038121°

then, DMS: = 51°50'25.37"

What a surprise. (See Illus. K4 and L1.) Was *this* what the Designer was saying? It seems to be a reference to the Pyramidon's apparent apex angle seen above (J5, J7). If we are looking for an accurate Great Pyramid reference, this center BSC dimension computed with the IDT height allows a much greater precision than that I made of the apex of the tiny Pyramidon triangle of the All Seeing Eye. *If* we can believe the precision in the measurement at the edge of the seals is as good as one thousandth of an inch at both ends, we then may distinguish this particular "slope" from the above pi-based form of BSC4. It would appear that the Designer *was fully able to use the simple, purely π-based pyramid form or 51°51'14".* But based on measurement using BSC2, he seems to have chosen *not to do so.* And this also demonstrates (but does not prove) that he *may not* be merely following a purely Davidsonian derived concept of the Pyramid. For some reason or other, he appears to match William Flinders Petrie's transit telescope-based observation of the North slope of the Great Pyramid. Perhaps the Designer's angle is derived from other ideas yet. (J4, J5 and J6).

Illus. L1: Profile of the
Great Pyramid formed
by Seal Centers and
height of LOR

$\tan^{-1}(h/b) = 51°50'25.37"$
*(A slope similar to the slope
observed at the Great Pyramid
by William Petrie.)*

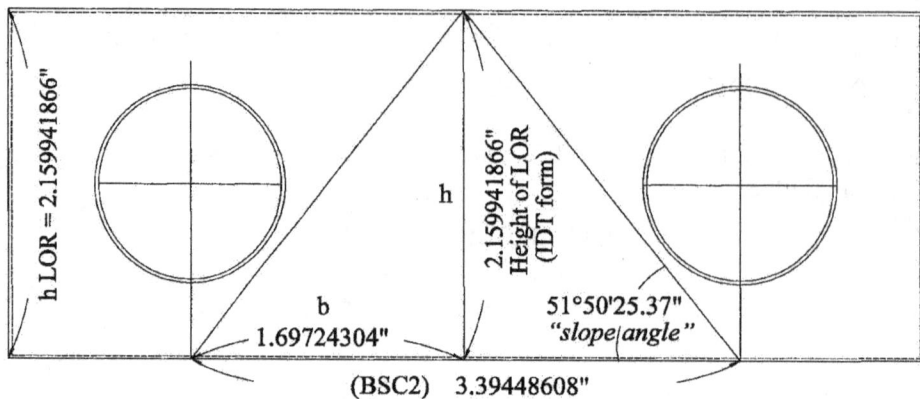

(BSC2) 3.39448608"

Notes:

[1] "Occam's Razor": William of Ockham (or Occam) was an English philosopher of the early 14th century. His (powerful) addition to modern or scientific thought was the idea that he would reject all complicated explanations of the unknown, in favor of the *simplest possible* explanation. The extraneous is to be pared away, as it were, by an unsentimental razor. Sometimes this concept is thought of as the Gold Standard of scientific and mathematical thinking.

[2] Also interesting, from the purely arithmetical point of view--being prompted by the reference for the number "360,"--is an interesting theoretical calculation possibly pertaining to the proportions of the Pyramid of Giza; 360; and the length of the dollar. I discovered this many years after this period of investigation. Suppose we took 360 as a height of a B-proportioned pyramid, and computed the length of one side of its base, then proportioned it down by 100 to make this length comparable to the length of the dollar:

Squaring of the Circle Formula: Twice the height 360 multiplied by pi (Steps 1 and 2 here):

(t)   (1) $360 \cdot 2 \cdot B = 2261.946711...$

this is the circumference a B-based pyramid that would have four sides:

(t)   (2) $2261.946711 / 4 = 565.4866776$

This side looks very familiar, numerically.   Then, reducing the proportion by 100:

(t)   (3) $565.4866776 / 100 = 5.654866776 = a$

It is obvious here that this number bears a close resemblance to the Independence Day Theory-length discussed earlier. Is this what the "360 thing" is about? Just how close is it to the IDT number-form?

(t)	(a)	5.654866776
(t)	Length, LOR (IDT)	5.654801219
		(-)--------------
(t)		0.000066

If this "a" above was the source of the Dollar's length, once again, any difference would obviously be unmeasurable, as a physical dimension, since this would round to about 7 one-hundred thousandths of an inch--a little more than a half of a ten-thousandth. But if anything can be made of this beyond coincidence, it is the startling similarity between these theoretical numbers--to the tune of one part in 86,000. (Those who apply the formula used in the

Independence Day Theory to this length will find that it will make a "date" of about seven and a half "days" later than the 4th.) So, might this be part of the reason for the Designer's choice for that pattern of fractions? Turning this around, *and using the IDT number times 100* will give 359.9958265, or, quite close to--but not exactly 360. But is any of this meaningful? It looks really quite coincidental, but cannot be ignored. Later I came to conclude that it was *probably not central to the issue,* and that other mechanisms were hidden in these three steps that were likely to be more important. (See Chapter 5 for Steps solution, pgs 169, 170.)

CHAPTER 4: ANOTHER LAYER

Section I

What was our Designer thinking about in 1935 when the new dollars finally appeared at the Treasury? I think by now, most people would agree that a large, mysterious puzzle had been built into the new, green side of the dollar. And yet, unless one looks very closely, there is hardly any hint of these complex puzzles. What an unusual person this must have been. He left the surface of the bill literally jammed with tiny geometric relationships. In time, the investigator will scarcely believe the number and complexity of things found here. What sort of person was the Designer? I wondered. Did he mark-off his calendar to retirement like other public employees? Or was he the busy all the time and never look back -type? Did he wonder how soon his patterns on the dollar would surface? How many others around him were aware of what he was up to? He had a team of engravers--were they included in on the plan? I like to imagine that on that autumn day in 1935, he smiled, wiped his bald head, hung up his printer's apron and would have happily driven to home and family in the suburbs of Washington D.C. (He was the "Superintendent of Engraving of the U.S. Bureau of Printing and Engraving", so maybe he was management and a suit at that point. Was the apron gone by then? No, I think we have good reason to think that he was very much a "hands on" type of guy. See Edward M. Weeks Biography, Appendix B.)

But what *did* he think? The Designer must have planned for investigators, and clearly laid trails for them. *The future investigators of the dollar* were intended to be lead by stages of discovery in slow unraveling of the dollar's design. New emerging facts would force earlier assumptions to be replaced by more accurate ones, by *successive approximation*. Increasingly subtle clues were intended to appear in later measurement and inspire new arithmetic, logic and construction of theories. Even though I was becoming accustomed to finding new things in the dollar all the time, I was not really prepared for the surprise and overall character of the next stage of discoveries. These were fundamentally different and a new "layer"in thinking. They had been just around the corner for a long time, and I hadn't taken notice of the evidence. This new evidence--the evidence found in this chapter--will be what allows us to "crack the code" of the deeper elements of the mathematical plan of the dollar.

From what I know now, from what I have seen of his clues, I probably wasn't following the Designer's logic exactly the way he had intended. What I did was to latch onto a few important oddities in the beginning, measure everything that looked promising, and slowly try to piece things together. But this is a long way from coming to understand the *motive* for all of this effort.

There was a gradual realization that I had *no satisfying explanations* for what I had found thus far. There were interesting relationships here and there, apparently strung together in intriguing ways. Most of these symbols seemed recognizable in one way or another through various traditions. But the Designer's purposes were elusive. This was beginning to look like a work of religious art, but in mathematics. But my intuition sensed *there was something still missing,* something not showing up on my radar. Was this whole design: "I make mathematical riddles and you solve them?" Or: "I make arcane relationships, and see if you can find out what they are?" There had to be some satisfactory pattern to all of this. I wanted to find a big picture, not a lot of *ad hoc* discoveries. I was still curious: the un-scratchable itches just wouldn't go away. "If the Designer was into the Pyramid idea, wouldn't he use the traditional number- forms like those of Piazzi Smyth?" I wondered. "Can it be that this whole design is only about the Date of American Independence? Is that all there is?" But that might well been all there was: after all, *it is* U.S. currency--and it may as well be patriotic in design.

Still unsatisfied, still spending a lot of time reworking these riddles, it (now) seems inevitable that the built-in logic of this design would finally force me--or anyone else for that matter--to dig and find the deeper elements. There *were indeed* many more hidden ideas. And the ones I did find, were found mostly by accident and rarely by a thoroughly reasoned hunt. And I should say that true satisfaction is still absent, even though--at long last--many reasons from tradition for these patterns did finally appear. Even now, I am convinced that I have passed over many basic design ideas: almost any time I re-investigate the basics on the dollar's design, new and important things of emerge.

****

Here I should back-up a bit. The end of the last chapter bore the fruit of several much later discoveries, such as those made from computer graphics and so on. Yet the telling of this story from where we left off would omit much of how I got there.

At the height of my excitement I really couldn't talk to anyone about the dollar. This research had gradually became a secret. With the exception of two or three close friends, I didn't mention it at all. Around spring of 1977 I felt the urge to announce my discoveries, and I composed a few pages of pencil drawings and equations which I showed around to some friends. To my amazement, there was little or no interest in my investigation, even from most of those who did find any of this credible. And, from those who evidently didn't, there was mostly uncomfortable silence and a quick shift to a safer topic. "So, what do you think about the Cubs this year?"

I bared my odd, little secret to a lady friend at some point, and I knew things weren't going well when she began to address me by both my full Christian name *and* surname. After thirty minutes of discussion about the dollar and measurements she said, "Of all the strange things that you have told me, *this* is easily the weirdest," or words to that effect. And I hadn't even gotten to the part about the itty-bitty steps on the little Pyramid. I think she said I should "get outside more," and when I mentioned that I already worked outside, in surveying, she then said that maybe I was "getting too much sun" and should go for an inside-type of job. (Maybe I should have worked in some charts and graphs here.) I don't really know why this astounded me then, and now I look back to this with a lot of amusement.

It was not this event especially, that caused me to keep mum about my little investigation, but the growing realization that this geometry-on-the-dollar-thing *was truly a peculiar idea,* perhaps requiring some caution. There was, of course, the total strangeness of it, judging from what certain people said, and a great unlikelihood that anyone would do this kind of thing with the dollar. *Why indeed?* After all, this was made by an older, no-nonsense U.S. Government of the 1930's. It is now a national symbol, the Great Seal of the United States, a government document, The U.S. One Dollar Note. It was an audacious concept to hide mystical mathematics in the etching of the dollar's public design. It was a fundamental image, and *some* people that I talked to turned out to *have very strong feelings* about the dollar--intense feelings that might otherwise have never surfaced before the question came up. Was this the sort of thing that brings forth villagers to the castle with the pitchforks and torches in the night? Igor: "Some people at the gate to see you, Doctor F." So, I decided to all keep all my little findings to myself for the time being.

****

David Davidson's Theory of the Great Pyramid:

Sometime late in 1977, I found a rare copy of <u>The Great Pyramid, Its Divine Message</u>, in a local library. This is the massive work by David Davidson and Dr. H. Aldersmith, running some 568 pages. It is completely jammed with detailed technical drawings, mathematics, tables, text, captions to text in margins, footnotes, huge foldout maps and historical tables: all on the subject of his theory of the Great Pyramid construction, prophecy and related Biblical prophecy. And what a strange book it was: an old fashioned and very British book in language and character; high quality old time engineering, together with classical history, astronomy, religion, mathematics and commentary. Not a book that you could just

81

skim through: a career could be made digging through it. I hadn't been able to locate this unusual book for years, it was always absent from libraries for one reason or other, and it was out of print.

In this library after work, I spent many hours studying Davidson's theory of the mathematics and design of the Great Pyramid. This was a real education in that sunlit, dusty old library, and I gained a great respect for the mathematical skill and research of Davidson and Aldersmith.

Davidson's work was widely ignored by professional scientists and academics of his time. They had to ignore it. A lot of his ideas were pretty far out. His work was technical as well as religious in tone and very much non-mainstream. Therefore he was destined to be sidestepped by most contemporary critical thought. The scholarly detractors of Davidson's work did not attempt to criticize the technical aspect of his work. After a short look through it, most wouldn't have tried to read very far in his book. Then as now, few who are literary are technically skilled. Most technically competent reviewers remained silent, while privately recognizing that the "numbers worked." As one reviewer said, "This has got to be completely wrong on all counts, somehow." There was dead silence in some quarters and elsewhere shrill criticism. Now, in our time, there is little or no learned comment on this author or his work at all.

The academic world can be partly excused for being closed to controversial questions of ancient mathematics, engineering or astronomy. Such things are not specifically the topic of any one of their professions. Beyond that, there will always be the delicate subject of where their next grant money is coming from. This is the curse of all professional scientists. Naturally, they will all want to continue to compete in the accepted, orthodox arena from where their paycheck comes. But if they say something that sounds foolish, the grant money dries up. Grant money people are often non-scientists, usually from a generation back in time. They are trusted, conservative and the bottle-neck to new science. They are at the choke-point. This doesn't really make the recipient academics dishonest, but they do not have true independence. They have practical realities that drive the choices for risks that they are willing to take with their reputation. So, much of this world is largely closed to their research. Yet in time, the geodetic evidence and patterns of ceremonial mathematics of the Pyramid shown by Davidson and others are thought provoking and will inspire new research.

The Pyramid as Designed and Built:

I was already somewhat familiar with the popular lore, the Pyramid data and theories in various forms through several books. These included Peter Tompkins' excellent Secrets of the Great Pyramid, (which includes Livio Stecchini's grand appendix on ancient metrology) and André Pochans' awesome The Mysteries of the Great Pyramids. But what I soon discovered in Davidson's theory, was that *even if only a small amount of what he shows is correct*, then the Great Pyramid shows us the remains of a vastly more sophisticated period of ancient Egypt than any accepted professional archaeologists now believe.

Pyramid as Prophecy:

Davidson was convinced that the Great Pyramid was intended to be an of imperishable book of prophecy, geodetic standards and other arcane data. Some of this might be true. But scientific research into the ceremonial or sacred geometry of the Great Pyramid is naturally forced to ignore the subject of "pyramid prophecy." Strictly speaking, this is Davidson's (and others') interpretation of the geometry and the measurements, and is an altogether separate question from technical considerations. Or I should say, not really so much a scientific question as a religious one--a matter of faith. Reason requires *doubt*, which is always betrayal to *faith*. I have always kept the prophetic side of his theory strictly at arms length, and I would strongly advise this course of action to others who want to avoid confusion and understand this field of study. In this book I will only touch on some aspects of his theory, and it appears that only a small amount of it pertains to the dollar's design. For perspective, we should note that Davidson may have believed from his research, as did many of his followers, that the world would come to an *end* on the special date of September 16, 1936. It didn't, of course, but this does not invalidate his splendid mathematical work, in which he did not cut any corners. He was a Structural Engineer, a hard discipline to rise to then, as now. His engineering work in locating these patterns is easily distinguished from religious ideas.

Later in the investigation of the dollar, we will see that there is unmistakable evidence that much of this special world of Pyramid theory also attracted the dollar's Designer. Several of the unique and arcane ideas of the Great Pyramid are taken directly from Davidson's work and are incorporated into the dollar's plan. One of the important elements is the above special date of September 16, 1936. Our Designer seems to have thought of this special date in a different light than Davidson's followers, but also, apparently, with great reverence. This *date* is likely to have had a very different significance to our Designer. He arranged the timing of the first printing run of the dollar to commence the first public issue of the millions of new dollar notes a year and a half before this date. It is as though the printing had been *made ready for that time,* at the special point in time shown in Davidson's Great Pyramid calculations, where the chronological timeline just enters the expanse of the King's Chamber. Perhaps this is more of a large scale symbolic announcement whose meaning was imagined to unfold later. It doesn't seem to me like something one would do if he believed the complete, apocalyptic end of the world was right at hand. Yet this special date in 1936 was still of great importance to the Designer for whatever reason--important enough to be made to play fundamental part in the dollar's design. And not only important enough to put on the dollar at all, but in several places and in different forms.

There were several clubs and societies in the U.S. and Britain that were aware of the technical details of Davidson's theory, and awaited the outcome of that day in 1936. One can not help but wonder if the Designer visited any of these groups, or had met David Davidson. His book gave a multitude of past and future dates and historical connections. The final date approaches soon at some point in September 2001. (This is at "The end of 6000 years", a computed length of 6000 Pyramid Inches ending a point at the center of a lentil called the Granite Leaf in the Antechamber. See Illus. M9 pg 99.)

****

The Historical Great Pyramid and Davidson's Theory:

Although I can not adequately abbreviate the relevant history, nor critique Davidson's whole theory here, I would like to show some pieces of his mathematical scheme of the Great Pyramid design, and early theories and background that relate to the Great Pyramid. (See Illus. M1.) Generally, we will be looking at various mathematical design elements that connect to features found in the dollar. The following ideas are thoughts on the plan and building of the Pyramid and historical background of the pre-Davidson world of ideas on the Great Pyramid of Giza. This evidence will lead us to see the origins of the composition of our dollar's design in the mid-nineteen-thirties. From what little we have seen, we can safely guess that the Designer of the dollar was deeply familiar with most of these facts and theories:

The Structure and Construction Logistics:

For those familiar with stonework and the building trades, the Great Pyramid is an exceptionally well built structure. It stands *alone* among ancient Egypt's ruined pyramids for quality of workmanship, and the structure still commands amazement among those trained in science and engineering, even in modern times. Perhaps nowhere in the rest of the ancient world, or even the *modern* world, has any stone structure been built to this level of care or precision, on this large a scale. Only the particle-beam alignment of big cyclotrons of modern times have anything like this need for large-scale precision. We should bear in mind that this Pyramid is an object that is nearly a third as tall as the World Trade Center in New York, covering 13.1 acres of area.

Illus. M1

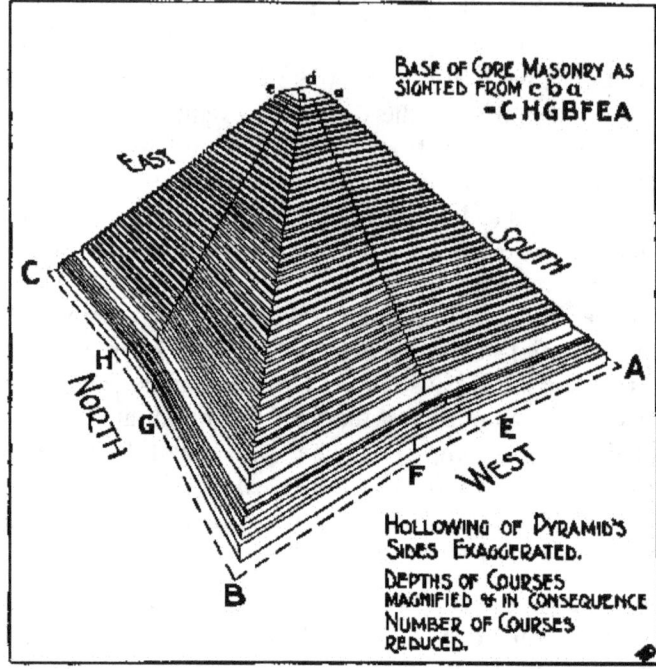

BASE OF CORE MASONRY AS SIGHTED FROM c b a
-C H G B F E A

EAST
C
NORTH
H
G
F
WEST
A
E
B

HOLLOWING OF PYRAMID'S SIDES EXAGGERATED.
DEPTHS OF COURSES MAGNIFIED & IN CONSEQUENCE NUMBER OF COURSES REDUCED.

Davidson's design diagram of the exterior theoretical geometry of the Great Pyramid

Note the Eight-fold symmetry above a break in courses (this is said to occur at the Thirty-fifth Course) and a Twelve-fold arrangement of lines at the Base, making for a total of thirteen planes above the Base, including the top.

From The Great Pyramid and it's Divine Message.

It was made without any artwork, ornamentation or religious writing inside, and shows no funerary murals or other evidence of having ever been the resting place of a king. It is *naked* of all markings that are typical of ordinary ancient Egyptian architecture. It is not even known for certain *how long ago* it was built or by *whom.* But it is *very old*--old enough that it is hard for us moderns to grasp just how ancient it is. The Pyramid of Giza was *already* not less than two-thousand three-hundred years old when Alexander the Great's armies conquered Egypt about 331 B.C. And it may well have been quite a bit older than that. So, it was at least as old to the ancient Greeks in their time, as those ancient Greeks are *to us in ours.* To the Ptolemaic Greeks it was as awesome as it is to us now, and in those days it was still covered with its original alabaster casement stone, described in those days as smooth and polished. There was much speculation among the Greeks about its origin, and the conquered Egyptians had no consensus of where it came from themselves. The period of Khufu (Suphis I) in the Fourth Dynasty is now thought to be when it was constructed, but this actual time period may have been anywhere between 2700 B.C. to 4700 B.C. The Greeks who were familiar with all the other pyramids in Egypt, called it one of the *Seven Wonders of the World.* Of the Seven Wonders, it is the only one remaining today.

Instead of being built of dry-fit stonework, as are all the other Egyptian pyramids, there was found to be a 1/50th of an inch of *fine mortar* binding the 2-1/2 ton stones of the Pyramid. This is a unique feature to all of Egypt's pyramids. This mortar is thought to have been made of crushed red pottery, which has mostly outlasted the Pyramid's stone blocks through the passing ages. To the amazement of modern-day engineering research, most of the builders' errors are in the neighborhood of 1 to 2 hundredths of an inch, and this very peculiar and careful work continues at this astonishing level of accuracy over the whole 13 acres of the base. In general, this range of precision continues to be evident in the internal stonework, all the way up the ascending passage ways to the chambers. The West Face base line at the foot of the Pyramid is aligned to about 2 and 1/2 minutes of arc West of true North. There is even some evidence that this tiny deviation from North may have been *intentional.* This is not the kind of work that to be expected of forced labor, as imagined by some. But this is the incredibly careful and pains-taking work of religious devotion--a devotion not unlike the cathedral builders of Medieval Europe.

David Davidson believed that a Semitic tribe called the Hyksos, ("The Shepherd Kings" who invaded Egypt and who seem to have introduced the wheel to Egyptian society), were the ones that built the Great

Pyramid and the Giza Complex. He proposed that after gradually infiltrating and subduing Egypt and finally building the Giza Complex, they later left Egypt and emigrated to the Palestine and the pre-historic British Isles. He implied that they were the Biblical group called the "Caphtor" by Moses, another earlier foreign group who were said to have left Egypt much as the Jews did (Deuteronomy Chap. 2). In this way, he speculated, (following Piazzi Smyth's theory) they would have taken their traditions, technique and equipment standards with their measuring units to Britain and the Palestine. All of these events were thought to have happened in pre-dynastic Egypt or as late as the Fourth Dynasty.

One of the few things that we moderns have discovered about the remote period of early Egypt, was that it once had an extensive canal system, some of which connected far out into what is now Libyan desert. There is much evidence that Egypt was a much wetter, grassland climate in those times. From recent research dating the erosion of the stone surface of the Great Sphinx, there is some evidence that Egypt's early civilized period may stretch several thousand years back, further than was once thought, and into a time of much greater rainfall. We have uncovered that the present Suez Canal was only *re-excavated* by modern engineers where it had *once been before,* and that there is even evidence of *another* great canal, once cut from the Nile River *to the Red Sea.* This huge, early canal system, curiously, was not fully maintained into later periods of history. My guess is that all of this was still present at the time of the construction of the Great Pyramid. Within the period of these canals, it is reasonable to conclude that there must have come the development of barge traffic. Probably through this method the great masses of stone for the Pyramid were most likely moved down the Nile from the quarry and then to the Giza site. And elevated canal locks are possibly connected with the placement of stone at the Pyramid itself.

The common theory of "ascending ramps of sand" for construction cannot have been the answer. They would have been even *more* massive than the Pyramid and ultimately would have to be removed later--a very impractical plan. But as for canals, there are, in fact, still the *remains of a stone canal causeway* up into the Giza site. Considering that this existing canal is substantially higher than the level of the Nile, we must wonder how water was raised to that level to make it useable. Several authors have proposed that some kind of ascending steps of water locks or wooden dikes may have been used for placing the stones on the Pyramid. Although this might not seem realistic at first, it is not so far fetched in a culture dominated by the practice of canal technology. In this theory, the water in the dike areas would have also provided a natural leveling standard for the builders, (a technique we do know the Egyptians used), as well as an easy way to manipulate the huge stones through attached floatation around the construction level. But the problem here is the source of water at that elevation--dikes or causeways would have had to have been flooded from a elevation source hundreds of miles up the Nile to the south. Water must be level in a dike, and the Nile is the only source of water. Just how was this existing causeway actually flooded to the height of the Giza Plateau? By an aqueduct?

A nice consequence of this theory is that the material requirement of wood would actually *decrease* over time, since the dikes would be getting smaller as the structure neared the top of the Pyramid. But all theories on the construction of the Pyramid seem to have major logistical problems.

From *Secrets of the Great Pyramid,* by Peter Tompkins, p. 226

Illus. M3   The Water-Locks and Crane Construction Theory
of Pyramid Building:

Progressive wooden locks to raise water levels and canal
transport of rock by barge; and the use of wooden rocking
cranes to hand off materials to upper courses.

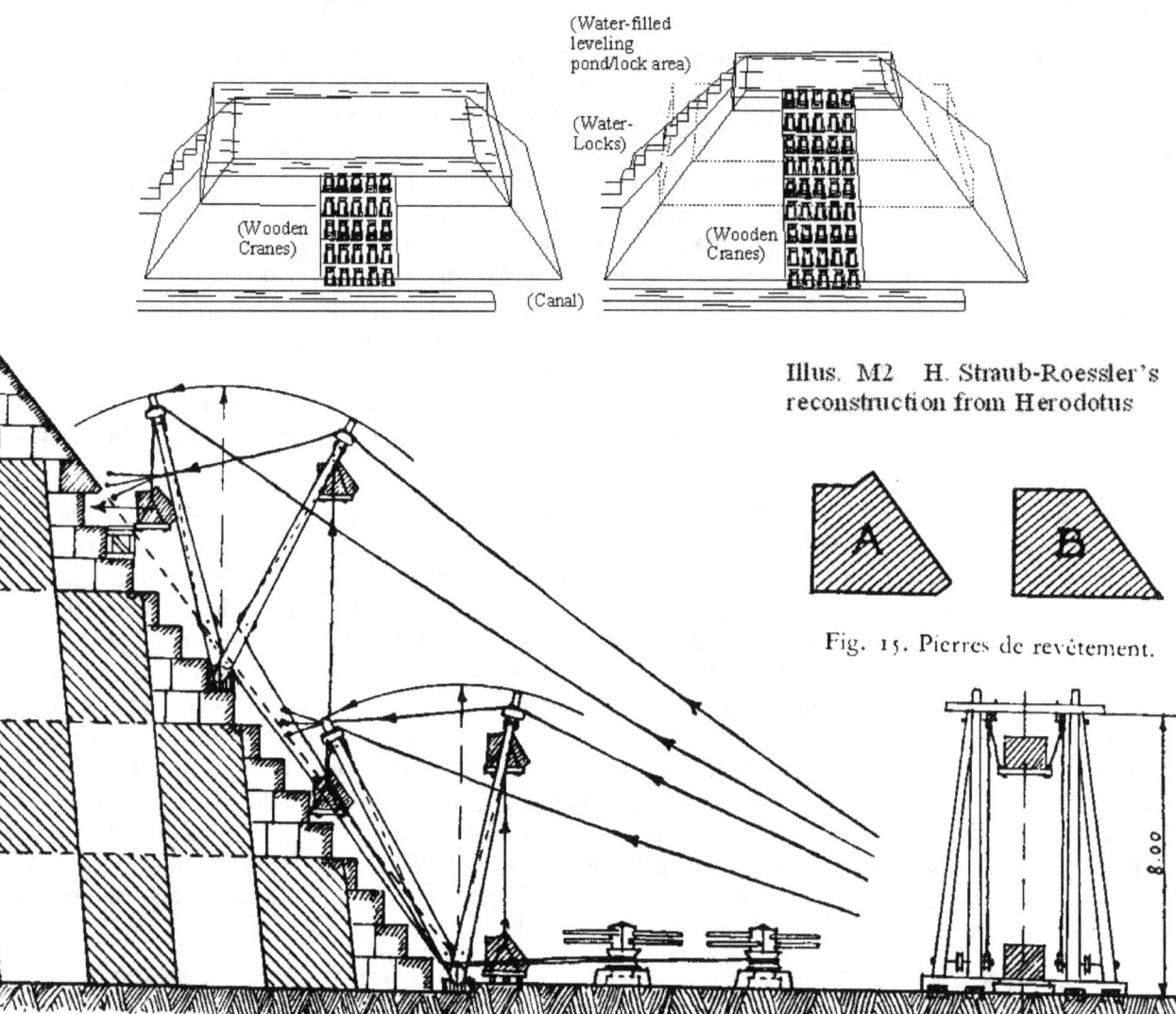

(Water-filled
leveling
pond/lock area)

(Water-
Locks)

(Wooden
Cranes)

(Wooden
Cranes)

(Canal)

Illus. M2   H. Straub-Roessler's
reconstruction from Herodotus

Fig. 15. Pierres de revêtement.

If we added to this theory an idea from the report given by Herodotus that these stones were moved up the slope by rocking wooden cranes, as the idea is expanded by Cotsworth and Straub-Roessler, thence to a leveling lock structure, then the question of the *necessary speed* might be answered (Illus M2). Any raising of stones by water locks would in all probability be very slow--ascending dikes or leveling pond would have to be filled and maintained somehow at high elevations. But there was a canal at Giza. If workable, dikes like this could have also supplied the water necessary for the mixing of the thin coating of mortar between the blocks, perhaps at a point just below the base of the top level of construction.

Both ideas together--dikes with cranes--might well answer some of the hard questions such as speed, but we just don't know how they did it. The Canal Period and the Pyramid construction period are not mentioned anywhere, even briefly, in any real historical record. We know of the system of canals only through unrelated archaeological work. From this evidence for this Period came the possible suggestion of locks, hinted at by *the strict leveling* found at every course of the Great Pyramid. Some have claimed that there are water line marks observable on the Pyramid's rock courses, so investigation of the dike concept may be approachable through archaeological evidence, if any exists. (See Illus. M3.)

No matter how it was built, speed would have been a *real* problem. Just looking at the numbers question shows that the planning and labor arrangements would have been an incredibly difficult project. I don't think the common estimate of 20 years for completion seems nearly long enough, but then again, we don't know the size or organization of their work force. Just to cut, move and place the 2.3 million huge stones over 40 or so years would have required *the finished placement* of about 57,500 stones per year *every year*. At the very least 157 stones a day would have had to be set, this being about one two and a half ton stone set every three or four *minutes*, if we assume an 8 or 10 hour day. This would be a great expense of labor and provisions in any ancient empire. Now, *Halving* that time frame to the above "20 years," would *double* the logistical requirement up to an unlikely average of one stone for every 2 minutes--every day, all year long, no matter what the Nile was doing. Hard to believe.

Additionally, the Nile itself would have been another big problem for the builders. Egyptians divided their year into three equal seasons: one of which they called "Flood." When it becomes flood season on the Nile it is also the time of planting, and it is likely that some of the builders must have had to go home to families and work their farms in the river Delta and elsewhere. This leaves an undetermined amount of labor leftover for cutting, moving and placing the on-going flow of big stones. A pyramid project of this sort would have *dominated* the economics and politics of the ancient Middle East for generations. This bespeaks of a very large base of prosperity and a solid civil organization, a very long time ago.

Although the surface of the Pyramid is now mostly removed, the original outside edges of the casement stone of the surface have left markings on an existing stone-paved surface below. These markings are still measurable today. This is called the Base Line, a crucial and much debated feature of the Great Pyramid. Many researchers have noted that much of the stone work of this particular historic period was unusually exact, "the work of an exceptionally fine hand." This Pyramid has attracted the technical attention and admiration of the British and other Europeans as far back as the Englishman John Greaves and his measurements in the 1630's.

****

The Pyramidologists' Theories

It is said that Davidson *originally set out to destroy* the prophetic chronology theory of Robert Menzies, but that over time "the more he attacked the data, the more he was obliged to assimilate it."[1] Most of Davidson's work relies on data from William Petrie's survey work. Sir William Flinders Petrie made a highly regarded and precise land survey of the Great Pyramid and the Giza Complex around 1881. Petrie was a surveyor by profession and later became a chief authority on the subject of the archeology of ancient Egypt of that period. He had studied pyramidology since his teenage years from the influence of his father, a pyramidologist, William Petrie. In an ironic symmetry, Petrie had developed an opposite turn of mind from Davidson, gradually becoming one of the foremost critics of pyramid prophecy theories, having originally favoring them in the beginning. But he remained greatly convinced of the Ancients' mathematical prowess at the Pyramid and their staggering building skill:

"Merely to place such stones in exact contact would be careful work, but to do so with cement in the joint seems almost impossible: it is to be compared with the finest opticians' work on the scale of acres."

Many archeologists have said that the unusual precision of this stonework found at the Pyramid and elsewhere from the period of the Fourth Dynasty seems to have reached a certain high-point in quality, at a time reckoned by Petrie to begin circa 4777 B.C. Subsequent Egyptian building arts were very inferior and said to be "in decline" throughout all later Egyptian civilization. It was just this extraordinary and weird ancient precision that caused a lot of interest and elaborate theories, not to mention the expensive and careful survey work of people like Petrie and James Humphrey Cole.

Davidson's new theory on the Pyramid that appeared in the 1920's shows that he accepted and used data from the skeptical Petrie's exterior location survey, as well as (believer) Piazzi Smyth's interior measurements of the Pyramid. Both of these elaborate surveys were recognized as high quality work.

Davidson began his study of the Pyramid with a general knowledge of pyramid geometry and prophecy theory, based on Robert Menzies and Piazzi Smyth. These sources proposed that the Great Pyramid's original base of four sides was supposed to be built to represent *the length of the Tropical Year,* or the Solar Year in days, as would be measured from vernal equinox to the next vernal equinox. The Great Pyramid's base was thought to be built according to a special natural units called "Pyramid Inches," a unit very slightly larger than British inches. The total Base length in these units was thought to represent an exact year if multiplied by one hundred. Which is to say, that the perimeter of the base of this building was supposed to be 36,524.2 of these "Pyramid Inches," or, representative of a one-hundred year cycle of *days* in these units. These inch units came to be thought of as "year" units in prophecy within the shaft lengths of the Pyramid. Robert Menzies and Piazzi Smyth believed they found specific parallels from Bible chronology and prophecy in the measured geometry passages of the Pyramid using these pyramid inches.

The life of Christ, for instance, was said to be depicted in mathematical architecture as a 33.5 Pyramid inch *slope length* along the Ascending Passage. The location for this event in the passageways is crucial to most prophetic theories. The proposed chronological position in time for this event in the Ascending Passage's slope was derived from a mathematical intersection of passage geometry. Where the projected floor and ceiling lines of two main passages join at the beginning of the Grand Gallery, an imaginary triangle is produced within the masonry (sometimes called the *Messianic Triangle*) having an hypotenuse of 33.5 Pyramid inches along the Ascending Passage. This has had the effect of being a logical "anchor" whereby the theoreticians could attach the recognizable event of the 33.5 year length of Christ's life to a place in their proposed historical chronology. The passages were reckoned to be a prophetically symbolic, historical timeline. In time, this place in the passages came to be thought of as a point representing a separation of the historical-prophetic path of the Elect with respect to the rest of the world, where the great opening at the lower end of the Grand Gallery is the beginning of the Christian Age. This chronological anchor point also has the consequence of making the doorways and rooms at the upper end of the Grand Gallery become important markers for prophetic chroniclers since they are in the neighborhood of 1,900 Pyramid inches distant. If the inches are prophecy, that area should depict present day events. (It seems likely that the above general theories are the original source of most ideas in dollar design. We should take note at this point, with respect to the dollar design, of developing abstract design concepts of (1) *specific units,* (2) *in specific lengths as messages,* and being (3) *embedded within a sacred mathematical geometry,* e.g., a special ascending angle. These ideas are important on the dollar, and we have already begun to see some of these elements appear in the dollar design, as shown in Chapter 2.)

Where did these ideas come from and why is the Great Pyramid special?

At least as far back as the 1600's, the Great Pyramid was of unusual interest in a technical sense, and it eventually became known in Europe to have very unusual proportional geometry. By 1859, the British mathematician John Taylor noted that it appeared to literally hold the ideal form of the classical concept: "Squaring of the Circle." Over many thousands of years, this idea had been a long-term, hallowed quest in mathematics within many cultures around the world. With the observed proportion for this structure, if one somehow built a perfect hemisphere centered around the Great Pyramid, *having a circumference made exactly the same as this Pyramid's square base length,* this particular Pyramid's apex point would *exactly touch* the top point of the hemisphere. (See Illus M4) A very unusual pyramid. If the purported slope angle of the sides is correct, this shape would *graphically demonstrate* a knowledge of the circle to square relationship, or the knowledge of the mathematical idea of "π" on the part of the ancient builders.

Beyond this, it shows their great *reverence* for this mathematical function. This Pyramid--for whatever else it may be--is a *religious structure,* a work of great devotion and precision to a Divine concept.

It is now known that this special number π (3.14159...) had been used in ancient Egypt and has been found in the approximate form of 22/7ths or 3.1428... in mathematical papyri. This is not quite correct as "π," but this, or some similar approximate number is an absolute necessity for computing circular areas with reasonable accuracy in any culture. The possibility that the Ancients might have known this function was sensational in Taylor's time. Not long after, people would argue about what range of precision the Egyptians had computed for π, or in our terms: "how many decimal places had they gotten to?" The number "π" was considered by most scholars as very hard to compute, and many were certain the ancient Egyptians were not up to the task.

Illus. M4   The π-Based Pyramid to Hemisphere relationship--the square and circular perimeters are the same

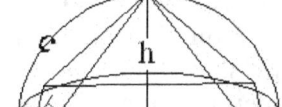
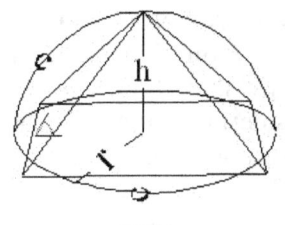
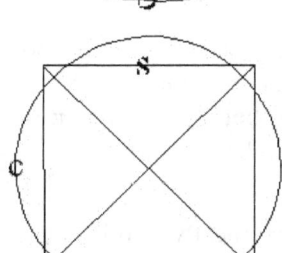

Circle and Square *are equal:*

Twice the height times π *gives the circumfrence:*

$2h \times \pi = c$

Four sides, divide π *gives twice the height:*

$4s / \pi = 2h$

h = height   *(2h = spherical diameter)*
r = radius   (Note: r *is the same as* h)
c = circumference
s = side
π = 3.14159265359...
△ = Slope Angle: 51°51'14.31" (θ)

It doesn't seem likely that the π-problem was a great hardship for ancient Egypt. The fact that it took the Europeans, Arabs, Hindus and Japanese only a few hundred years from mediaeval arithmetic to develop exact formulas for π makes this an empty question. The Egyptians had much more time historically, and vastly more interest in *numbers* than the above cultures have had. There are literally hundreds of very simple number series that will generate perfect π-ratios. Many of these are discovered and re-discovered accidently by high-school students *all the time.* To the ancient Egyptians, mathematics was of very serious and sacred interest. It was the instrument of their religion and magic, and the "good stuff" such as π would have been the object of their continuous, primary focus. Due to secretiveness, they might have lost and re-discovered an exact formula for π dozens of times in the course of their civilization--even though all we can document is the scribes' "22/7ths" practical value. And yet, unlike practical mathematics, this Pyramid seems to be a *consummate* and expensive celebration of a clear knowledge of this value--a structure showing their knowledge of the universal and fundamental importance of π.

And this is a *very unlikely shape* for any pyramid. For practical reasons, there are many *much* more easily constructed slopes than this. (The Second Pyramid's slope is also quite odd; it however, follows the slope of the hypotenuse of the sacred "3,4,5 rectangle" as discussed earlier. This is a similar slope angle of 53°07'48.37", being a little more than a degree steeper than that of the Great Pyramid thought to be about 51°51'14.31". The 3,4,5 angle would be lots easier to build--and if nothing else, we can see that these pyramid builders at Giza were *very much* into architecture strongly symbolic of revered landmarks in geometry.) Math was a large part of their religion. Here Horus conquers evil:

An illustration from the Book of the Dead Horus killing the serpent Apep with a spear.

This drawing secretly shows the Right Triangle--but first, it must be measured by the student to identify it's dimensions.

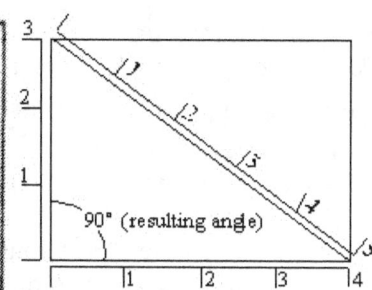

Geometric proportions of the religious illustration. Note the element of power that conquers the serpent is a *diagonal line* apparently symbolic of the unspoken diagonal length or secret number 5, completing the 3,4,5 triangle.

Original illustration at left from <u>The Book Of The Dead</u> by Wallis Budge

Great Pyramid researchers of the Eighteenth Century found that the measured square base length divided into twice its height appeared to give this special π-ratio, or very close to it. Alternately, as others found (by way of Herodotus), its *surface area to height proportions* also seemed to give (instead) a golden ratio connection by means of the square root of phi ($\sqrt{\varphi}$), a number quite similar to one quarter of pi (4/π) of Taylor's theory. (Mathematically, any direct relationship of $\sqrt{\varphi}$ to 4/π is only approximate. But there is a relationship between the two functions involving a form of Davidson's Displacement Factor that we will look at later in Appendix A. David Davidson, we should note, *never mentions the golden ratio any-where at all.* This is an interesting fact for us, considering that our Designer--who turns out to be quite Davidsonian--was obviously very taken with the φ-proportion, using it as a *basis* for the dollar design.)

Beyond the general π-proportion, the rough numeric height/base length proportions of the Pyramid of 116.5/366, caused John Taylor in the 1850's to wonder if the *denominator* of this fraction might have been intended to represent the *length of the year*, in addition to being part of a π-ratio. The solar year is about 365 and a quarter days in length, pretty close to 366. The question then arises, *what is the real length at the base of this Pyramid?*

Oddly enough, In real numbers of standard *British inches*, the Pyramid's base was estimated to be *right around 100 times* the number of the days required for a year: about 36564" +/- 5" in the data available in 1850. (For this to match the true Tropical Year, it would have had to have been about 36524 units around the base of four sides, and about 5813 units in height, or a proportion of 11626/36524.) So some-thing like a unit about the size of an inch *did indeed* become reasonable to Pyramidologists as a symbolic length, were it to be taken as one day in a symbolic one hundred year cycle at the pyramid base.

Polar Inches:

It had already been noted by Eighteenth Century geodeticists, that there were very nearly 500,000,000 (five-hundred million) *British Inches* between the North and South poles of the Earth. This is the Polar Diameter, which is distinctly smaller than the Equatorial Diameter, and is a length which is still used today as an important fundamental measure of the Earth (See pg 196). This measurement interested many researchers, since in inches the polar diameter came very close to *a completely round number* of inch units. "A very curious numerical coincidence," said Piazzi Smyth. In time, this round-number idea gave rise to the notion of the unit called the Polar Inch, or the so-called Pyramid Inch among Pyramid-ologists. This ideal relationship defined the "Pyramid Inch" (PI) as *exactly one 500 millionth part* between North and South poles. The actual British Inch length was later found to be closer to one 500,550,000th part of the polar diameter--a little smaller than the ideal. Could anyone in ancient times have known the true Polar Inch? Was the curious base of the Pyramid a multiple of these as a Great Year of some sort? The ancient Greeks had computed fairly good values for the circumference of the Earth, using the geometry of solar shadows at different latitudes. One might imagine a genius computing Original illustration at left some arbitrary, even units using the Earth's size as a base. It was widely believed among Eighteenth Century scholars of metrology that the Ancients preferred to use natural, or Earth commensurate, even units in their various temple units, geodetic units and even trade measures.

The Intellectuals of Revolutionary France (ca.1789) also felt a deep need to find just such "natural units" to replace the old French units associated with the aristocracy and Church, and this ancient natural-unit concept was the same general ideal behind the original creation of the present-day Meter. But the Polar Inch was also a unit that was naturally occurring or "Earth commensurate" between the poles, and just a little bigger than the British Inch: approximately 1.0011 inches in length. Both systems of units found their champions. (See Illus. M5: Inches vs. Polar inches.)

Illus. M5

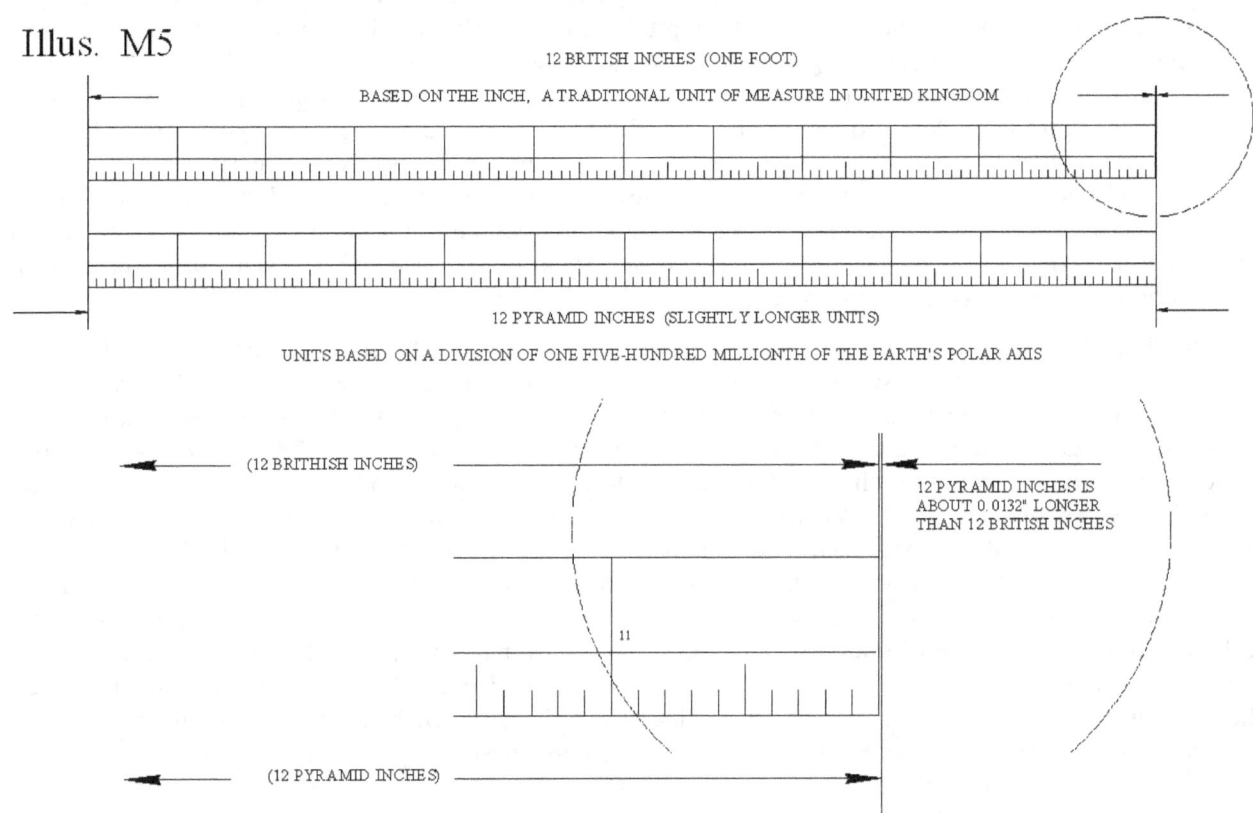

12 BRITISH INCHES (ONE FOOT)

BASED ON THE INCH, A TRADITIONAL UNIT OF MEASURE IN UNITED KINGDOM

12 PYRAMID INCHES (SLIGHTLY LONGER UNITS)

UNITS BASED ON A DIVISION OF ONE FIVE-HUNDRED MILLIONTH OF THE EARTH'S POLAR AXIS

(12 BRITHISH INCHES)

12 PYRAMID INCHES IS ABOUT 0.0132" LONGER THAN 12 BRITISH INCHES

(12 PYRAMID INCHES)

The Polar Inch unit, strangely enough, was promoted to the scientific community by the famous British astronomer Sir John Herschel, against the then pending international scientific adoption of the Meter. He was not a Pyramidologist. To Herschel, it seemed more rational and scientific to make a division of a line going straight between pole to pole for a natural basis of scientific measurement, rather than to have ten-millionth division of the irregular surface of the Earth, along the elliptical arc from equator to the pole, as the French Meter partisans had preferred. The Pyramidologists also thought that they had uncovered a special *Sacred Cubit* of 25 Pyramid Inches--a favorite of Piazzi Smyth and oddly enough, Sir Isaac Newton. Isaac Newton's metrology is said to have been based on a study of Biblical archaic measures. This unit was thought by many to play a special part in the Pyramid's design, as well as certain other ancient structures. Under other historical circumstances this 25 Polar Inch unit might well have become competition for the Meter, a unit that would produce an *even polar radius of 10 million* of these Pyramid Cubits. This cubit-like unit was actually proposed later in France in 1795, and was once known as the Collet Unit, being about 25.0275" or 0.6357 meters.

The Polar Inch and Cubit were quite convenient for the pyramidologist. They made theoretical sense out of the Pyramid, if used together with the apparent height and π-proportion. With the Polar Inch unit, many concluded that the Pyramid's base clearly fit the numerical "year-length" from survey data. British (and some French) Pyramidologists felt this must have been the *natural unit* the ancient Egyptian engineers had used, and that it would naturally symbolize the days of the year through "a wonderfully natural, ancient religious and metrological system." The Egyptians, Hebrews, British, and certain other

European cultures were thought to have inherited their many similar inch-like units from the ancient world, through the descendants of the emigrating Hyksos to ancient Europe. And then, through traditional use over many generations, to have carefully preserved this inch-dimension into modern times. And if true, in the British Inch case, the error would be smaller than one part in nine-hundred.

The Astronomer Royal for Scotland, Piazzi Charles Smyth, refined the pyramid theory, a development of earlier ideas by Robert Menzies in 1865, and John Taylor before him. Their idea proposed that the shape of the Pyramid, through its slope angles of the sides was built exactly to symbolize π, where the internal angle to base would be exactly $\tan^{-1}(4/\pi)$, or 51°51'14.31", or John Taylor's original computation. Piazzi Smyth's theory proposed a precise *100-fold year base circuit* in this unit, based on the evident proportions of the sides and its apparent height. Smyth thought that the presence of the ancient inch unit in British Isles marked the *English People* as the inheritors by blood of a common tradition between the Builders of the Pyramid and the Hebrew Tribes. Thus began the concept of *British-Israelism*. Smyth believed that there was a Providential connection linking the Hebrew and British peoples' destiny by ancient inherited ethnic and tribal ties from the Old Testament, together with this natural unit, Earth-based metrology linking both time and space.

Smyth stated that this Pyramid was designed by a scientifically superior, ancient culture as a monument to be discovered, having superior mathematics and natural geodetic units. "Written" in the universal language of mathematics, this pyramid was to have been built to survive long after their decaying civilization had collapsed and language gone. It would have been *devised for discovery* by our present day people, a people with sufficient science to measure and decode its mathematical language. Smyth believed mathematics to be *the language of God.* The Pyramid was to be evidence of their ancient science and faith, and the Pyramid Inch, the dimensional key to the prophetic time line in the geometry of the passage ways in the Great Pyramid. This idea is a direct development of Robert Menzies' original concept of the passage ways as a chronological representation of prophecy where *one Pyramid Inch* was to be taken as a scale value of one year of time. (Note that for the dollar, the Designer appears to have expected *the same type of discovery process* for his work. Davidson, in his development of the prophecy theory, came to see the "Primitive Inch"--the Polar Inch or Pyramid Inch--as the key length symbolizing (1) one year or day at the base; and (2) one year on the base diagonals for the Precession of the Equinoxes; (3) one historical year along the ascending passage way; and (4) in another symbolic form, one Egyptian 30-day month along the floor surface South of the feature called the Great Step. This last 30-day Pyramid inch idea appears to have had its origin from a Canadian, William Reeve, in 1909, finding support from a passage in the Book of the Dead. Davidson speculated that the Pyramid mathematics connected the height of the floor, beginning at the Great Step, with the *lunar month* of 29.53059 days. Davidson believed all of this information and as well as astronomical relationships were incorporated into the Pyramid by the builder through Divine Inspiration.)

To Smyth, in 1865, the hundred-year-base-circuit idea seemed a compelling length and self-evident. He had no doubt in his mind about this, *but it was really not true.* Smith could never accurately measure *the Pyramid's base*, during the time he lived at the Pyramid at the Giza Plateau, and its base mostly remained covered with sand and rubble until just after the turn of the Twentieth Century. So he was really making *an estimate* to which he freely admitted. The validity of his survey is mainly confined to the measurement of the inside passages: the rooms of the King's Chamber, Antechamber and Queen's Chamber, Grand Gallery and Ascending Passages, areas for which he was evidently careful and thorough.

Roughly at the point where David Davidson enters the scene, as the rubble and sand was beginning to be removed around 1909, it was becoming known that the Great Pyramid's base was somewhat smaller than this ideal length. The technical basis of Smyth's theory was then becoming open to ridicule. Soon, the total base course of the four sides was found to be very close to 921.453 meters, from the Cole survey (done by John Humphrey Cole in 1925, for Borchardt). Nearly identical figures were found using the earlier W. Flinders Petrie's survey who had actually dug down to the Pyramid's corners. The Cole data made a total of 36237.7 Pyramid inches. Close to the ideal Solar Year length, but not close enough--it was too short. The consensus was that this was curtains for the Pyramid inch theory--it was to become almost universally spoken of as having "been discredited." But this base still showed some very interesting relationships for the Pyramidologist.

The relationships, as shown by Peter Tompkins in the <u>Secrets of the Great Pyramid</u>, (pg 209) revealed the newly determined base length actually worked out as fairly *even* divisions, fractions or multiples of *many ancient measurement standards* found all around the Mediterranean area. This discovery further suggested that Smyth's theory was way off base, and led some to a conclusion that the original side lengths had now been determined with some exactitude. But how could this be? How could *all of these* lengths turn out to be even fractions of this base? This would be a truly enormous and unlikely coincidence. One direction of logic leading from this discovery is compelling. Even though we know nothing about the historic origins of the Great Pyramid, we are still left with a startling conclusion: since most of these *unrelated* ancient standard units were *evenly divisible into this base*, it seems likely that this whole multitude of ancient standards *may have been originally taken from this Pyramid's base* as their source, several thousand years ago. To propose a theory: it seems as though several distinct cultures and probably at different times, somehow borrowed from the overall physical length of the base at the Great Pyramid, making for themselves convenient, even divisions for their own purposes. Perhaps this may have happened by way of an ancient student body--such as foreign neophytes to priestly keepers of the Pyramid at Giza. The students perhaps returned to their own countries, such as the Greek investigators were to do much later. Metrology in this form is likely to have become a corpus of sacerdotal information to the neophytes. Perhaps over time, the units used in the neophytes' measuring equipment became the local standards in their various countries. Although we cannot know any of this for certain, *there is no other known single length like this* from the ancient world, in which *all* of these archaic measuring standards find a general, divisibly even fit. It is not at all likely that this base was contrived in reverse to accommodate all these dissimilar units from different cultures. It is far more likely that the concordant standards were somehow derived from *this particular structure*. There must be more to this, even beyond Tompkins's geographic fractions argument.

Quite apart from this discovery--Davidson's calculations from Smyth's data and Petrie's survey, scaled to the now doubtful theoretical Pyramid inch Units, raised two enigmatic and important facts. (These facts were also evidently of interest our Designer of the dollar.) The following is another "very curious numerical coincidence," as Smyth would have said:

> (1) *The centerline* of the passage ways of the Great Pyramid *were all offset toward the East* with respect to the centerline of the base by somewhat more than 23 feet and 10 inches (7.275 m), *or about 286.1 Pyramid Inches*, and

> (2) The measured base length as reconstructed, by producing intersections of the lines of the casement edges, by Petrie (as well as J.H. Cole and others) was, oddly enough, the same number 286.1 of Pyramid Inches *short* of being 36524.2 Pyramid Inches--the ideal number to represent the day length of the Tropical Year.

Something that gradually became noticeable after the exact figures were published was that this weird offset *was all that was necessary* to add to the base to make the special Tropical Year number. So, the year/Polar inch idea had surfaced once again, but this time in a very different way. Strangely enough, this "year-form" is quite precise. It was as though the original builders had chosen to veil the year length through the use of the number, "286.1", as a *secret offset length.*

Illus. M6

(1) The Ideal Base
(exterior lines) and the

(2) Stone Base as Built
(light shading)

*Exterior lines said to follow
subtileties in Pyramid geometry.
These are believed to represent the
astronomical year lenghts by
direct measure:
Tropical (ABCD)
= 36524.2465 PI,
Siderial (AEFBGHCJKDPR)
= 36525.6471536 PI and
Anomalistic (AaBcCdDa)
= 36525.997317 PI*

*(Illustration and figures by
Davidson.)*

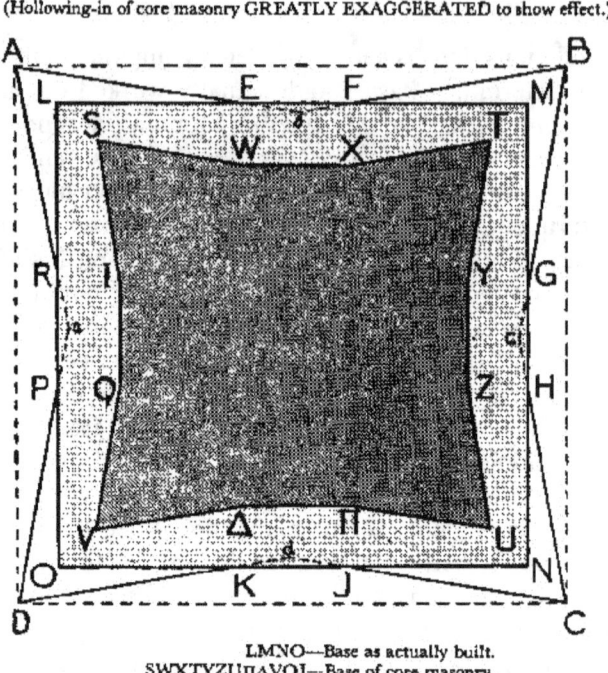

CONSTRUCTION OF THE GREAT PYRAMID'S BASE.

(Hollowing-in of core masonry GREATLY EXAGGERATED to show effect.)

LMNO---Base as actually built.
SWXTYZUΠΔVQI---Base of core masonry.

Davidson's concept of the Pyramid is much more complex than four equal triangles. Some of this can be seen from the following illustrations (See Illus. M1). Davidson stated that Petrie had *omitted the hollowing* evident in the sides for the finished casement stones in his final survey report. He argues that this is answerable for the over-all length discrepancy of base course length measurement between Smyth's work and Petrie's survey, which Davidson said would have their corners at ABCD in Illus. M6. This could change the reckoning of the base course length if true. However, Davidson used the ending "socket corner stone points" from Petrie's survey (which are about three feet further out from the pyramid), and almost every other researcher has used the extended or "produced edge lines" for their theoretical base of the Pyramid as built. (In Illus M6 the "produced edge lines" are between points LMNO: the lighter shaded area and the existing base where most of the outer casement is missing, is in the darker shading. These are the *three fundamental astronomical year lengths* that are based on theoretical mathematical extrapolation of his theoretical proportions. Note that these lengths would differ by less than two inches in length over all. In this diagram point L would be about 35 PI from line AB--or an 1/8th part of 286.1 PI. Davidson used the above figures in his calculations, and these results are fairly similar to modern astronomical computations. These dimensions are not measured dimensions being considered at the fourth or seventh decimal place, but are based on his theoretical mathematical proportions. These may be derived from part of Davidson' general theory of the Pyramid's geometry. Modern astronomy gives 365.24219 solar days for Tropical; 365.25637 for Sidereal; and 365.2595844 for Anomalistic Year durations.)

But *whether or not* Davidson's reconstruction of the exterior is correct is really apart from the point: *even if* Davidson's choice of outer starting points was wrong, and even if we were to use the modern, smaller measured length of the Base Course (LMNO) by J.H. Cole, a clear case can still be made that this base could represent the ideal hundred year circle anyway, *if you had in your possession this key number length of 286.1 PI.* If added to Base Course LMNO, the 100-fold Tropical Year number dramatically "appears"--a self-demonstrating hypothesis that forces the observer to notice the Builders' reverence for the "286.1" shaft offset. We should note that this again supports the notion of *a secretive character* in the beliefs of the Builders, who were clearly given to hiding the details of their ideas in architecture. (This is not the only kind of offset clue here. Many internal chambers have various other kinds of offsets, such as shown by Smyth in the King's Chamber floor with respect to volume. There is even a very small offset on the northerly base near the centerline, 0.0355 m westerly of centerline, found by Ludwig Borchardt that we will look at later. See Appendix A.)

286.1 Pyramid Inches:

This "286.1 PI" offset has turned out to be a special theme in the Great Pyramid's design. Davidson found other methods to independently derive this special hidden offset number through fairly straight forward geometry (See Illus. M7, PLATE XXIII) from the Pyramid, and also through an independent numerical formula:

$$\sqrt{(\pi \cdot 25 / 128)} \cdot 365.242465 = 286.1022156$$

(The above value of 365.242465 solar days, that Davidson used in calculations, is an *average* Tropical Year computed over several ages of astronomical precession.) He called this resulting number the "Displacement Factor." (His number, "286.1022156" is capital "D" in formulas shown later.) He used this special number to compute the size of the legendary "missing capstone" through geometry on his reconstructed, theoretical Pyramid design, (See Illus. M7 and M8) where the horizontal displacement from the centerline defines edge lines of the "unfinished" area at the top of the Pyramid. This number had deep religious significance to Davidson, as shown his geometry of the "Stone that the Builders Rejected" capstone concept. He saw it as a pre-Christian symbol for the promised Messiah, perhaps like a universal form of the *name of God* in mathematics. Davidson believed this was a crucial piece of prophetic and inspired knowledge, known to the ancient engineers who built the Great Pyramid.

Illus. M7   Davidson's scheme for geometry demonstrating 286.1022156 PI (Fig. A, between F1 and k1) and the internal passages' easterly offset (Fig. B between lines YW and ZU).

And it is quite true that this 286.1 length (and the number itself as a fraction) appears in various forms and geometric configurations throughout the Pyramid, even beyond those that Davidson found. It is unmistakable and certainly earns its position as a major pattern in the ceremonial mathematics of the Great Pyramid. I think it is fair to say that this special number must have also had great religious importance to the ancient engineers who designed the Great Pyramid. With many tables and formulas, Davidson compares this number and many of his other discovered numerical patterns to the mathematics of Biblical prophetic symbolism, all in elaborate detail. (In the design of the dollar, this number figures prominently, and in several forms, leaving us little doubt of the connection to this tradition.)

PLATE XXIII.

THE GREAT PYRAMID'S EQUAL AREA GEOMETRY DEFINES DISPLACEMENT OF PASSAGE SYSTEM.

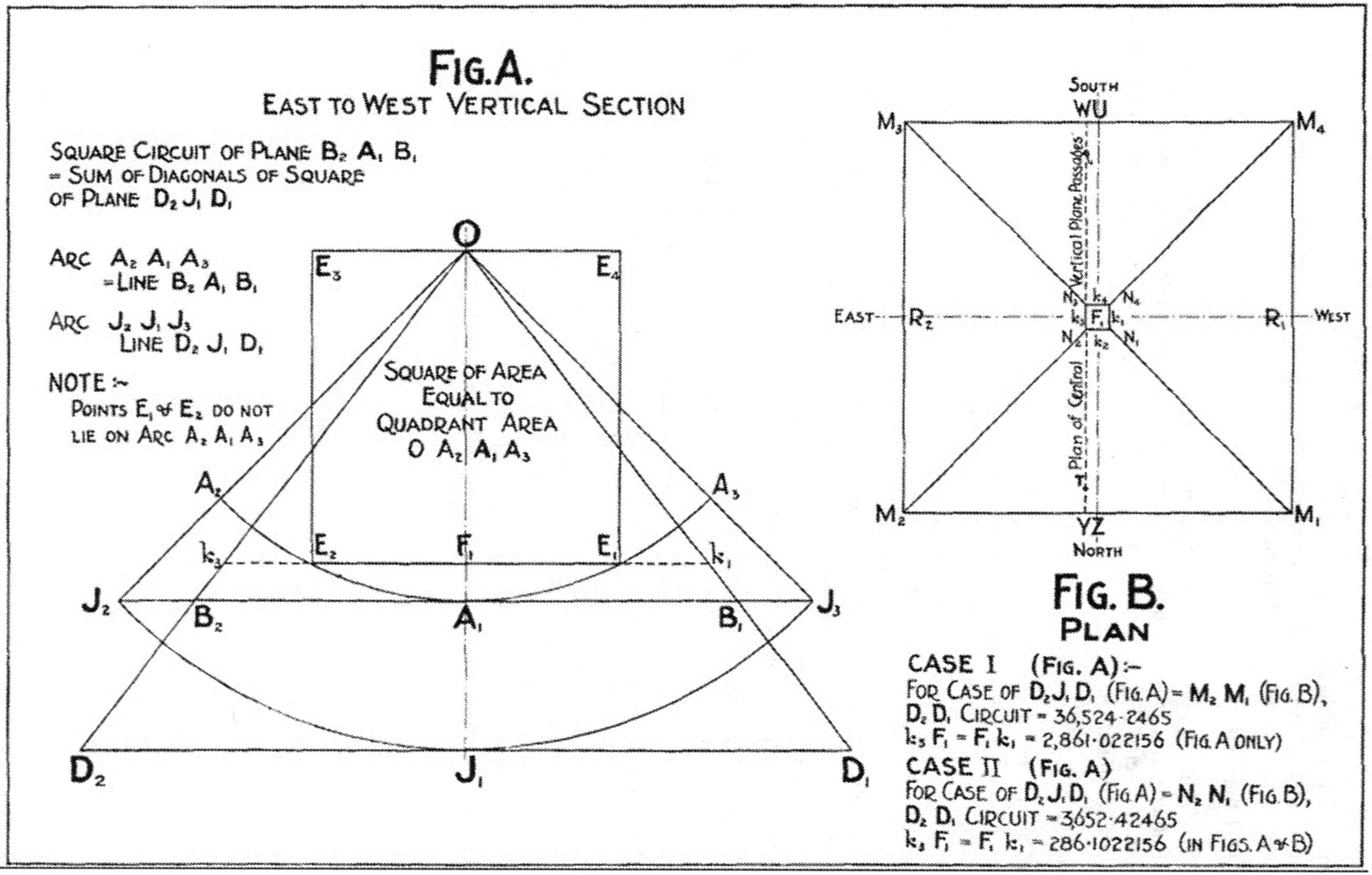
FIG. A.
EAST TO WEST VERTICAL SECTION
SQUARE CIRCUIT OF PLANE B₂ A₁ B₁ = SUM OF DIAGONALS OF SQUARE OF PLANE D₂ J₁ D₁
ARC A₂ A₁ A₃ = LINE B₂ A₁ B₁
ARC J₂ J₁ J₃ LINE D₂ J₁ D₁
NOTE :- POINTS E₁ & E₂ DO NOT LIE ON ARC A₂ A₁ A₃
SQUARE OF AREA EQUAL TO QUADRANT AREA O A₂ A₁ A₃
FIG. B.
PLAN
SOUTH
NORTH
EAST
WEST
Vertical Plane Passages
Plan of Central
CASE I (FIG. A):-
FOR CASE OF D₂ J₁ D₁ (FIG. A) = M₂ M₁ (FIG. B),
D₂ D₁ CIRCUIT = 36,524·2465
k₃ F₁ = F₁ k₁ = 2,861·022156 (FIG. A ONLY)
CASE II (FIG. A)
FOR CASE OF D₂ J₁ D₁ (FIG. A) = N₂ N₁ (FIG. B),
D₂ D₁ CIRCUIT = 3652·42465
k₃ F₁ = F₁ k₁ = 286·1022156 (IN FIGS. A & B)

Illus. M8 Davidson's scheme for the exterior geometry of the Pyramid with respect to the passage offset of 286.1 PI. At top is the "Missing Capstone" area, as a tenth of the overall Pyramid. (See previous Illustration.)

PLATE XXIV.
THE GEOMETRY OF THE PASSAGE SYSTEM DISPLACEMENT.

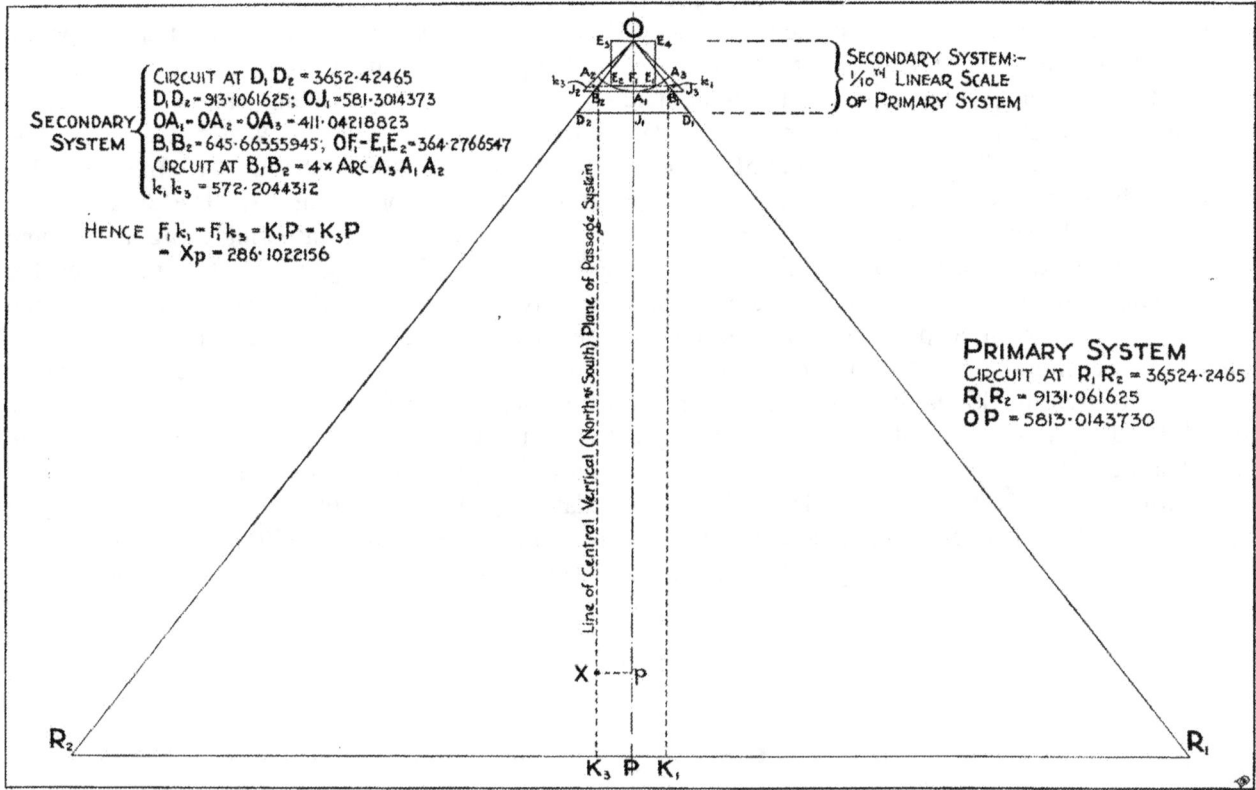

SECONDARY SYSTEM
{ CIRCUIT AT $D_1 D_2$ = 3652·42465
$D_1 D_2$ = 913·1061625; OJ_1 = 581·3014373
OA_1 = OA_2 = OA_3 = 411·04218823
$B_1 B_2$ = 645·66355945; OF_1-$E_1 E_2$ = 364·2766547
CIRCUIT AT $B_1 B_2$ = 4 × ARC $A_3 A_1 A_2$
$k_1 k_3$ = 572·2044312 }

HENCE $F_1 k_1$ = $F_1 k_3$ = $K_1 P$ = $K_3 P$
= Xp = 286·1022156

SECONDARY SYSTEM:- ⅒ᵀᴴ LINEAR SCALE OF PRIMARY SYSTEM

PRIMARY SYSTEM
CIRCUIT AT $R_1 R_2$ = 36524·2465
$R_1 R_2$ = 9131·061625
$O P$ = 5813·0143730

The above Illustration M8 seen from the north, is a side view of Davidson's proposed Displacement alignment. The position of the central passages in the Great Pyramid when seen from this view are also shifted by this amount easterly, or toward R_2 on the left. This axis shift has been a puzzle for students of Egyptian architecture for centuries, and this is Davidson's geometric explanation for it's lack of central symmetry. (We should note in passing Davidson's choice of letters "Xp" in his geometric design denoting this special offset. We know that he regarded this offset as a specifically symbolic of Christ, and these letters have a striking resemblance to the Greek letters Chi Rho (*XP*), used as an abbreviation for *Christ* in Christian iconography. This is called an "ellipsis" or a traditional Greek abbreviation taken from the first two letters of "Christos" which is sometimes seen on vestments and altars in the form of a ligature of the two letters fused together as a symbol.)

The above illustration is a geometric method for deriving the Displacement Factor (D) from nothing but the geometric form of the Great pyramid itself. The "Square of Equal Area/Quadrant Area" system is numerically equivalent to the $\tan^{-1}(4/\pi)$ formula seen above, or for a pyramid whose face has the special angle of 51°51'14.31". This is the mathematical construct atop Davidson's Plate XXIV, above (Illus. M8). Fig B shows this as an offset applied easterly to the center axis of the Great Pyramid from the top view. (The exterior entrance to the Pyramid, discovered by Khalif Al Mamoon in Fig B. would be on the "Y" line approximately at "T.")

Davidson's theory was that this was an empty area or a square platform made at the top of the Pyramid having a side of 572.2044312 (2 · 286.1022156 or K_3 to K_1 in Illus M7). This was thought to be the final level of the Pyramid as built, having a circumference of 1144.408862. This was the area and dimensions for "The Stone that the Builders Rejected": the place in Masonic symbolism where no stone was perfect enough to be placed, where the perfect one was to be tragically cast away. (See pg xv.)

Prophetic Lore and History:

There is an awesome amount of detailed material on prophecy in Davidson's book. There is quite a bit of large scale interpretation of historical events from mathematical patterns found in the Pyramid and similar patterns found in Old Testament prophecy. The part of Davidson's (complicated) prophetic theory that captivated his audience in the 1930's was that the position of the outside edge line of the door at the entrance to the Antechamber, (at the South end of the Grand Gallery) when a sum length is scaled according to Davidson's data, gave the numerical date of the beginning of the First World War, August 4-5th, 1914. The opposite inside edge line of this lentil gave the end of the Great War, November 10/11th 1918. ("The Great War" was the name of World War I, prior to World War II.) These dates were developed out of *Pyramid inches with fractional lengths* of Pyramid inches. Continuing along in the same direction, the inside edge line of the entrance to the great room, called King's Chamber, produced the special date of midnight September 15-16th 1936. (See arrow, far right in Illus. M9. The line of the inside edge of this room as projected down to the floor marks the instant on the surface that the prophetic timeline is measured on. In this part of the Pyramid inches are said to mark *months* rather than years.) From the apparent markings of the previous momentous events, this entrance way appeared to announce something at least as important for the world as a great war, if not more so, based on the great size of the next chamber. The entrance to the great room was hoped to be symbolic of the advent of the end of the world, Messiah, or the Kingdom of God. (This second date September 15-16th 1936 is the special date discussed earlier that appears in mathematical form in the dollar design. It too, will be found as result of a *fractional length* on the dollar, and also hidden on the dollar by means of a mathematical function.)

Illus. M9

Internal geometry of the Pyramid by Davidson (facing east) to show prophecy as linear stations in the alignment of the central passage.

The end of the Ascending Passage is to the left, the beginning and ending of the Great War (WWI) are the vertical lines extended down at the First Low Passage and the special date of September 16, 1936 (see arrow) is at the right, at the end of the Second Low Passage--or entrance to the King's Chamber. (All units area in PI, prophetic "months" begin at the Great Step at 2 Aug 1909.) Though one may argue with prophecy, these spaces are in mathematical harmony with the overall structure of the Pyramid. PLATE LXVc.

The General Chronology System of the Great Pyramid's Passages is given to the scale of 1 Pyramid inch to a solar year, and defines a period of 6000 years, commencing from 4000 B.C. A *connected* Special Chronology System, however, as shown on the diagram below, applies horizontally from the Great Step to the South Wall of the King's

Chamber. This Special Chronology System is given to th scale of 1 Pyramid inch to the month of 30 days, and c Final Period of the Great Pyramid's prophecy as apply interval from 2 August 1909 to 20 August 1953 A.D., from Step dating to the South Wall dating of the King's Chambe

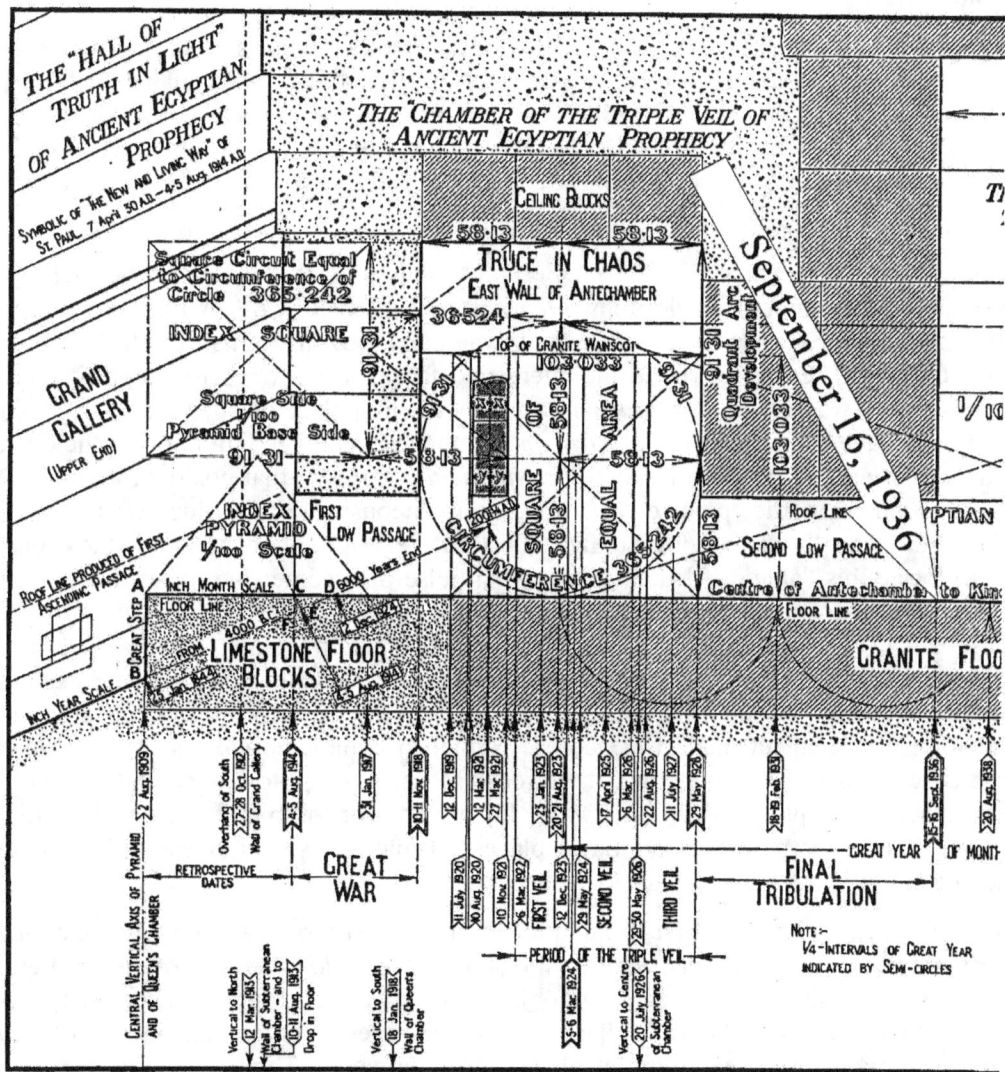

From the beginning of the 1st Low Passage (4-5 August, 1914) to the end of the 2nd Low Passage (15-16 September, 1936) is symbolised as a *natural* period of Chaos. Into this period the Ante-chamber Symbolism is inserted as an intervention—a "Truce in Chaos" from 10-11 November, 1918, to 29 May, 1928.

The Egyptian Messianic prophecy and the Pyramid's Messianic

prophecy indicate that the symbol of the year-circle relate Messiah.

The Ante-chamber and its position and purpose are and indicated bv the year-circle. It is therefore defined as lising a period of "Truce in Chaos" due to Divine Inter The insertion of the Ante-chamber Symbolism into the *natura*

The Ancient Egyptian and Pyramid Messianic prophecies refer to the Messiah as "The Lord of the Pyramid," "The Lord of the Y In the course of time these prophecies became paganised by the substitution of Osiris (the god of Vegetational Decay and Regenerati traditions which led scholars to conclude that the essential basal elements of Christianity had been borrowed from previous religions.

Ken McGrath

Geography:

Multitudes of astronomic alignments, solar reflection data, geometric configurations, and historical items of interest in and around the Great Pyramid are shown in his work. The Bible was believed to go hand in hand with the Pyramid. The words of the Biblical Prophet Isaiah were used to show the Great Pyramid as having an especially sacred and unique geographic position:

ביום ההוא יהיה מזבח ליהוה בתוך ארץ מצרים אצל-גבולה ליהוה:

והיה לאות ולער ליהוה צבאות בארץ מצרם

"In that day there shall be an altar to the Lord in the midst of the land of Egypt, and a pillar at the border thereof to the Lord. And it shall be for a sign, and witness unto the Lord of Hosts in the land of Egypt." Isaiah 19:19-20.

This is a puzzling statement even for Isaiah. Davidson understood this oracle by way of a scholarly argument made by Smyth. He believed that this section of the prophecy of Isaiah 19:19-20 specifically defined the Pyramid of Giza, due to the peculiarity of Egypt's *paradoxical topography* and because of the translation of this Biblical word *"pillar"* from Hebrew as a pyramid. At first glance, it is difficult to see what this passage *could mean*, since the wording seems to describe something that couldn't be.

Pyramidologists state that only the *Northern Nile delta region* is supposed to be "Egypt," technically, from Biblical Hebrew. (*Mizraim* מצרים). Gesenius' classic *Hebrew-Chaldee Lexicon to the Old Testament* seems to bear this out by defining *Mizraim* as *"lower Egypt."* or northern Egypt. (As the probable Biblical authority for his theory, this book was published in English in 1847, and would have been available to Smyth.) Gesenius raised the interesting fact that this word for Egypt is plural ("-im") and that the other part was called *"Pacharom"* (פחרום) when spoken of specifically in singular form. *"Pacharom"* would have had to be south of *Mizraim,* or upper Egypt. Geographically, the Great Pyramid of Giza stands very nearly at the exact *apex of the Nile delta*. So from that position, it would be both the "mid-point" *and* southern border point of *Mizraim*, simultaneously. (See Illus. M10. The word Giza itself is also said to mean "border" from Arabic.) Curiously enough, after the word study, one would be hard pressed to locate this "altar" of Isaiah 19 anywhere else than somewhere near the Giza site. Apart from the above, the word "pillar" is convincing in it's own right as meaning "pyramid." In Hebrew this word is "matsabah" (מצב), that the King James Version and others translate as "pillar." *(The ninth word from the right in the top line of Hebrew text above.)* Gesenius gives this word as "a monument, a pillar" and shows that it was also used this way in ancient Phoenician inscriptions. It also appears as the word in Hebrew used for the monument made by Jacob at Bethel to commemorate his dream of a ladder and ascending and descending angels--made out of the rocks he used as a pillow. But this word is practically identical to the ancient Egyptian word *"mastaba,"* for pyramid or tomb and this word is still in current use in Egyptian Arabic for pyramids or burial places. Isaiah's text could easily be talking about a pyramid, and at this particular place in Egypt.

John Greaves appears to be the first to have noted ancient writers' traditional association of pillars, oblesks and pyramids as the being *identified as the same general idea as a pyramid*, and as having *an astronomical nature,* in his book <u>Pyramidographia</u> [2] written about 1646. There may be a deeper symbolism behind the idea of "pillars." Pillars are a crucial element of sacred furniture in *any* middle eastern temple, especially the dual pillars at all temple doorways. The *dual pillar concept* may also abstractly connect to the functions and fine structure of the Great Pyramid. See Appendix A.

Illus. M10

Position of
the Great
Pyramid at
Apex and
Mid-point
or
Centerline
with
respect to
the
traditional
boundary
of Egypt

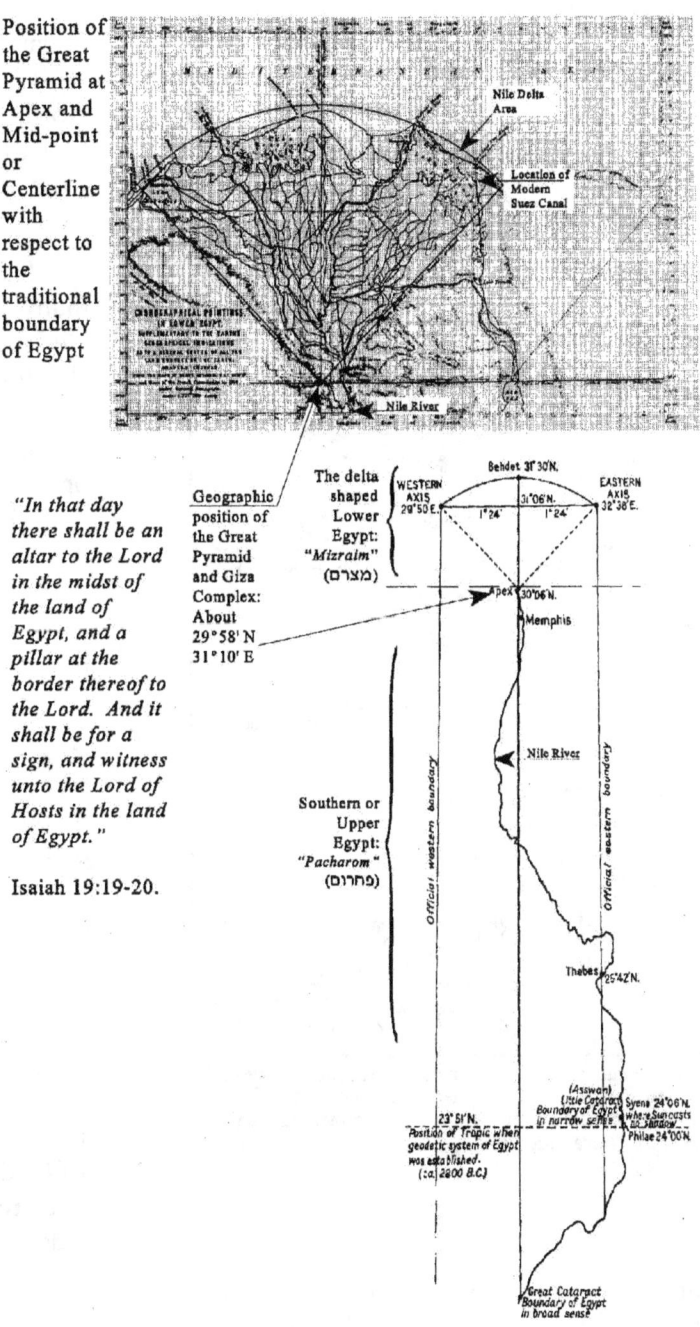

*"In that day
there shall be an
altar to the Lord
in the midst of
the land of
Egypt, and a
pillar at the
border thereof to
the Lord. And it
shall be for a
sign, and witness
unto the Lord of
Hosts in the land
of Egypt."*

Isaiah 19:19-20.

Geographic
position of
the Great
Pyramid
and Giza
Complex:
About
29°58' N
31°10' E

Illustration adapted from the Secrets of the Great Pyramid, by
Peter Tompkins, pgs 46 and 181

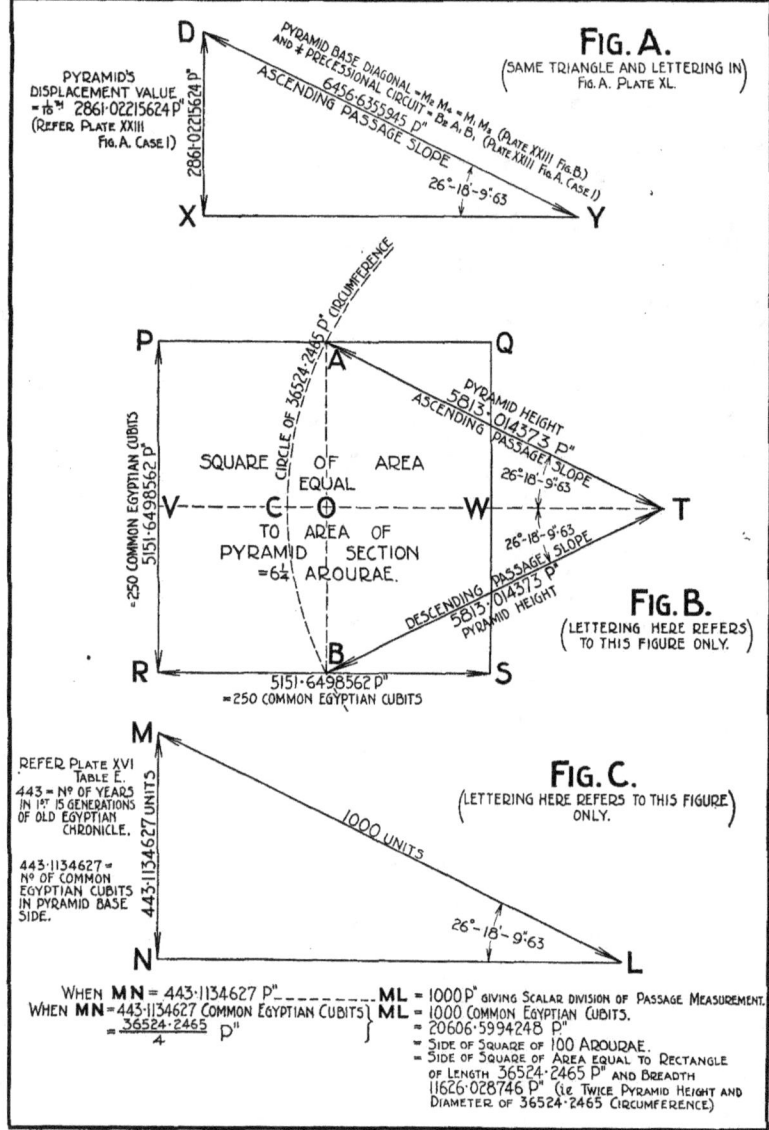

PLATE XXXVIII.

GEOMETRY OF PASSAGE SLOPES.

Illus. M11   Davidson's geometry to demonstrate the connection between the Displacement Factor (here multiplied by 10) and 1/4th of the Pyramid base, with the Ascent Angle of the primary internal passage, or the "Ascending Passage."

Geographic Alignments based on the Pyramid:

Other map based prophetic lore: In a geographic alignment, Davidson provides a map that illustrates the internal *Ascent Angle* of the Great Pyramid passage ways transformed into a special North-based *geodetic azimuth* toward the direction of Israel, from the location of the Pyramid.

This angle is thought by many to be an astronomical elevation angle in the northern sky pointing at the star Alpha Draconis at a chronological alignment, (Thuban in Arabic, or Tanin in Egyptian and Semitic languages). Davidson however, showed that he could derive this angle directly, from nothing more than his pyramid-based mathematics.

Davidson said this Ascent Angle was 26°18'09.63" and that it was based on the theoretical Pyramid's base course diagonals and the Displacement Factor. This is one of the stronger architectural arguments for the clear use of the number 286.1022156 (D) by the ancient engineers of the Great Pyramid. ("Arourae" in M11 are pyramid-related measures of land area not unlike acres or hectares, of 2000 B.C. or older.)

Davidson's Ascent Angle in the Pyramid is taken from the arcsine of half the diagonal of the theoretical Pyramid divided by *10 times the Displacement Factor*:

(t) $\sqrt{(9131.061625^2 + 9131.061625^2)}\,/\,2$ =   6456.6355 PI = a

then: =   Half of the Pyramid's theoretical diagonal length,

(t) $\sin^{-1}(10 \cdot D\,/\,a)$ = θ

(t) $\sin^{-1}(2861.022156\,/\,6456.6355)$ = θ

(t) $\sin^{-1}(0.443113469)$ =   26°18'09.73"   (See Illus. M11 above.)

                                                              =   The theoretical Ascent Angle of the Pyramid. (This is 0.1" larger than Davidson's figure, which is negligible.)

This theoretical Ascent Angle closely approximates the observed, internal angle of ascent in these passage ways, falling within an observational error of 10 or 15 seconds of arc. This would amount to a vertical difference of about 1/2 an inch at the King's Chamber. Davidson wrote that this observed difference in height at the level of the King's Chamber from his ideal computed angle was due to the 45 centuries of interior settlement through compression of the stone structure. (It would, however, be interesting to compute and compare which angle is closer to what would have been the true slope--this mathematical angle, or the angle of the Lower Culmination of the star Alpha Draconis in simultaneous alignment with the star Alcyone at the Vernal Equinox in 2750 B.C., as theorized by Smyth. The astronomical alignment at this instant in time is supposed mark *the beginning* of the over-all prophetic chronology of the Pyramid--an alternative source as a chronological "anchor"point. See pg 88.)

This angle was also proposed as a prophetic alignment as *a geodetic azimuth on land.* In Davidson's map (Illus. M12) we see one of the more celebrated and frequently cited of his ideas. The ascent angle is taken as an Initial Heading (HI) of a Great Circle course based on the passage way's ascending angle (See Illus. M11 and M12. This would give a North azimuth: 90° minus 26° 18'09.73" = 63° 41'50.27".) The angular alignment beginning at the Pyramid crosses the North end of the Red Sea and proceeds to Bethlehem (and further on, south of Jerusalem) in Judea some 230 miles away. Davidson stated there was a place on this line where the Hebrews originally crossed the Red Sea in the Exodus period, (at "Pihahiroth," a shallow, the famous narrow point in the Sea of Reeds, North of the Bitter Lakes). He suggested that there was a continuing, underlying prophetic path, which finally ended at the location of Christ's appearance in Bethlehem later in history. This geographic point, as noted by some Pyramidologists, would be similar by analogy to the position of King's Chamber in the Pyramid at the end of the Ascending Passage.

Curiously, a *trial by a passage through water*--not unlike the Hebrews crossing the Red Sea, the mystical *yam suf*--is paralleled by pre-existing Egyptian mythology found in the Book of the Dead and elsewhere in Egyptian lore. This suggests the actions by the Hebrews at Exodus had a deeper meaning than we are now aware. Perhaps to those aware of ancient lore among the Hebrews, this was a ceremonial procession with an inner meaning. The Book of the Dead ("coming forth by day") is often cited as being a *literary form* of the passageways in the Great Pyramid--where the "Pit" in the Grand Gallery is identified as this "trial." We can see a trial of water at the infancy of Moses (and his name), and Christian baptism as well are related symbols of rebirth. Is the idea behind *Exodus* the same as "coming forth by day"?

Several sources state that the Tabernacle in the Wilderness, Ark of the Covenant, sacred equipment, tent organization, and so on, carried by the Levites into Canaan bear a sort of symbolic similarity to the pattern of inner passages of the Great Pyramid. Many have said the Holy of Holies was originally hidden in the Granite Coffer of the King's Chamber until the time of Moses. The Egyptians shared with the Hebrews the idea of a contract with God kept in a box and carried around ceremonially by priests, guarded by two winged images. The *unit volume* of the Coffer in the Great Pyramid, as computed by Smyth is approximately the same as the Ark of the Covenant, as given in the Bible. (We will return to the importance of the volume of the Coffer later, as a 2861-fraction of the King's Chamber. See pg 202 and 203.)

PLATE LXIII.
THE ROUTE OF THE EXODUS

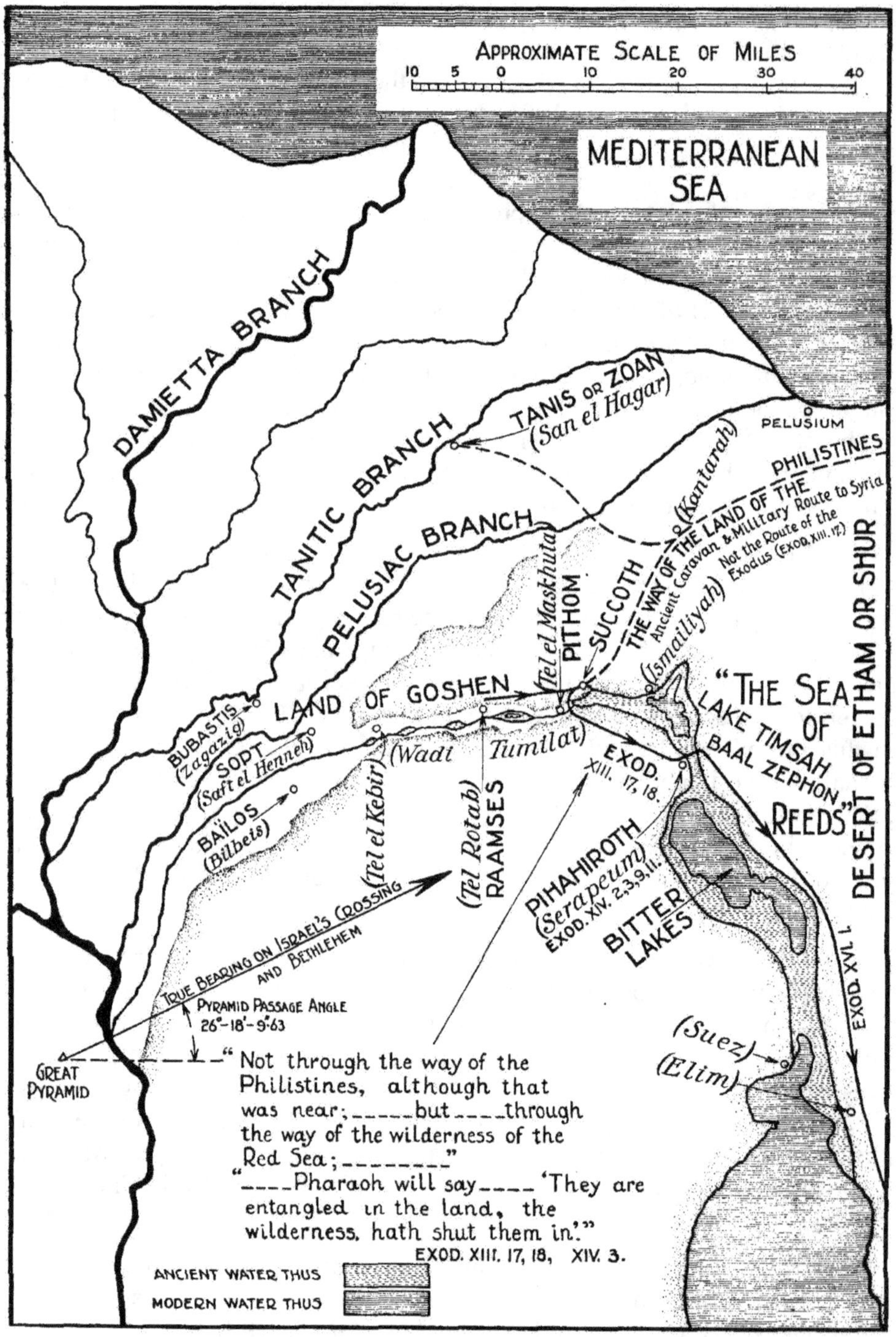

 APPROXIMATE SCALE OF MILES

MEDITERRANEAN SEA

The Extension of the Gulf of Suez shown is from Sir Hanbury Brown's "Map of Goshen." In view of Sir Gaston Maspero's theory of the Exodus Route, it was surprising to find that his map ("Dawn of Civilisation," p. 349) shows the Extension above given.

Illustration from Davidson. Note Ascending Passage angle.

" manifested in the light "

"pert em hru"
Source E.A. Wallis Budge
The Egyptian Book of the Dead (1985)

The Book of the Dead, the *Pert em Hru,* the all important *sacred book* of Egypt has another curious connection to the Great Pyramid. (*Pert em Hru,* means, "coming forth by day"or "that which was hidden brought to light.") The Egyptian chronicler, Manetho, wrote that the ancient Pharaoh Suphis I (Khufu) *built the "largest pyramid,"* disagreeing with Herodotus, who said that Cheops built it. He then goes on to say, revealingly, *that he also wrote the Book of the Dead:*

> "He was arrogant towards the gods, and wrote the sacred book; which is regarded
> by the Egyptians as a work of great importance." (Piazzi Smyth, The Great Pyramid)

Evidently, Akhenaton was not the first, nor only "rebel-innovator" of Egypt, as we so often hear. If Suphis I (Khufu) did indeed build the Great Pyramid, this passage might make a certain amount of sense--for whoever did build it had many unique ideas for an Egyptian and sealed up the Pyramid with no evidence of any interest *whatever* in the standard panoply of gods and other funereal trappings. Considering the possible ritual/design connection seen above, it is interesting that Manetho credits Suphis I with both building the Pyramid *and* writing the fundamental Book of the Dead. But he couldn't have written it all: some parts are rites known to come from a time much earlier than his, and as a book, existing examples vary in composition. The only known Egyptian tradition that devalues the gods in any fashion is the Egyptian-Greek literature called *Hermetic* writings, (e.g., the many gods "as a human creation," as seen in *Asclepius*, III, 38). These writings are also *strongly monotheistic* in character. Wallace Budge, the classical translator of the Book of the Dead, said that the *Pert em Hru* was indeed part of the "body of Hermetic writings". It is tempting to draw a possible connection that the Great Pyramid, the *Pert em Hru* and other Hermetic literature are representative of an underground Egyptian religious movement from very remote times.

The fact that these writings are monotheistic, and share many important concepts with early Biblical and Greek philosophy, may suggest an early causal tie to many historical developments--such as that of the Akhenaton/Moses period of monotheism and also certain recognizable ideas seen in Zoroastrianism (e.g., the concept of the "Good Mind" etc., of the Gathas.) If this Pharaoh, Suphis I (Khufu), did have a special involvement or knowledge of the Great Pyramid concurrent with the Pyramid's construction, he may well have written something to commemorate ideas and ceremony associated with it's religious beliefs or rites. This then might have become a part of the *Pert em Hru.* The body of the so-called Hermetic writings are not thought to be very old however, (ca. 200 B.C.--200 A.D.), but these may well reflect a much older, monotheistic, and inner tradition of Egypt.

But what do we really know of that ancient time? Here, taken from the Hermetic writings, the Ancients themselves seem to have known that we wouldn't know all that much about them--except from their remains where they carved in stone:

> *O Aegypte, Aegypte, religionum tuarum solae supererunt fabulae, eaeque incredibiles posteris tuis, solaque supererunt verba lapidibus incisa tua pia facta narrantibus. Et tunc taedio hominum non admirandus videbitur mundus nec adorandus.*

> "O Egypt, Egypt, of thy religion nothing will remain but an empty tale, which thine own children in time to come will not believe; nothing will

be left but graven words, and only the stones will tell of thy piety. And in that day men will be weary of life, and they will cease to think of the universe worthy of reverent wonder and worship."

(The Poimandres of Hermes Trismegistus, Libellus I
Latin, c. 200 AD, Translation by Walter Scott.)

This is a brief look into the complex, arcane world of the pyramidology of Davidson's time, and one for which there is an enormous body of related literature. As we will see shortly, the Designer of the dollar was clearly interested in at least some of this lore as well as the technical details of the Displacement Factor. Deeper aspects of the Pyramid's unusual geometry, uniquely Davidson's work, soon surfaced in my research on the dollar.

****

Section II--Related Discoveries in the Dollar's Geometry:

The Three Dimensional Dollar Uncovered

During the time I was studying the mathematics of the Pyramid, I happened upon a small similarity between the numbers of the Pyramid's formulas and a part of the dollar's math. This related to the length of the Small Inner Rectangle--the first golden rectangle of Chap 2 (SIR, see pg 45.) This quickly unfolded into something much more significant.

I noticed--I can't quite recall just how I got onto this idea--that the reciprocal of the number 0.86 was quite similar--ignoring the decimal place--to twice Davison's theoretical height of the of the Great Pyramid if divided by 10,000 or, 1.162790698. (The actual theoretical number from Davidson is 11626.02875.) This similarity did not appear to be especially important. But when I happened to divide 0.86 by the "39.37 1/16ths" number for some reason or other, the number 2.861 appeared:

(t)   ("39.37 1/16ths" = 2.460625")
At first I found--

2.460625 / 0.86 = 2.86119186   *(Davidson's displacement factor: 286.1022156.)*
Then, later trying to do the math again, I got it backwards--

0.86 / 2.460625 = 0.349504699   (Therefore: 2.86119186 1/x = 0.349504699)

Entering the "39.37 1/16ths" number into the calculator would have been by force of habit. In the first case, I recognized a similarity to Davidson's special Displacement Factor number, and thought I should dismiss it as one of those coincidences that happen occasionally with numbers. But ignoring that for the moment, there was something about the *opposite number in the reciprocal form* that rang a bell for me. I remembered something about this from earlier somewhere in my research.

Later, after many and varied searches through related numbers, I tried Davidson's number, multiplying 2.861022156 (or, D/100) by 0.86. This gave 2.460479054 which was very close to 39.37 1/16ths, of the Small Inner Rectangle or SIR rectangle. The square of this number (6.05397176) was no more than 6 ten-thousandths of an inch longer than my original computed diagonal dimension for the Large Outer Rectangle or LOR rectangle, and was similar or nearly the same as my original diagonal measurement of 6.054". This seemed important. Earlier I had noticed there was an intent to make the length of SIR the square root of the LOR diagonal (See Chapter 3). This stood out somewhat in my thinking. (The "0.86" aspect of the trail, however, grew cold and didn't appear to shed any more light on matters.)

It was the *other number* out of this: the "349" number, that really stuck out. I knew I'd seen it somewhere before--and sure enough, there it was back in my original scribbly notes. It was among my various experiments many from years before on the dollar's dimensions using φ as a divisor:

(m, t)	5.655 / φ=	
(t)	5.655 / 1.618033989 =	3.494982206

This number was *ten times* the earlier reciprocal 0.349 number. This meant that 5.655 might be related in a mathematical way to Davidson's "286.1" number. After searching around the dollar with this 3.49 number length some years before, and having found nothing, I moved on to something else. But it did appear to be reasonable search at the time. At the time I thought *some* graphic feature on the dollar ought to align to the 3.49 number length: it was, after all, the first, or *primary division by φ of the dollar's length.* Yet I couldn't see any alignments arising from it then, or anything else of value. I even wrote down the reciprocal, 0.2861 in my notes but didn't see anything in it at the time. Because it was *a reciprocal,* it didn't seem very important. By the time I had begun to see the potential importance of *any reciprocals* in the Designer's scheme, I had forgotten all about the little 0.2861 number.

Shortly after this discovery and review of my notes, I tried using just the pure, theoretical numbers themselves. I found to my astonishment that the *length of the dollar might just as well be the simple relationship of Davidson's Displacement Factor and one thousand times φ*--giving a new possible source for the LOR geometry:

1000 · φ / D	=	DFT (Displacement Factor Theory length)
1618.033989 / 286.1022156	=	5.655440261

Why couldn't this 5.655440261 be the length of the dollar? Here again, like the Independence Date Theory length, we have a number that *if it was* actually the intended inch length of the dollar, no one would be able to measure the difference to make the distinction (See Independence Date Theory in Chapter 2 on pgs 29 to 31). And yet the tiny difference is twice the size, and in the opposite direction. How much difference is there? Here is the original measurement to IDT theory subtraction, showing a possible 2 ten-thousandths difference, compared to measurement to the DFT length above showing a 6 ten-thousandths difference:

(m)	5.6550"		(m)	5.6550"
(t) (IDT)	5.6548"		(t) (DFT)	5.6554"
	(-)----------			(-)----------
	a=0.0002"			b=-0.0004"

Not a lot of difference in either direction. How about the difference between the two theory lengths?

(t) (DFT)	5.655440261	
(t) (IDT)	5.654801219	
	(-)----------------	
	0.0006	Not much of a difference in any of these three cases.

In terms of measurement this is not an observable difference. It amounts to a little more than half of a thousandth on the long side--when compared to the earlier theory--but this subtly new figure of the DFT theory will put a whole new slant on the dollar design.

It soon began to dawn on me that this length could be intended to represent *the golden ratio in its three dimensional form viewed edge on.* This "3.49" number would be it's *third dimension as a depth.* This was really a *box*--an exact form of the Golden Cuboid. As the new idea slowly crept up on me, I sat in stupefied amazement:

Consider now *why* the dollar's Square of the Golden Ratio *shape* (1: 2.618033989...) was chosen. Was it the most convenient classical shape that the Designer could think of: "fits better in the wallet?" Was it a little more economical for ink and paper than the earlier larger $1 note? No to both. Something new has become evident here. Something more than a two dimensional, flat dollar. The Designer might very well have been able to trot out some easy explanations of lengths and proportions if need be, such as the hypothetical answers offered in the Introduction and they would all be *true*. Yet it's likely that he would leave the next part unmentioned. I don't think that he would have offered to explain that this length could be said to be a deeper, arcane number, *a composite number*. This is a number composed of one thousand, the golden ratio *and* Davidson's Displacement Factor. And the Displacement Factor has been "folded backward" mathematically as a reciprocal. The "286.1" had been veiled by arithmetic. More than that--the dollar's face *is the top of an invisible box*, or the top side of a *virtual* Golden Cuboid as we saw in Chapter 1. I am sure that this too, would have gone unmentioned. Could this "depth" be a mathematical coincidence? No it couldn't be: in a single step *it's reciprocal is the naked form* of Davidson's special number, locked "inside" the LOR rectangular face by the golden ratio shape.

Illus. N1
The Dollar as the top face of a Golden Cuboid

The Dollar Rectangle (LOR) provides the depth (z) for the top of an invisible Golden Cuboid

The riddle of the height and length of the Dollar is found through the discovery of the virtual depth: 3.495254302"

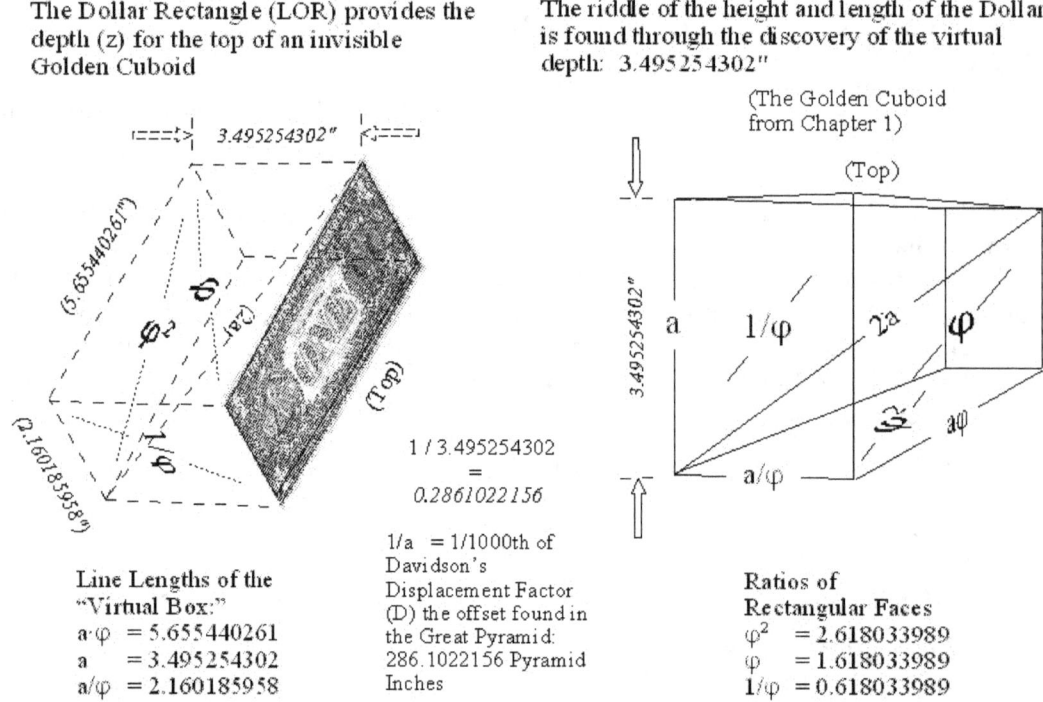

1 / 3.495254302
=
0.2861022156

1/a = 1/1000th of Davidson's Displacement Factor (D) the offset found in the Great Pyramid: 286.1022156 Pyramid Inches

Line Lengths of the "Virtual Box:"
a·φ = 5.655440261
a = 3.495254302
a/φ = 2.160185958

Ratios of Rectangular Faces
φ² = 2.618033989
φ = 1.618033989
1/φ = 0.618033989

This box would be the classic be the Golden Cuboid shape: x = 5.655440261...", y = 2.160185958..." and *the hidden dimension, or invisible depth* is z = 3.495254303..." being the exact reciprocal of 0.2861022156..." or D/1000 (See Illus N1). A clear announcement, once seen, from a mathematical point of view. (J1, and J3, J4 J5. See Chapter 1, pg 13.) How did I miss this? Added to this surprising virtual shape, are the many unusual and unique properties of this Displacement Factor Theory box. One of these is the diagonal from the inside corner to inside opposite corner. The cubic diagonal of this shape as shown in Chapter 1, is exactly *twice* the height, or twice the invisible z dimension. And so then it must be also *the spherical radius* of an invisible orb--as we saw in Chapter 1-- where all eight points of this golden box would exactly touch its surface, whose basis is a *radius of z*. The *z*-side and 2z diagonal proclaim its basis only for those who venture into this inner, mathematical space.

The following is the formula for the cubic diagonal of any regular Cuboid. But if it happens to be a Golden Cuboid, the diagonal will always be *exactly double* that of the z dimension (2z):

(t) $\sqrt{(x^2 + y^2 + z^2)}$ =   the cubic diagonal of the golden cuboid
(t) $\sqrt{(x^2 + y^2 + z^2)}$ = $2 \cdot z$
          Or in real numbers:
(t) $\sqrt{(31.98400469 + 4.666403392 + 12.21680264)}$ =
(t) $\sqrt{(48.86721063)}$ =   6.99050861
(t)   *which is the same as:* $2 \cdot (1/D \cdot 1000)$ = $2 \cdot 3.495254303$

Also, the *area* of the dollar face itself is unusual: the height times the length is *the square* of the hidden depth, z. In other words, the dollar's face area is *exactly* same as another piece of paper, made as a square measuring exactly 3.495254303" by 3.495254303" or, z squared ($z \cdot z$):

(t) $x \cdot y$ = z^2
(t) $5.655440261 \cdot 2.160185958$ =   12.21680264
(t) 3.495254303^2 =   12.21680264

After all this, would it surprise anyone to learn that *the volume* of this box *is the cube* of the new hidden width, z? In other words, this imaginary rectangular cuboid has exactly *the same volume* as a cube-shaped box whose three basic sides are made exactly 3.495254303" in length, or, z cubed ($z \cdot z \cdot z$):

(t) $x \cdot y \cdot z$ = z^3
(t) $5.65440261 \cdot 2.160185958 \cdot 3.495254303$ =   42.70083207
(t) $z^3 = 3.495254303^3$ =   42.70083207

Also notice, that if you compute a new cubic diagonal of the z^3 cube: it is the *same diagonal length as the flat diagonal on the face of the dollar* using the DFT lengths (See Illus. N2.) That is to say, the diagonal corners of the flat dollar face would just exactly touch at the *cubic diagonal corners* of the z^3 cube:

(t) $\sqrt{(z^2+z^2+z^2)}$ =   6.053958037
(t) $\sqrt{(x^2+y^2)}$ =   6.053958037 (DFT theory *diagonal*, Large Outer Rectangle)

What an intoxicating find. I went around for days open mouthed, and amazed as the parts fell into place. I was surprised that I hadn't found this connection earlier. But who would have thought the Designer could build a design *into* the dollar as a three dimensional depth?

It should be noted that these odd, self-referential and amazing relationships will not happen with any other ratio than φ. These are obscure and perhaps unique properties of the golden ratio.

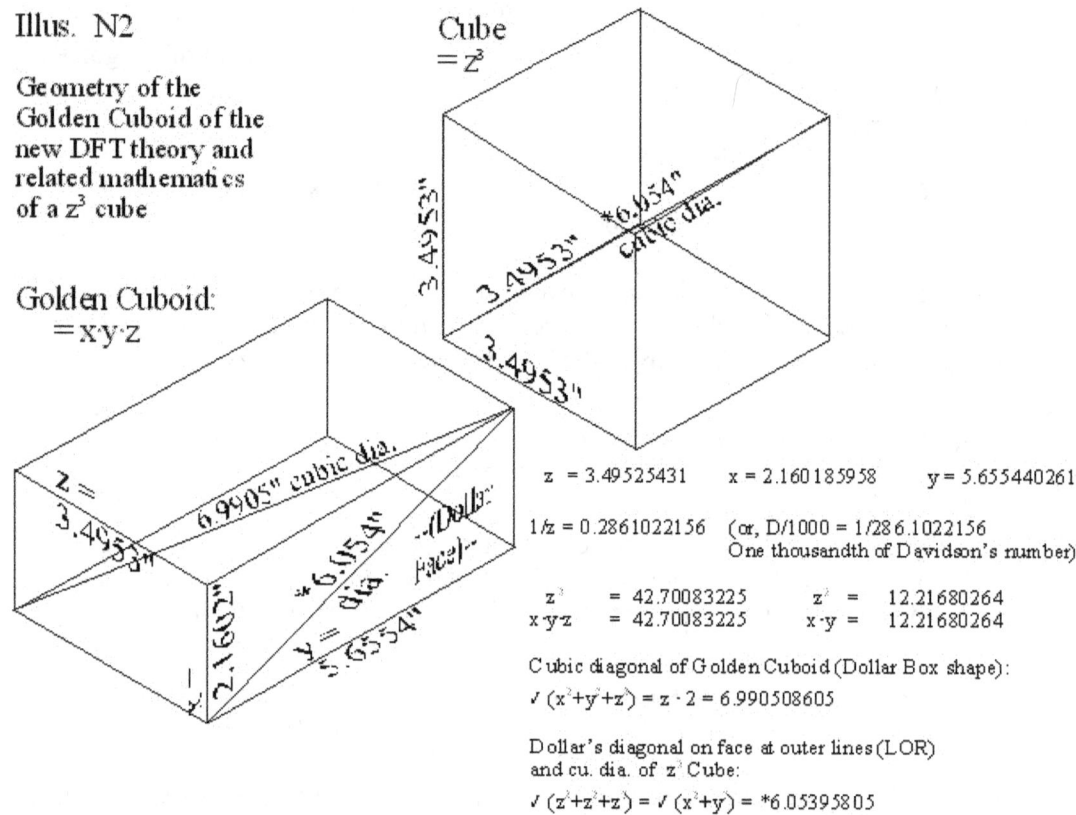

Illus. N2

Geometry of the
Golden Cuboid of the
new DFT theory and
related mathematics
of a z^3 cube

Cube $= z^3$

Golden Cuboid: $= x \cdot y \cdot z$

$z = 3.49525431$ $x = 2.160185958$ $y = 5.655440261$

$1/z = 0.2861022156$ (or, D/1000 = 1/286.1022156
One thousandth of Davidson's number)

$z^3 = 42.70083225$ $z^2 = 12.21680264$
$x \cdot y \cdot z = 42.70083225$ $x \cdot y = 12.21680264$

Cubic diagonal of Golden Cuboid (Dollar Box shape):
$\sqrt{(x^2+y^2+z^2)} = z \cdot 2 = 6.990508605$

Dollar's diagonal on face at outer lines (LOR)
and cu. dia. of z^3 Cube:
$\sqrt{(z^2+z^2+z^2)} = \sqrt{(x^2+y^2)} = *6.05395805$

What is truly fascinating is that of all these three numbers, x, y and z, *the only number that we don't know* from the dollar by direct evidence is this new 3.495254303. The elegance of the design is that *it doesn't need to be said anywhere in the graphic evidence, at all*. Its *hidden character* is its charm and its graphic *absence* most revealing. Yet when you do find it, and work it out mathematically, additional evidence is not given or needed. It explodes in your mind as an astonishing *fait accomplis*. The weight of certainty and simplicity of this finding become enough on their own. It was made so that the hidden z must be located by digging in the math, and only then is the invisible world of the Designer revealed. And I don't believe that this number occurs in any measurable form in the dollar design, or by any clear indication in the printing of the's lines. Yet to the researcher it becomes the unmistakably *fundamental number* of the design. (J4)

By the placement of this number 3.495254303 in the theoretical box shape as its *middle* variable z, the Designer forced the shape to trumpet this numerical identity in an avalanche of related mathematical forms due to the odd properties of the golden ratio. This is the "definiteness" of demonstration--the object of Rule 3, writ large, like added exclamation points.

All of these hidden forms stem from the bare choice of the original height and length for the dollar, and the specific dimensions. From only these bits of evidence, do numerical identities, diagonals, angles, proportions and three-dimensional shapes come forth. We have seen this kind of care earlier in the SIR/SOR shapes, as well as Pyramid of Giza and Pyramidon angles. For the investigator who had gotten this far, the Designer has still not been found using the Displacement Factor with a *clear*, graphic dimension. (Okay, he *does* in one spot--but I should say, not with any definitive proof of sharp markings. There is almost an appearance of coincidence when found.) The 2861 number never appears with any clear length in the geometry--only by hints or alignments that you have to determine on your own. Until you divide this proposed, discovered depth by the *number One*; until you look at *the reciprocal* of z, you won't see a real proof of its presence.

****

I could see at this point that the Independence Date Theory was obviously not the whole story. From the clear, many fold pattern seen here, the Independence Date idea soon paled in significance and began to seem *coincidental*. This is an example of the "clustering" problem. I spent quite a bit of time thinking about this. I gradually decided to shut out the old, beloved IDT theory from serious consideration in the various mathematical routes I was then following. I chalked it up to a fortuitous choice by the Designer for the theme of the dollar's design, and left it at that. Another strange *coincidence*. It was a big lesson, finding intent by interpretation of data. Afterward, many discoveries on the dollar appeared to show that the new Displacement Factor Theory (DFT) concept was the controlling idea. (But in fairness to the Independence Date Theory (IDT) idea, the "date"*does appear independently* in crucial places elsewhere in my research, and it seems evident that the Designer was completely aware of, and probably affected by both sides of these numerical constructions.)

They *both* seem like really good ideas, the IDT and DFT. It didn't seem likely at that point, that any other fundamental pattern would offer another serious, alternative explanation for the length of the dollar. (Wrong again.) I began to re-think the dollar's design, and saw that this element in the Designer's mathematics clearly identified him as a partisan of at least some of the Davidsonian picture of sacred mathematics. At this point, all I had was the Displacement Factor and the exterior slope angle of the Pyramid. Soon, there was mountains of evidence in this direction.

****

Davidson's Number tucked away in other places: Letter Serifs and the Thirteenth Line

With greatly renewed interest, I went back to measuring the dollar, an activity I had neglected for quite some time. Soon enough, I discovered a 2.861" length (D/100) in various forms all over the face of the dollar. Some right out in the open, and other forms in quite sophisticated arrangements.

The easiest 2.861" length to find appears right at the center, as the apparent distance between the intersections at the top of the SOR line and two mirror-opposite curls of leafy shapes on the right and left sides. Just where *this line slips out of sight*. Somehow, that seems appropriate. (See Illus N3). Notice, though, that from a geometric standpoint, these are very "weak" intersections geometrically. It would be very difficult as noted above, to *prove* the exact dimension of this line by measurement in the strict sense, to our limit of one thousandth of an inch. This is because these lines meet *obliquely,* with internal angles of about 20° *that gradually feather off,* as the leaves curve, to an even finer angle. This appears to make incontrovertible measurement--closer than one-hundredth of an inch--impossible.

Where, precisely, is either intersection? Try it yourself on a dollar with a dial caliper. After that, set the dimension 2.861" in the caliper and compare the points to this line segment. ("Okay," you might say, "it's pretty close," but in reality you will only convince a believer.) Depending on just what place on the line's width you find the intersection to measure from, quite a range of possible lengths might be found. And not just *one* weak, uncertain intersection to measure to, but *two*--and they are from non-standard curves (the leaf edges) crossing a straight line at an sharp angle. Two living objects against a ruled line. We can not be sure if these points were intended to be calculable intersections from the graphics, as most of the other graphic relationships are. This length seems to rest entirely on the dollar's graphics, or your best estimate from measurement. This is an "obscured ends" -thing, and this idea appears as an element elsewhere among the dollar's graphic puzzles. Perhaps this is a self-imposed rule by the Designer.

111

Illus. N3

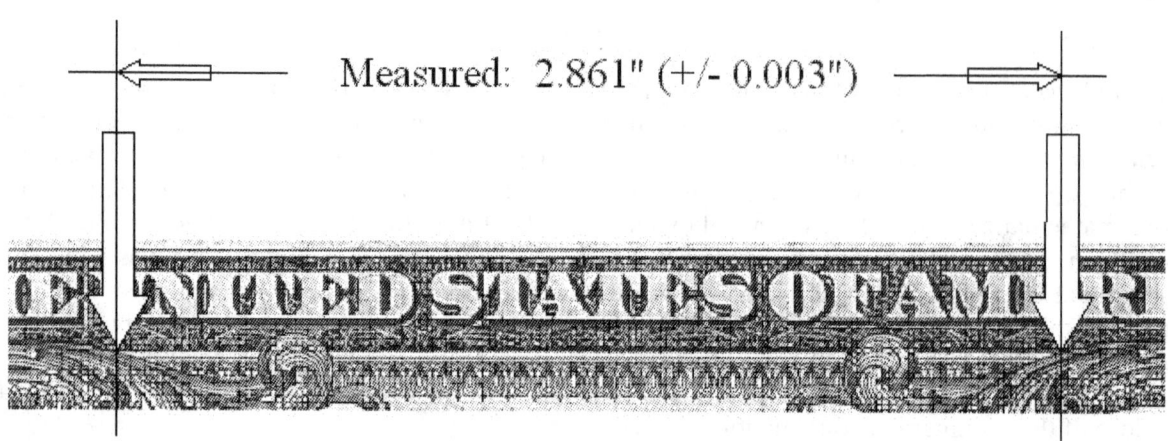

Measured: 2.861" (+/- 0.003")

But here lies the *problem*: You wouldn't have known to look for "2.861" there, unless you already *knew* about this special number in its reduced decimal form. On the other hand, you *wouldn't likely be persuaded of this dimension*, just from the measurements to these frustrating, vague intersections. But it is conceivable that you might have gotten a start there, *if you knew something of the lore* behind the Pyramid. These ends are not a good pair of points to take measurements from. Yet by laying a rule on the dollar between these points, even the casual observer will see that it is at least close to 2.86", plus or minus a hundredth of an inch (0.01"). So it provides a good confirmation of this number *if you already knew this number was going to be important*, but it doesn't help you if you were just looking around for good shapes to test with an exact measuring tool. A line marked obliquely by irregular curves is not a "good" shape for measurement. But the Designer says "yes" to this number, yet did not make it an obvious, clear, or an easy identification for the ignorant or the doubter. It's there, but it's *not:* how would the edge of a leaf mark a line? you might say he "did not so impart it." Should every pair of leaf/line intersection like this be measured and scrutinized? There are quite a few on the dollar, even within this line. (Although artfully "obscured," the above *end points* were not wasted by the Designer on this one idea of *length*--they have several interesting geometric connections elsewhere on the dollar.)

Returning to The Sum of the Four Elliptical Chords and the Fourth Elliptical Chord Length:

Back on pages 58-61 of Chapter 3, (See Illus. H4, H5, H6, and H7), we saw in Illus. H5 that the longest elliptical chord was 0.2865" a now familiar number. The sum of all these chords equaled 1.2217". Strangely enough, we have seen this second number *earlier* in this Chapter as a form of z^2. The largest chord measured in the CADD program was 0.2865", a chord length farthest from symmetry (J1). It gave us a number very similar to D/1000, but seemingly +0.0004" or 4 ten-thousandths *too large*. The total sum of the chords produced a number that initially suggested the height of the semi-major axis of the Left Face of the Great Seal. If instead, we multiplied this number by 10 and took it's square root, we get 3.495282535. The reciprocal of this is 0.286099905--a more sharply familiar number, being "0.2861" if rounded to four places. (The raised fractional exponent of "1/2" is another way to write "square root"):

$$(1.2217 \cdot 10)^{1/2} \quad = $$
$$(12.217)^{1/2} \quad = \quad 3.495282535 \text{ then,}$$
$$3.495282535 \quad 1/x \quad = \quad 0.286099905 \quad \text{or very nearly 0.2861.}$$

Notice that the Displacement Factor Theory $x \cdot y$ and z^2 values shown above are practically indistinguishable from the sum of the elliptical chords if multiplied by 10, or 12.217 and 12.21680264 (J3). These figures and this section are reproduced here from above:

(t) $x \cdot y$ = z^2

(t)	5.655440261 · 2.160185958	=	12.21680264
(t)	3.495254303^2	=	12.21680264

This sum (multiplied ten times) compared to z^2 has an error as small as one part in 62,000. This strongly suggests a non-chance relationship. Why then is there a 0.0004" excess in the fourth chord, when the sum is so closely recognizable? I believe that this is a "nested problem"--and that the Designer's puzzle here is still not fully solved. *Maybe* he wants us to shift the right elliptical face straight *down* a tad. This would be to diminish this chord length to become exactly 286.1/1000--to affect another relationship, such as to slightly *enlarge* the top tie distance of 0.440" to the LIR line by a tiny bit. We will look at this later in the "date number" calculations on page 130. (See Illus G5).

The Central 2.861 Circle

The next clear place we find is a radius of 2.861" made from the center of the dollar, at the point where the diagonals cross, within the body of the "N" in the middle of "ONE" (See Illus N4). Much that follows here is of the J6 class, the "kissing" or tangent alignments as we saw the Designer doing with serifs in Chap 2.

Illus. N4

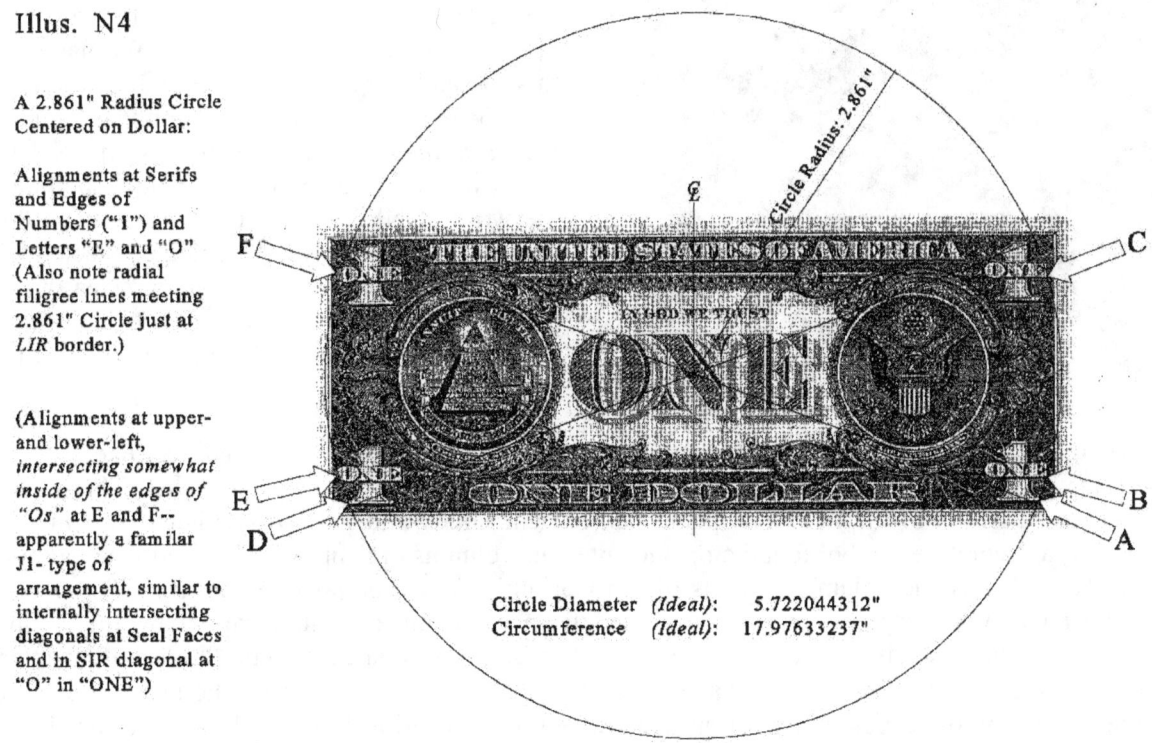

A 2.861" Radius Circle Centered on Dollar:

Alignments at Serifs and Edges of Numbers ("1") and Letters "E" and "O" (Also note radial filigree lines meeting 2.861" Circle just at *LIR* border.)

(Alignments at upper- and lower-left, *intersecting somewhat inside of the edges of* "*Os*" at E and F-- apparently a familiar J1- type of arrangement, similar to internally intersecting diagonals at Seal Faces and in SIR diagonal at "O" in "ONE")

Circle Diameter *(Ideal):* 5.722044312"
Circumference *(Ideal):* 17.97633237"

Starting on the right side, the 2.861" radius circle is exactly tangent to the farthest point at the *lower right corner point* of the serif of "1" in the white, unshadowed part of the letter (A). Next, further up, it is exactly touching the lower right farthest point (at the unshadowed white area) of the "E" in the word "ONE", the text superimposed over this numeral "1." (B) Okay, so this might be coincidence. But further up along the circle, it appears once again. The circle is grazing the *lower right point* of the white corner area again, yet in the upper right letter "E" of the word "ONE"(C). (See illus N4 and N5.)

On the left side of this circle, the lower left edge of the serif of the numeral "1" is also *exactly* tangent, (D). Here again, an exceptionally unlikely relationship. So, theory scores four, contextually related, exact hits with this dimension taken as a radius from the center point of the dollar. Since we

113

Illus. N6

Decorative *acanthus* planted at the base of Greco-Roman styled coluums The Palace of Fine Arts. (From the 1915 Worlds Fair, now a park near San Fransico, California.)

already familiar with the idea of "intersected ellipses," we might count this as six hits, even though there doesn't appear to be an exact intersection relationship. It may be interesting that the outer left edge of the "O" on the "ONE" at the upper- and lower-left sides fall slightly outside of the circle, perhaps suggesting a similar pattern to the one we have seen with the diagonals with the Seal Faces (E and F). If so, there may be some exact shift indicated by this clue. On the upper left corner, no serifs on the numeral are available to be touched. If the Designer is rigidly holding to the J1-form here, this area seems suggestive of a puzzle element. I have no solution to this puzzle yet.

We should notice, too, in the symbolic sense, that the *only places* where this special circle touches *anything distinctly,* is on something called "ONE," and then, just on the "Os" and "Es," or the number "1" itself. (We have seen this O- and E-relationship before in Chapter 2 pgs 23 and 24). If you also count the circle's *radius point* at the middle bar of an "N," this might be said to complete the whole spelling of the word "ONE" in various ways. This 2.861" number then, appears to be being connected with the previous message of "oneness," idea that we saw earlier in alignments having to do with φ-angle based diagonals. (J6, J7 See Chapter 2.)

Leaves and balls

At the right and left ends of the dollar there are two curly leaf structures near the Seal Faces. There are *two series of balls* connected like a spine or a stem forming the center vein of the leaves. These leaves are thought to be leaves of the *Acanthus* plant [2] an ancient Greek and Roman architectural symbol of sacred beauty, a decorative symbol found atop ancient stone columns in temples. This is also found in the ritually associated Masonic column symbols of the Corinthian and Composite Columns. The original Corinthian Capital was supposed to have been cut from stone on a column capital by the sculptor Callimachus. (Callimachus is said to have been a rival artisan of Phidias, the same Phidias of Chapter 1 who's name gave us the letter "φ." It was said that Callimachus was taken by the sight of a basket surrounded by acanthus leaves, a part of an ancient religious offering in Corinth, around the time of Pericles. See photo, Illus. N6.)

Many have noted that *thirteen balls* make up this leaf vein at each outer side of the Seal Faces. This is numerically in keeping with all of the other known Thirteens on the dollar: Stars, stripes, olive leaves, olives, arrows, steps of the pyramid, the letters of "ANNUIT CŒPTIS," the letters of "E PLURIBUS UNUM," the (stretching poetically) letters of "IN GOD WE TRUST" (12) plus "ONE" just below it--another symbolic play on "ONE." (Note that the phrase "IN GOD WE TRUST" was not in the original issue of the 1935 Dollar, but is an addition to the design called the "1935 A Dollar." Also note, that in many descriptive accounts, the Eagle's "tail feathers" are also said to be thirteen in number, but there are *actually nine.* Some traditions [3] state that these stand for the Nine

Supreme Court Justices, in the sense of being symbolically a rudder of State. It is interesting to note, that if Franklin D. Roosevelt had had his way (just around the time that the new dollar was appearing) we might well have had twelve or thirteen Supreme Court Justices after FDR's efforts. It was called "packing the Court" at the time. (Is it conceivable that additional tail feathers might have wound up on the Eagle?)

I say "known Thirteens," above, because there is at least one more group of thirteen, previously hidden, that we will come upon shortly.

****

The Thirteenth Line--The Line that Names Itself:

If we lay a rule on the dollar, *exactly at the centers of the final balls* of the leaf vein, the thirteenth balls on either side of the dollar, a line will be laid out that is slightly lower than the horizontal centerline of the dollar. (The line's position is about 0.0244" below the exact centerline at the center of *WB*.)

Illus. N5   2.861" Circle from Center: Alignments at "O"s, "E"s and "1"s

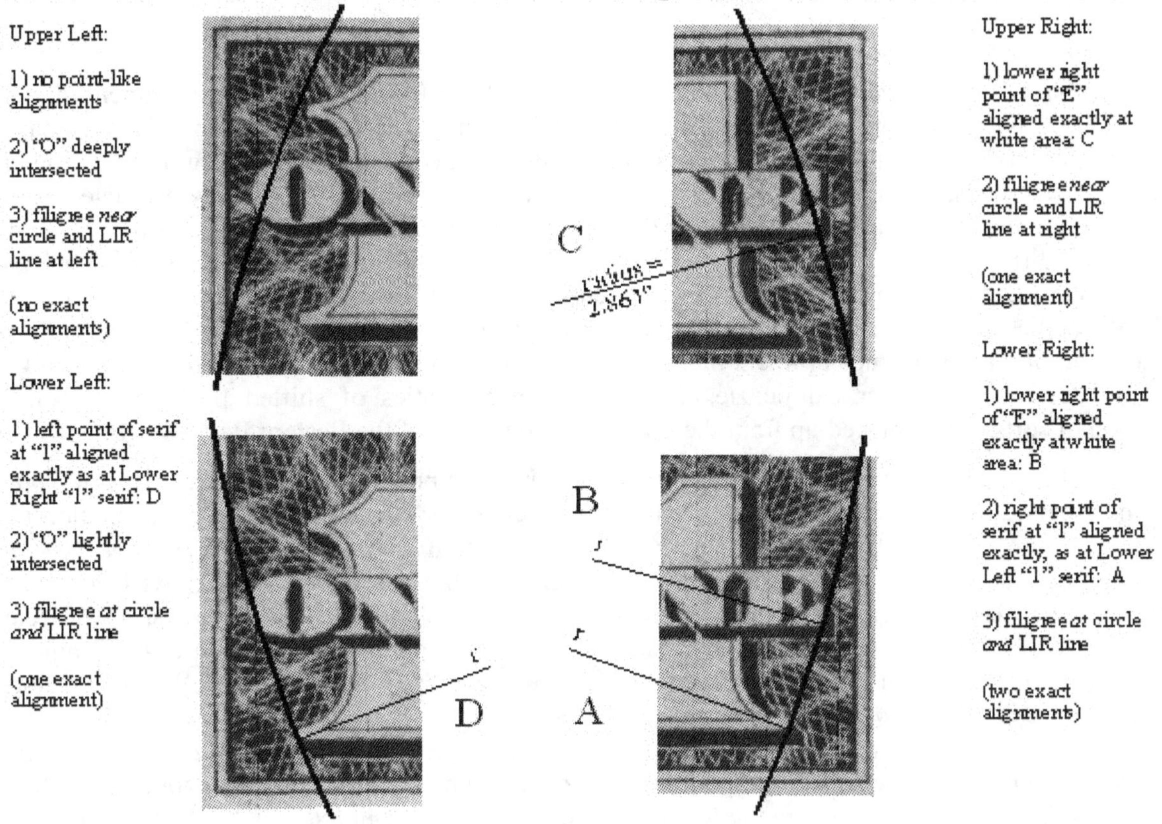

Upper Left:

1) no point-like alignments

2) "O" deeply intersected

3) filigree *near* circle and LIR line at left

(no exact alignments)

Lower Left:

1) left point of serif at "1" aligned exactly as at Lower Right "1" serif: D

2) "O" lightly intersected

3) filigree *at* circle *and* LIR line

(one exact alignment)

Upper Right:

1) lower right point of "E" aligned exactly at white area: C

2) filigree *near* circle and LIR line at right

(one exact alignment)

Lower Right:

1) lower right point of "E" aligned exactly at white area: B

2) right point of serif at "1" aligned exactly, as at Lower Left "1" serif: A

3) filigree *at* circle *and* LIR line

(two exact alignments)

Illus. N7 The Thirteenth Line, beginning at final balls on stems of Acanthus leaves, outside of Great Seal faces.

EB and WB =
Eastern and Western Balls

In a moment we will see why this came to be called the "Thirteenth Line." This alignment, like many found in the dollar, is an "imaginary line." In this case, it is imaginary two ways: not only is it imaginary vertically, but also horizontally. Its beginning and ending points are imaginary, beginning at the *empty centers* of these balls, and not starting at any sharp point or line. Since we can only estimate beginning and ending points, there is a real difficulty in determining the exact length of a line like this. One might ask: "Why is this so hard to measure: why couldn't we just measure from the right sides of the balls, or to both left sides, and divide by two to get a real dimension?" As it happens, however, the line-work of the final balls at the end of these leaf-veins is *not identical and are irregular* so this trick won't work. This is another kind of Obscured Ends problem that we saw earlier. Later, we will see that these open balls hide not only horizontal and vertical puzzle elements but peculiarities of shifted position as well. An indication of sorts can be picked up from the oddities of the shape of the Eastern Ball. (See Illus. N7)

The ball on the left, the Western Ball (WB) is regular; it is a practically *round* circle, *easily separated* from the other 12 balls, and totally *empty* of markings or engraver's hatching. Okay, that shouldn't be so hard to measure to. But the ball on the far right, the Eastern Ball (EB), *seems to have been made irregular intentionally,* with inner border at right covered by the previous ball #12. As if to say, "The other round ball is the precise clue, this oval ball indicates shifts and other possibilities." Apparently it was made to be hand-drawn in appearance, and slightly overlapping, shaded with a small amount of hatching as continuation of the blobby style of all the other balls of both groups. We should note the differences: it is (1) *elliptical,* and (2) its *major axis seems to be cocked to the right about 25 or 30 degrees,* (3) and it was *hatched* as noted. (*Hatched?* Five or six suspicious little *dots* within EB--See right scan in Illus N8.) On close inspection, the same overlapping by an inner border appears at WB. The big problem is that the inside line on the left side of the EB is *indeterminately behind or merged* with the right border of ball #12, *so that no certain average dimension can be clearly ascertained horizontally* from all four edges of the two balls WB and EB. You *can* make an average, of course, but it will not be rigorous or correct per Rule 1. It is almost as though any precise clues have been suppressed by the Designer, forcing the investigator to make educated guesses. The vertical-width information of the WB ball might have been *substituted* in measurement for the missing evidence for a horizontal length of the Thirteenth Line (since it is a circle and both dimensions would be the same) but this cannot really be done for EB.

Illus. N8   Differences between EB and WB: Attempting
to center circles vs. ellipses, clear vs. vague borders

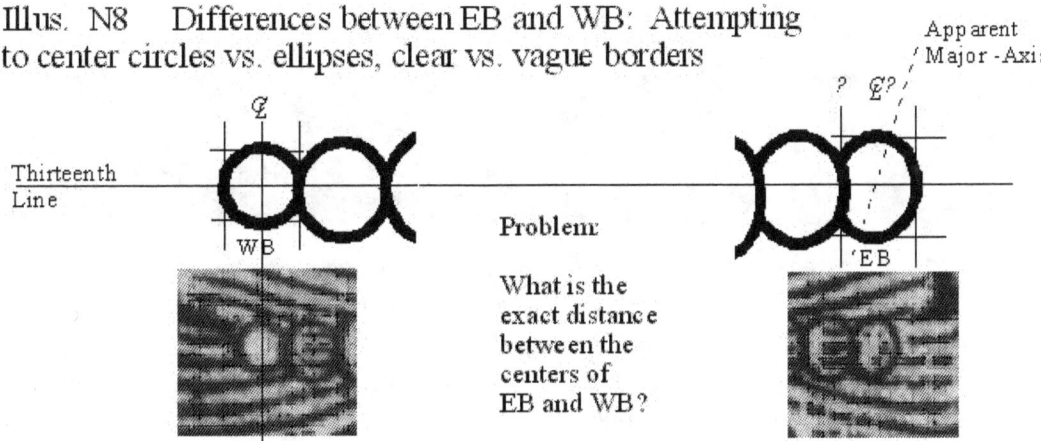

Problem:

What is the exact distance between the centers of EB and WB?

So from where is one supposed to measure? The ball at left, WB, does seem to have been made to stand out in an especially clear way for some reason--but EB has been partially obfuscated. This is a J1-puzzle arrangement. Since both end balls are not truly regular with respect to each other, measuring for a true average to exactly determine the line's length seems to be ruled out. But the balls' center points could be closely estimated through apparent symmetry of these circular shapes by eye starting points to analyze this line. This method will allow approximate beginning and ending points precise as the neighborhood of one thousandth of an inch or so at either end. (In the work below, strict conformance with Rule 1 will be exchanged for a reliance on Rule 2, and demonstration under Rule 3. See Chapter 3.)

****

Interesting Things Encountered Along the Thirteenth Line:

Illus. N9   Beginning at center of Western Ball (WB), the line next grazes 7th ball at (a), and passes just above

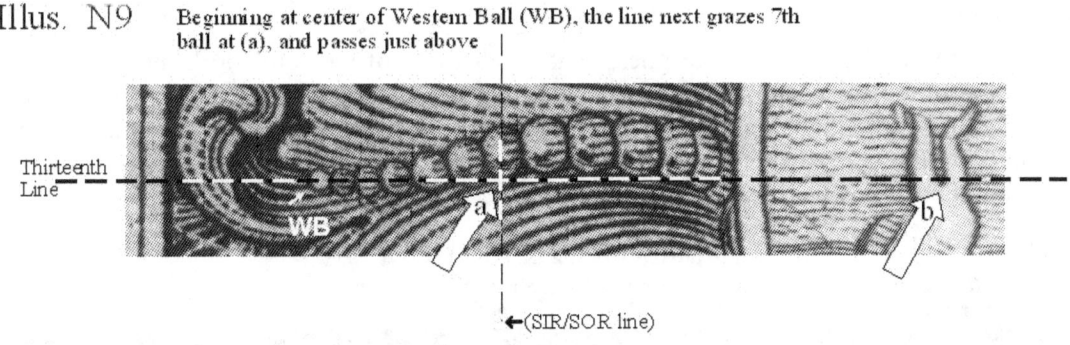

the "v" point (b) in flag tails of scroll.

Proceeding along this imaginary line, beginning at the center of WB from left to right, this line *just grazes* the lower edge of ball number seven at a point (arrow a) just where the line ending SOR/SIR crosses this line. It then passes just above the inverse apex, or "V" shape at the bottom of the left forked tails of the scroll (arrow b) that carries the message "NOVUS ORDO SECLORUM." (See Illus. N9.)

Then, following the line across the little Pyramid's face, (Illus. N10) it hovers parallel and a little bit above the top line of the Sixth Step, (arrow c) and this together with the bottom points of the "V" (arrows b and d) seem to be intended to draw attention to the top of *step number six*, which we already know is of great interest, as shown in Chapter 3. (There could be a little more space above the line at the point of the right "V" of forked tails, than on the left.) Does this indicate a hint of a *downward* shift? We saw the bottom point (d) of this fork in an earlier alignment in an SIR diagonal in Illus. E of Chapter 2, the diagonal projected from the dollar's centerline.   The line then passes (apparently uneventfully) through the middle section of the bill, until it reaches the right face of the Great Seal.

117

Illus. N10   Thirteenth line just above top line sixth step (c) and above "v" point between scroll flags, once again (d).

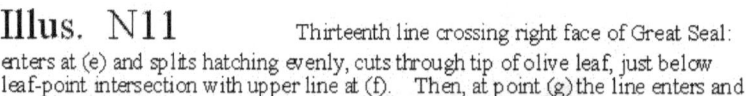

← The shifted *Whirling Squares* line, projected from centerline of dollar. (This meets the pyramid at the edge, just at the top of the Sixth Step. See Illus. E).

On the Right Face there is mostly regular, *formal*, horizontal hatching--unlike the Left Face, which seems more of a naturalistic, perhaps *informal* artistic hatching with complex shading, clouds, topography, plants, odd horizons, diagonal rays, etc. Perhaps this is another reciprocal J3 element, with respect to informality/formality.

When the line enters the hatching of the right face of the Seal, (See Illus N11) it falls *exactly between two lines of the horizontal hatching* at (e). Proceeding onward, it *cuts through the tip area* of the topmost olive leaf, just below its uppermost point at (f). This leaf tip might also be seen as a *solid object "V" shape pointing up*--as opposed to the *negative space "V" shape pointing down* in the forked tales of the scroll on the other side. Does this indicate an *opposite* vertical shift? J2 again, and J3 for the inverse directions of the "V's," as well as the oppositeness of the full and empty spaces in the "V's." (Why "V" shapes? As marking arrows, are they directional in more than one way? Recalling the centerline of the dollar, an offset to the center is marked by the inside, right hand "V" shape of the "W" of WE of "IN GOD WE TRUST." There are, in fact, other geometric alignments in connection with this leaf tip not shown here.) The actual *point* of the leaf tip exactly intersects the line of *hatching* immediately above. Is the hatching *spacing* special? This is once again a tiny portentous difference--and determining hatching width might be a useful puzzle clue of some kind.

As for these "V's," I am inclined to guess that at least *one* of the scroll-fork inner points is exactly *reciprocal* to the olive-leaf tip position--that is, I think that this scroll-fork point on the left end of the dollar probably lines up to the lower line in the *hatching* in the Right Seal Face, while the leaf tip inverted "V" at (f) marking the upper hatching line provides an opposite limit. In other words, exactly between these two V-point positions lies the Thirteenth Line.

Illus. N11   Thirteenth line crossing right face of Great Seal: enters at (e) and splits hatching evenly, cuts through tip of olive leaf, just below leaf-point intersection with upper line at (f). Then, at point (g) the line enters and

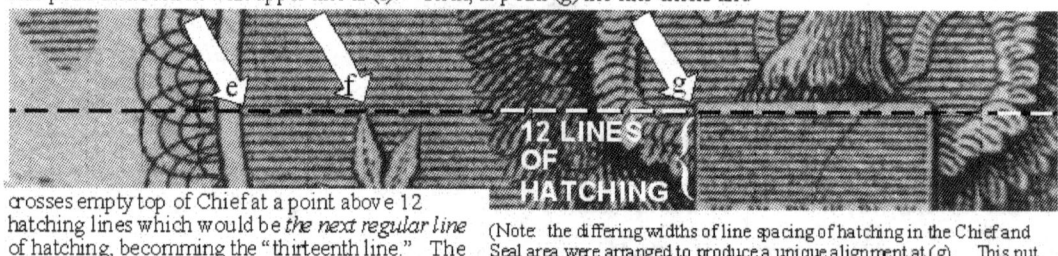

crosses empty top of Chief at a point above 12 hatching lines which would be *the next regular line* of hatching, becomming the "thirteenth line." The line names itself.

(Note: the differing widths of line spacing of hatching in the Chief and Seal area were arranged to produce a unique alignment at (g). This put the mid-point of larger lines exactly at the next line interval of the Chief.)

The line then proceeds to cross through *a white space* at the top of the Chief of the Shield on the Eagle's breast. This happens at a place (g) precisely at the height where the *next line* of hatching on the Chief of the Shield *would be,* if the Designer had continued hatching upward. The lines of the hatching just below, making up the shading of the Chief area of the Shield are *twelve in number, leaving our line above these, the thirteenth line* of the Chief. If there actually *were* a thirteenth line of shading there, this small space would not be white in appearance, and it would fit right in. It would be positioned exactly and evenly at that space in the Chief. Compare the visual appearance of this place in Illus. F1 of Chapter 2 and Illus. N11 above. He wants us to *think hard about this line.*

Maybe we can say that the Designer forced the researcher to complete the job of *shading the Chief,* just by *discovering* this line. And if one used a very sharp pencil and delicately drew a fine line between the two final balls, it would materially have this effect. From a "statutory" symbolism point of view, since this Chief is supposed to be blue, or azure in color as described in Federal law (See Thomson's "Remarks and Explanation" adopted June 19, 1782, see pgs 186 and 187) , it would appear that one would have to be making a "blue line" maybe, in a virtual sense. It is interesting that old conventions in heraldic etching used horizontal lines for "blue" and vertical for "red". The this line apparently has several vertical "sliding" possibilities: J2-type relationships with the hatching in the right ellipse in an "up direction" and perhaps the top of the Sixth Step of the little pyramid in a "down direction," and many other possible subtleties involving "V" points on the scroll, and other features not mentioned here. But in the unmoved position, it apparently fulfills the Designer's objective of being the "thirteenth line" of hatching the top of the Chief. This appears to suggest that something important is intended for this line as design element-- this invisible line ensconced as a central feature to the whole dollar design.

Illus. N12 Thirteenth line grazing 7th ball at (h), and ending at center of Eastern Ball (EB)

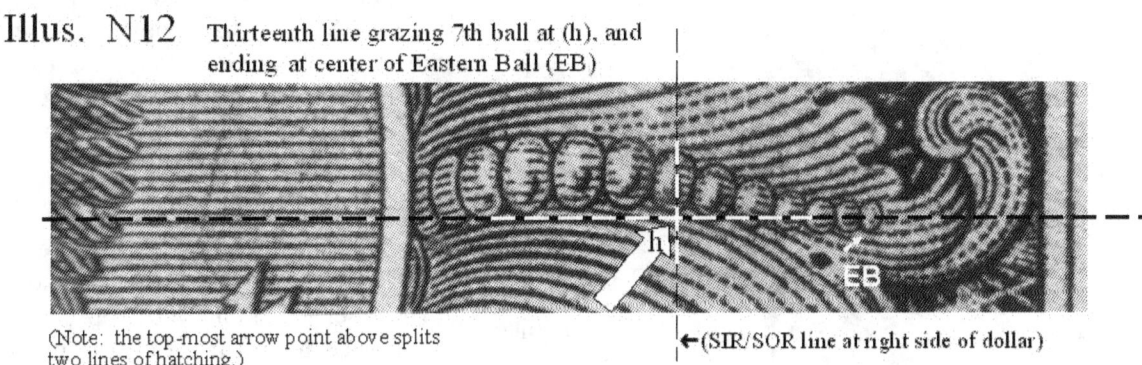

(Note: the top-most arrow point above splits two lines of hatching.)

←(SIR/SOR line at right side of dollar)

Following the top of the Chief, the line crosses the right-hand SIR/SOR vertical line. As on the left side, all intersecting together again as was the case on the left end SIR/SOR line, exactly at the lower edge of the opposite ball number seven on this side (See Illus. N12).

The ideas behind this line are much more complex than just a graphic alignment. The graphic alignment merely allows us to find its location in an approximate way. The true nature of this line will only appear through working out the meaning of other clues that are dependant upon the measured positions of WB and EB. The Thirteenth Line will turn out to be very interesting for an large number of reasons, and we will return to it shortly after looking at some of the dependant clues.

In the next section, the reader might find a draftsman's divider useful for their own investigation of the arc intersection points. A divider is exactly like a compass, but has two needle points rather than having on one side a pencil lead or a drafting pen. This is the classic tool of boat navigation, commonly found at drafting equipment suppliers. Using a precise machinist's scale to separate the points to the correct length, some of the small details of the following discussion can be seen with the help of a 5x or 10x magnifying glass. (Please *do not* mark your bills, either by scratching with points or poking holes at radii. No one wants to have a collection of "defaced federal securities." If any markings are necessary, place fresh, unwrinkled bills (secured by drafting tape) under a sheet of 5 mill acetate or other hard transparent plastic and make whatever marks are need on this. Most of the important discoveries here can be verified through calculation from supplied dimensions.)

* * * * * * * * * * * *

Arcs and Diagonals, having their origins from the WB and EB end points of Thirteenth Line:

Projecting another arc of 2.861", this time from the center of WB gives several interesting results of the J6-type, or alignments at edges of objects. Starting from the top down, the circle misses intersecting the centerline of the dollar at the upper line of LIR by a small distance horizontally, about 0.023" left at (A). Next, it touches at the upper left tip of the "T" in "STATES" at (B) and then the lower left serif at the bottom of this letter (C). Next, it appears to repeat the performance with the *next* letter "T" in "TRUST," but it is a tiny bit too far to the right (about 0.003") of those exact points on this letter (D). The arc also meets the lower left serif as it did earlier but closer to the tip at (E). Once again, a tiny difference

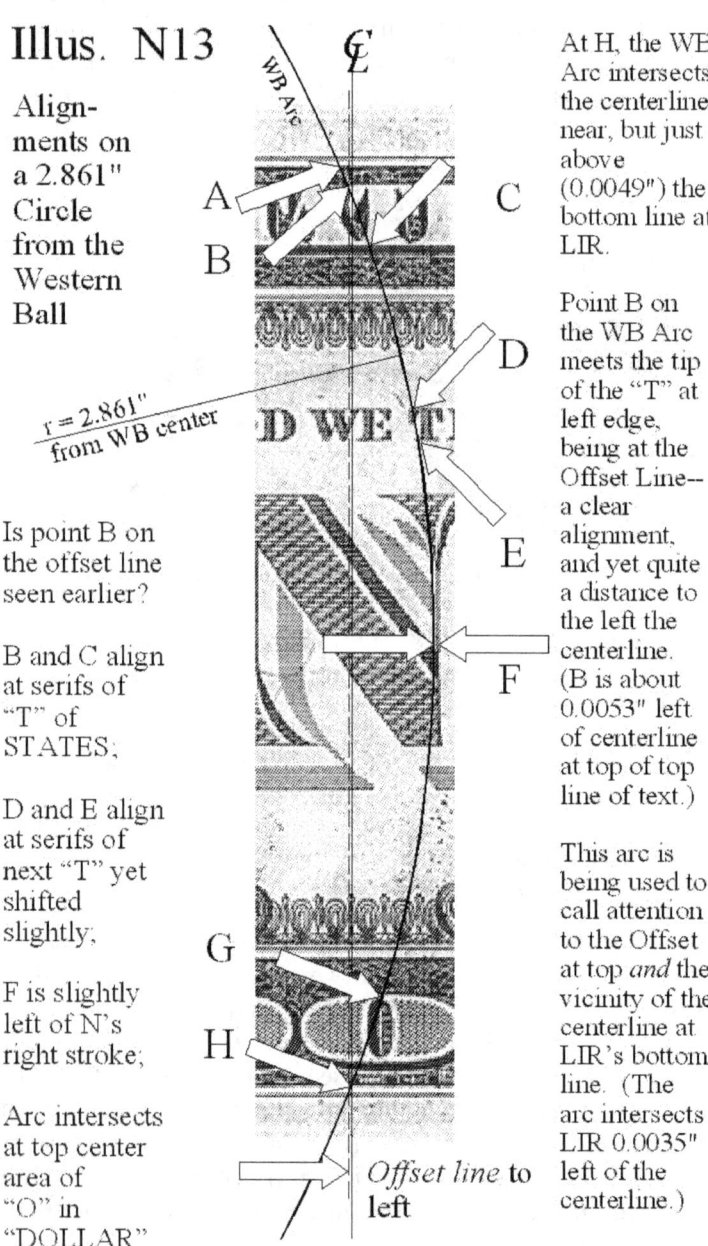

Illus. N13

Align-
ments on
a 2.861"
Circle
from the
Western
Ball

r = 2.861"
from WB center

Is point B on
the offset line
seen earlier?

B and C align
at serifs of
"T" of
STATES;

D and E align
at serifs of
next "T" yet
shifted
slightly;

F is slightly
left of N's
right stroke;

Arc intersects
at top center
area of
"O" in
"DOLLAR"

Offset line to
left

At H, the WB
Arc intersects
the centerline
near, but just
above
(0.0049") the
bottom line at
LIR.

Point B on
the WB Arc
meets the tip
of the "T" at
left edge,
being at the
Offset Line--
a clear
alignment,
and yet quite
a distance to
the left the
centerline.
(B is about
0.0053" left
of centerline
at top of top
line of text.)

This arc is
being used to
call attention
to the Offset
at top *and* the
vicinity of the
centerline at
LIR's bottom
line. (The
arc intersects
LIR 0.0035"
left of the
centerline.)

apparently signaling something of the J1-category. Then, it passes slightly inside (about 0.008") of the thin, right-hand upright staff of the large "N" in "ONE." (Is this yet another tiny difference to investigate? This would seem to require something like another "N" for comparison like the two forms of "T" above. Maybe the opposite arc seen in the next illustration could be used for symmetry clues to another puzzle. J1, J2.)

After this, the WB Arc crosses a point on the top edge of the "O" in the word "DOLLAR" where the centerline of this "O" appears to exactly intersect it at the inner area at the top. No tiny difference here--this "O" was moved to that place to do this act, and it appears to be another one of "the center of the O" - alignments found all over the dollar. Next, the arc *intersects the lower line of LIR very near the centerline of the dollar but not on the Offset line.* The alignment displacement of the WB Arc at Point B at the tip of the "T" serif is not the same as the intersection at LIR at point H. The difference at top is the offset line width from the centerline, about one half hundredth of an inch (0.0053") and a little more than a third of a hundredth of an inch at bottom near LIR, (0.0035") This last dimension will loom large in importance as we head into deeper mysteries of this unusual arrangement. Both intersection areas are of interest--but now the one at the LIR bottom line appears to require careful study (See Illus N13).

It seems reasonable to assume that *the lower intersection* at or near LIR at H is being identified as important, both in position and in relation to the leftward offset shift at top at the tip of the "T." It is indeed important, and this be confirmed next in the illustrations of the EB Arc having the same radius.

Another 2.861" arc made from the opposite easterly ball, the EB Arc:

This opposite arc has only two points of real interest. It intersects the WB arc *exactly at the same intersection* at the upper left tip of the "T" shown above at point B. *But* the opposite (and not vertically symmetrical) intersection occurs *slightly above the bottom line of LIR very near the centerline, being about a half-hundredth above LIR.* (The intersection is about 0.0049" above LIR and 0.0017" left of the centerline. See inset for Bottom Area.)

The top line of text at B is 0.044" below LIR. (Recall that the distance from LIR to the top of the Right Face outer ellipse of the Great Seal as shown in Chapter 3, is *ten times this*, being 0.440". This value is very important and we will return to these numbers later.) There are (at least) four elements of evidence for geometry that intersect the left tip of the "T" serif: the WB Arc, the EB Arc, the Offset Line and the *top line* of the borderline of the text string crossing the "T." (See inset for Top Area, Illus N14.)

Illus. N14

Align-
ments on
a 2.861"
Circle
from the
Eastern
Ball

At top:

The WB Arc
and EB Arc
intersect at
at the Offset
line *and both*
at the top line
of text at
point B.

At bottom:

The WB Arc
and EB Arc
intersect *at a
point about
0.0049"
above the
bottom line* at
LIR,
but *also* very
near the
centerline,
*about 0.0017
left of true
center.*
at the same
point H.

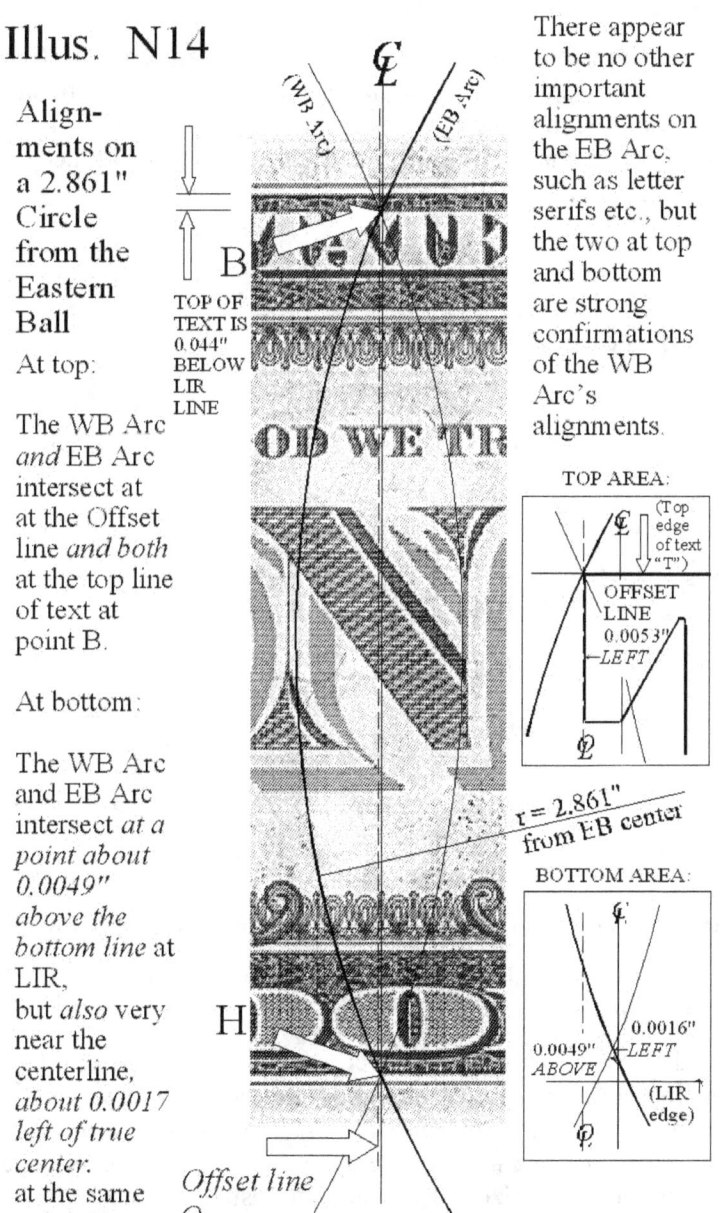

There appear
to be no other
important
alignments on
the EB Arc,
such as letter
serifs etc., but
the two at top
and bottom
are strong
confirmations
of the WB
Arc's
alignments.

Notice that both arcs, as well as the tiny offset line, have been arranged to exactly intersect. There is good reason to believe that the Designer did this to make a special point. Again, the geometry demonstrates an interest in signaling something about the intersection on the tiny offset line, with respect to the centerline. Both intersections confirm of the importance of the *limiting lines,* and some leftward shift. The *top* intersection doesn't seem to connect to anything of great importance, beyond the top of the top line of text. But the "top of the top line of text" has it's *own* significance, which will appear later. These *vertically skewed intersections* also say something about the nature of the end points, and the dependant 2.861" circles position on the Thirteenth Line itself.

Below at H, the arcs are in a kind of conjunction with the centerline, but *also just above the bottom line of the fundamental LIR rectangle.* This, I believe, is the crucial area for which the Maestro wants us make special note. He wants us to *do something* in this area in our own theoretical work. Were we to only look at this area visually, this point seems to be right at the center of the dollar. But not quite.

To the analytical students of geometry, the arrangement will make their minds cry "tilt". The geometry of these circles has been carefully constructed in order to accomplish the feat of making apparent intersections at both places--with the top intersection, being *on a special offset away from the centerline*. After the surprise from skewed intersection points wears off, we might wonder if the bottom intersection is or is not actually touching the centerline. We will need to apply mathematics to what solid clues have been provided in order to learn just what the Designer really had in mind, since measurement and observation will not allow us to follow much deeper. We know from experience that exact mathematical statements seem to continually parallel this sort of graphic display. But just from the logic of the arrangement, just from what we already know of the Designer's *modus operandi*: reciprocal action is certainly indicated. The crucial question must be about the true *position* of these two intersections produced by circles having a radius the sacred dimension 2.861" from WB and EB.

First, perhaps, the investigator is asked to analyze this peculiar arrangement. Then, the question arises: "How would the intersection at the bottom line look if the circle intersections *were not* vertically skewed? Would this happen if the lower intersection touched the LIR line instead the upper touching of the top of the top line of text?" And, "Would this require that one of the circles shift? If so, how much?"

****

Illus. N15

GEOMETRY OF HORIZONTAL CIRCLES AND THEIR INTERSECTION POINTS:

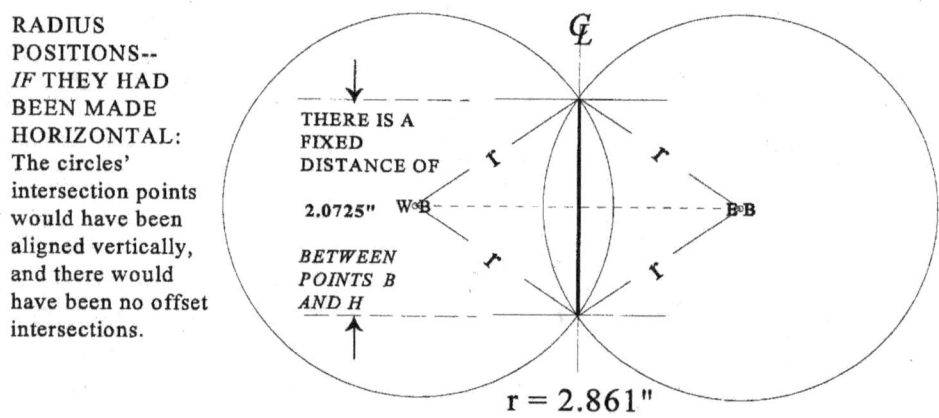

RADIUS POSITIONS-- *IF* THEY HAD BEEN MADE HORIZONTAL: The circles' intersection points would have been aligned vertically, and there would have been no offset intersections.

THERE IS A FIXED DISTANCE OF

2.0725"

BETWEEN POINTS B AND H

$r = 2.861"$

Observations: What can be said about the odd arrangement of circles and offsets seen above?

(1) If two of the circles do not have intersections exactly above and below each other, vertically, their origin radius points (i.e., at WB and EB) *must be at different heights above the LIR line.* (See Illus N16)

(2) Relative ball heights: If we observe a difference between intersection points of about one half hundredth of an inch (0.0053") to the left, at the top *horizontally,* and also about the same *vertically* above the bottom (0.0049), one or both balls would have to have been *shifted counter-clockwise* from *some point on the Thirteenth Line.* If it happens that *both balls are shifted,* one up and one down, the average difference vertically above LIR will be about equal to the horizontal Offset Line distance at point B, on either side, since the Offset and either ball are about the same distance from a center point on the Thirteenth Line. *If only one ball was shifted* (for instance, stretching the Thirteenth Line upward at radius

122

point EB) the shifted distance at intersections will be about twice that of the Offset line, since it is about twice as far from WB. This then, appears likelier: the evidence more closely fits a higher EB.

(3) Only *one* intersection of circles can touch either boundary at one time: Due to the radius and their distance apart, there is a *restraining fixed distance* between circle intersection points B and H (about 2.0725". See Illus. N15). The top boundary: The top line of the top of the text where we found the first intersection is 0.044" below the top line of LIR. The lower boundary begins at the lower edge of LIR. Since the whole height of LIR is 2.1213" (√4.5) we are left, after subtracting 0.044", with a *wider space* equaling 2.0773", bigger than that of the fixed distance. This means *only one intersection point can touch either limiting line at one time.* The fact that intersection point H is about a half hundredth of an inch (0.0049") above the bottom edge of LIR and almost directly above the centerline point at LIR, suggests that a vertical shift would have to come from shifting *EB downward* a distance perhaps twice the Offset width. The difference between ball heights should be easily measurable to the bottom line of LIR as a verification (See pg 125).

Illus. N16

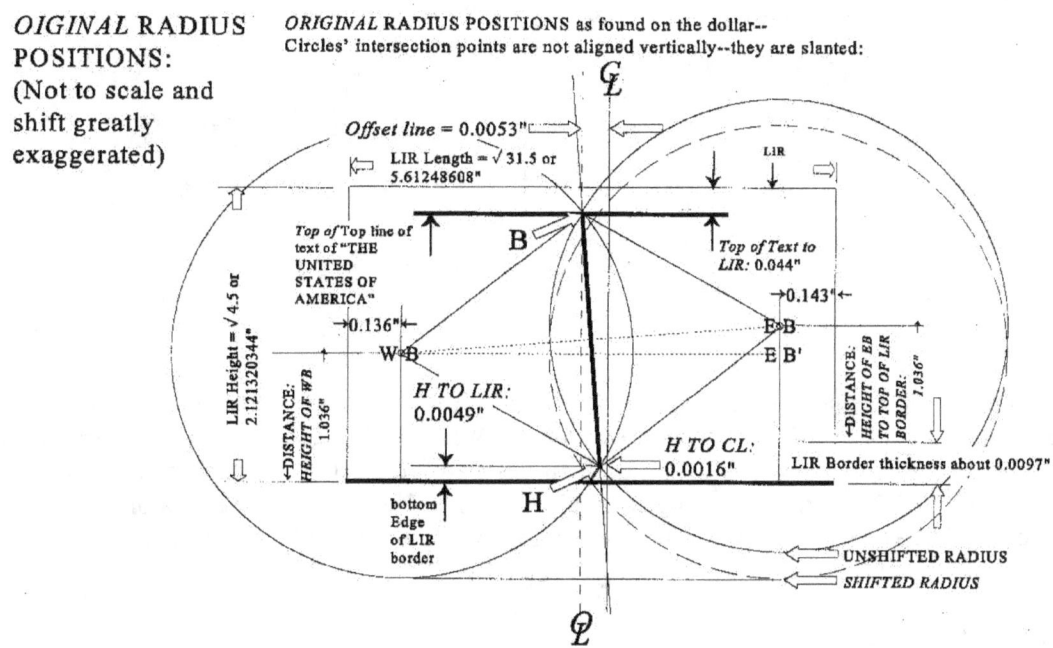

OIGINAL RADIUS POSITIONS: (Not to scale and shift greatly exaggerated)

To conclude from the above observations: it appears that the only angular shift for the Thirteenth Line that makes sense would be *clockwise, which is to say bringing EB downward.* This would have the effect of *bringing both intersection points B and H into balance at an equal distance left of the centerline, top and bottom.* We already know the Designer's motif of reciprocal exchange, (J3) and it would only make sense that an intersection "at the top" would be traded for an intersection "at the bottom." An angular shift of the ball positions from a centered mid-point on the Thirteenth Line (sliding one up and the other down about a half-hundredth) would not appreciably change the height of the intersection at point at H above LIR. An angular shift made from EB seems to be ruled out since both intersections would then fall even further from the centerline. Only a theoretical shift of EB downward makes any sense.

What needs to be done? Moving the EB radius point downward some distance *along a line parallel to the left end of LIR,* where it's arc would then intersect the WB Arc at a new point on LIR. This would be at a point opposite and reciprocal to the original arrangement above. The *height at left* of the WB ball then, is a crucial distance to know, as well as the *horizontal distances* of both WB and EB from either end of the LIR rectangle.

As we will see, *both* arcs must then intersect at the *same distance left* of the centerline on the unchanged WB Arc seen above: 0.0035" left. Meaning, that both arcs meet at the *same point,* at LIR's lower, outer edge. This will bring the radii into balance, which can be seen in the fact that the horizontal displacement of both WB *and* EB are mutually 0.0035" to the left of an average distance to the opposite ends of LIR. (See Illus. N17 and N19 ending dimensions: These are 0.136" and 0.143". Subtracting either number from their average, places the center on a line between them 0.0035" toward the *left.*)

Illus. N17

NEW RADIUS POSITIONS: (Not to scale and shift greatly exaggerated)

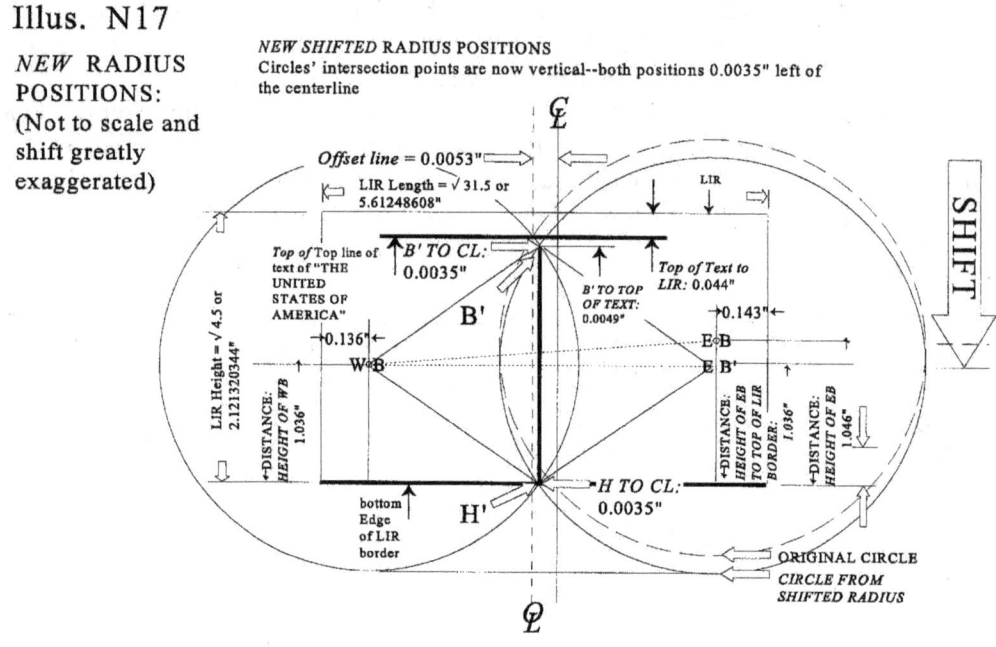

We can consider the EB position adjusted to the hight of the WB since this is consistent with the rules of symmetry. Our attention is drawn to the fact that the WB position *is unmoved and that its ball was made circular like a precise standard.* Now, both circles will intersect at the same place on the LIR line at H' in Illustration N17. After providing several clues and leading the investigator to perform this geometric exercise, the Designer has given notice that this new point must be important.

****

The Vertical Dimension of 1.036 Inches:

We can see from the two positions of EB and EB' that a reciprocal "exchange" of dimensions is seen in the height differences *above and below the printed LIR borderline* at the bottom line of the dollar. This sends us several important messages (See Illus. N18):

(1) The only length *common to both sides* is the vertical 1.036" at the faint *inner edge* of LIR at right and outer edge of LIR at left--bringing this dimension to our attention *twice.*

(2) The only *unmoved form of this number* is from the center of the circular *WB ball,* and

(3) This length from the center of WB ball goes to the *fundamental Outside Edge* of the printed border of LIR.

Illus. N18 Height differences measured between WB and EB centers and LIR border edges--*both inner and outer edges at the lower LIR border.*

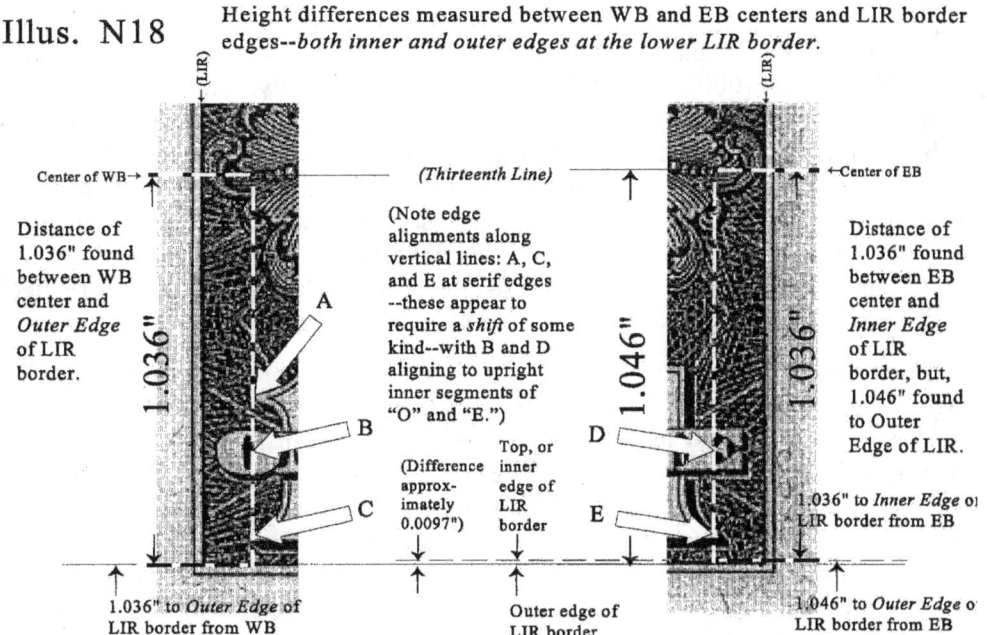

Center of WB →

(LIR)

(Thirteenth Line)

(Note edge alignments along vertical lines: A, C, and E at serif edges --these appear to require a *shift* of some kind--with B and D aligning to upright inner segments of "O" and "E.")

(Difference approximately 0.0097")

Top, or inner edge of LIR border

← Center of EB

(LIR)

Distance of 1.036" found between WB center and *Outer Edge* of LIR border.

Distance of 1.036" found between EB center and *Inner Edge* of LIR border, but, 1.046" found to Outer Edge of LIR.

1.036"

1.046"

1.036"

A

B

C

D

E

1.036" to *Inner Edge* of LIR border from EB

1.036" to *Outer Edge* of LIR border from WB

Outer edge of LIR border

1.046" to *Outer Edge* of LIR border from EB

It seems clear that this *1.036" length* is being highlighted. Since no adjustment is required at WB, a preference for the *Outside Edge* of the LIR border is clearly shown. In this case, we can see that the Designer is sharply aware of the distinction of the two edges, inner and outer, and uses them to show a preference. The use of the outer edge of the LIR border was successful in many later discoveries. And it would make sense: this is the sharpest reference line that the Designer could have chosen. As an *edge* it has no line width and is ideal for geometry. We have already seen a general preference by the Designer to use the edge at "white areas" contrasting dark areas as alignment elements. And the space between LOR and LIR surrounding the bill *is just such an area.* In the next illustration, measurements were taken from this outside edge of the LIR rectangle to the centers of WB and EB. This will be the typical of any later measurements I make "to the LIR line."

End Points and the Thirteenth Line:

As can be guessed from the above intersections taken from equal circles at the ends WB and EB, the ends of the Thirteenth Line are *not symmetrical* to the centerline, (See Illus. N19). The center of ball WB is about 0.136" to the right of LIR's left edge, and EB is about 0.143" to the left of LIR's right edge. With this information and the information in the above illustration, now a simplified mathematical hunt can begin by means of right-triangles.

Illus. N19 The Centers of WB and EB measured to right and left edges of LIR

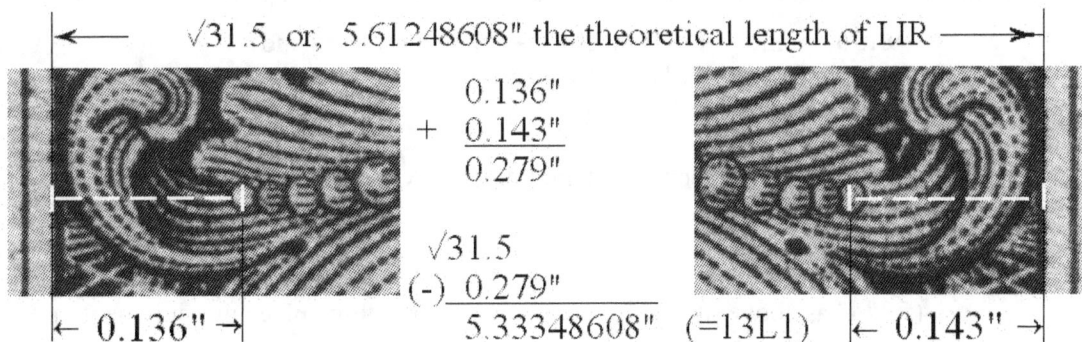

← √31.5 or, 5.61248608" the theoretical length of LIR →

$$\begin{array}{r} 0.136" \\ +\ \underline{0.143"} \\ 0.279" \end{array}$$

$$\begin{array}{r} \sqrt{31.5} \\ (-)\ \underline{0.279"} \\ 5.33348608" \end{array} \quad (=13L1)$$

← 0.136" →

← 0.143" →

The Tiny 0.0035" Offset:

The number 0.143" oddly enough, is *half* of 0.286" and we might say, in one sense, that "this line is offset by half of 0.286", or (D/1000)/2. (Perhaps a reference to the Pyramid Displacement of 286.1"? (J4). Is this a clue, a base of another pyramid? *Yes* it is, as we will see later.) We should also note the total *difference* by subtraction between these numbers amounts to 0.007", apparently the same as the *vertical height difference* between the semi-major axes of the right and left Great Seal faces (See pg 55). But again, the interesting thing here is that the actual absolute offset to the West will be *half* of this difference, that is being 0.0035" also taking us into the estimated fourth decimal place by logical deduction, rather than by strict measurement. (To simplify explanations, all *fourth* decimal place information in illustrations above were supplied by previous calculation.)

Notice that the reciprocal of 0.0035 is 285.7142857, which is suspiciously similar to 286.1022156, and even having the proper number of three digits to the left of the decimal. Since the Designer has laboriously taught us to look for and investigate *reciprocals of tiny distances*--he can then make small numbers such as this unfold and explode in our face as large numbers. The true, theoretical reciprocal of 286.1022156 would be 0.003495254, the same number we have seen before in a various ranges of decimals. And as a true *measure*, this of course, is not very different from the estimated three and a half thousandths. Obviously, this is something the Designer is highlighting for our attention. But although it is very tempting to suspect that the 0.0035" is really intended as the reciprocal of 286.1022156, we will we will stick with it as it appears: 0.0035" (See pg 134).

$$************$$

How long is the Thirteenth Line?

What we need to know is why the Thirteenth Line is found in such an important place on the dollar. After learning its length, why is it important? It would also be nice to know how this line fits in with the rest of the mathematics elsewhere on of the dollar. To begin with, we have only the starting points (WB and EB) and only a rough idea of the length of the line suspended between the balls at the ends of the leaves. But I think the Designer left us sufficient clues to extrapolate a trial length for a start. Since LIR is 5.612" long, (or theoretically, $\sqrt{31.5}$) this length is going to be in the neighborhood of 5.334" by subtraction of the two end distances. Measuring between the centers of these balls over the length of the whole dollar will be completely unreliable, since it is a long distance over the paper. But we might safely *subtract from the outside lines* if we believe we know what the whole outer length LIR is supposed to be. This is the same procedure used for the Between Seal Centers #2 dimension of the last Chapter.

These ending points were made intentionally vague--and it is my present thinking that this was to cause an investigator search through the many variations of possible lengths for this line. (There are, in fact, quite a large group of very interesting relationships a several thousandths either side of this 5.334"-range, not shown here.)

Following the previously successful practice of using measured dimensions to the even thousandth inch, subtracted from the theoretical LIR length we will get a theoretical Thirteenth Line Length ("13L"):

(m)	0.136" (Center of WB to LIR West line, at the *left* end)			
(m)	0.143" (Center of EB to LIR East line, at the *right* end)			
	(+) --------			
(m, sum)	0.279"			
(t)	5.61248608" (theoretical length of LIR, $\sqrt{31.5}$)			
(m)	0.279 "			
	(-)--------------			
(t)	5.33348608" =	13L1	(Thirteenth line #1)	
(m)	(5.334" as above, if we were only considering three place measurements.)			

Notice that the horizontally shifted position of this line allows for at least three obvious "takes" on the data of the J2-type, right away. There is a horizontal length that can be found from *the right* to the centerline; another quite different horizontal length from *the left* to centerline; and both considered together as one length. Suppose, for instance, that we looked at the left line length as "half of the length of LIR minus 0.136":

(t)	5.61248608"	(theoretical length of LIR)
	n / 2 =	2.80624304" (half LIR)
(t)	2.80624304"	
(m)	0.136 "	(Center of WB to LIR line, left end)
	(-)--------------	
(t)	2.67024304" =	1PL13 (1st Partial Length of the Thirteenth Line)

If this length is multiplied by φ the result is a number very close to *twice the height of LOR*; a tie to the mathematics of the other large fundamental rectangle just outside of LIR. (We will recall here, that the length of SIR was very nearly the square root of the diagonal of this LOR rectangle--a similar pattern.) The difference is a little less than a ten-thousandth of an inch with an error of about one part in 25,000 for DFT theory, or one part in 6500 for the IDT theory:

(t)	$2.67024304 \cdot \varphi$	=	4.320543998"	where
	$2.67024304 \cdot \varphi / 2$	=	2.160271999"	

This is well within the plus or minus one thousandth of an inch Rule 2 region for the length of the height of LOR (2.160185958 DFT or, 2.159941866 IDT). If this new length is then multiplied by φ^2, for instance, we get 5.655665519", for comparison (5.655440261 DFT or, 5.654801219 IDT). So, this is yet *another* possible version of the long dimension of LOR. A tie-in of sorts to LOR has apparently been made, *another* link from the interior design to the exterior LOR scheme.

Suppose we went the opposite direction, in from the other measured distance from the opposite side at right, 0.143" and made another length to the centerline point. :

(t)	2.80624304"	(half of LIR length, from above)
(m)	0.143 "	(Center of WB to LIR line, right end)
	(-)--------------	
(t)	2.66324304" =	2PL13 (2nd Partial Length of the Thirteenth Line)

After the close hit above on the height of LOR, and also the use of (half of) 0.286 as an offset (two seemingly intentional numbers) it is unlikely that this 2PL13 number could be a precise clue to anything else. That is because (barring chance) out of *any three such interconnected relationships only two can really be intentional since the third has to be dependant--and can't be intentional.* (But I will look anywhere since I don't know what was intended. In time, this 2PL13 became an important number: see page 137.) When I tried 2PL13 used *as half* of a side of a pyramid, (i.e, when multiplied by 2, then applying the "Squaring of the Circle" formula, times 4, and divided by π and divided by 2) an approximate Between Seals Center (BSC) number emerged, (about 3.39). This number-form foreshadowed a more exact relationship that appeared when *the whole length* of the Thirteenth Line was used instead, as one regular side of a base of a pyramid, or a base having a side of 13L1.

Suppose we thought of this sum, 13L1 "as one side of the base of a π-proportioned pyramid," and then we determine its height:

(t)	$(13L1 \cdot 4) / \pi / 2$	=	3.395402694" (= BSC5)

Isn't that amazing. This number looks a *whole lot* like the group of Between the Seals Center lengths seen in Chapter 3. *But now,* it is the *height* of a pyramid we are looking at, shifted by 90° to *vertical,* whereas before it was the bottom line or *base* of a pyramid, (See Illus. N20). *Another* new and different reciprocal J3-type relationship. What does this mean?

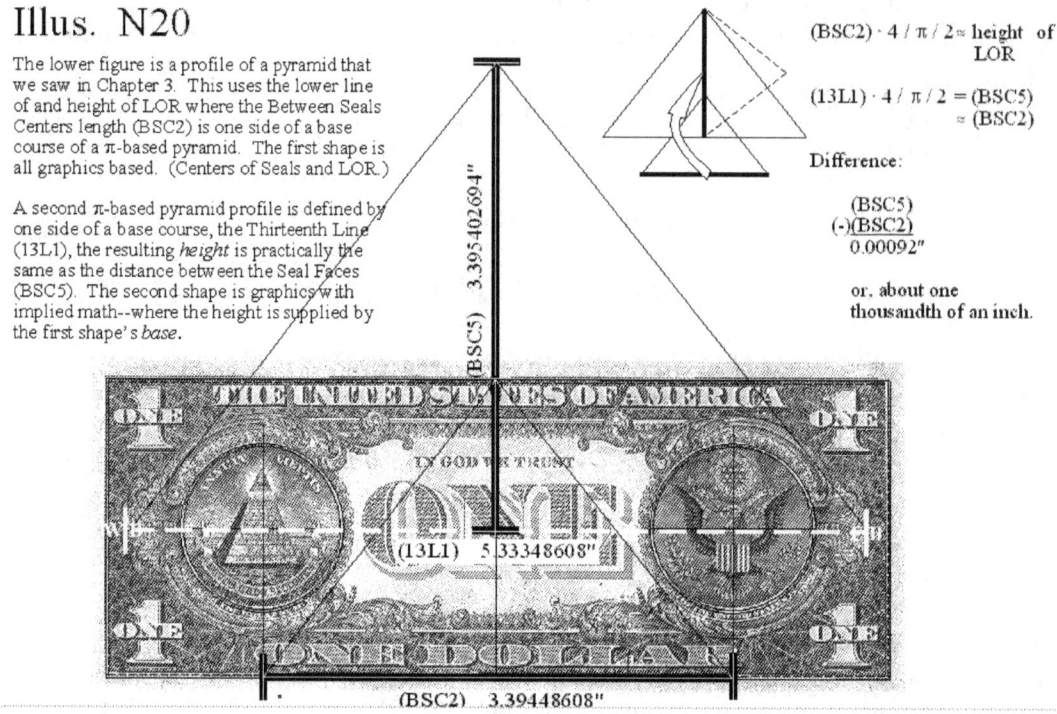

Illus. N20

The lower figure is a profile of a pyramid that we saw in Chapter 3. This uses the lower line of and height of LOR where the Between Seals Centers length (BSC2) is one side of a base course of a π-based pyramid. The first shape is all graphics based. (Centers of Seals and LOR.)

A second π-based pyramid profile is defined by one side of a base course, the Thirteenth Line (13L1), the resulting *height* is practically the same as the distance between the Seal Faces (BSC5). The second shape is graphics with implied math--where the height is supplied by the first shape's *base*.

$(BSC2) \cdot 4 / \pi / 2 \approx$ height of LOR

$(13L1) \cdot 4 / \pi / 2 = (BSC5)$
$\approx (BSC2)$

Difference:

(BSC5)
(-)(BSC2)
0.00092"

or, about one thousandth of an inch.

(BSC5) 3.39540269.4"

(13L1) 5.33348608"

(BSC2) 3.39448608"

Earlier in Chap 3 we saw that the height of LOR compared to the Between Seal Centers #2 (and other BSC numbers) carried possible mathematical footprints of the Great Pyramid in at least two possible forms: a π-proportioned (51°51'14.31"), and a similar proportion (about 51°50'25"). These profiles differed only slightly from each other. In the first case the height of LOR was the y, or up and down axis of the pyramid, and BSC was the x, or left and right axes of the base. But *here*, a BSC-type length can be seen now as the *vertical height* against a horizontal 13L (a Thirteenth Line, of some length) and *the Great Pyramid proportion reappears in a reciprocal relationship*. (J2 and J3.) So, in this new pyramid profile, we make a new connection to the other relationships seen earlier. And the 13L line was offset in such a way (using the Partial Length 1PL13) that a pyramid can be made to closely approximate *the same LOR height of the earlier pyramid profile* of Chapter 3, through the use of π as another kind of linking connection.

So, it looks as though 13L was chosen as a length to further display the Great Pyramid proportion, and an offset to tie-in the LOR height. Is that all? I don't think so--that can't be all there is to this. It seems that there must be some other connection or relationships for the Designer to have gone to this much trouble, and to have so centrally located this special line across the dollar.

It appears that these two lengths are begging for a comparison. Both are starkly unique, and both have clear graphic or graphic and mathematical definitions. One has the effect of defining the other by means of the *height* of one pyramid being a *single side* of the other pyramid. And you wouldn't have any basis to compute this second pyramid, without discovering the first--so the trail was arranged to go from *certain to implied*. This is one of several Rule 2 cases for puzzles and number trails on the dollar.

Here the J3-plan is the forward as opposed to inside-out pyramid equations: (1) A length as a side times 4, divide π, divide 2 gives the *height* of a pyramid; whereas (2) a length as a height times 2, times π, divided by 4 gives a *side* of a pyramid. But now we know both, what can they mean?

Let's look at the length BSC2 subtracted from the fundamental 13L1, both of which are the closest to being true, measured dimensions:

(t)	5.33348608"	=	13L1	(Thirteenth line #1)
(t)	3.39448608"	=	BSC2	(Between Seal Centers
	(-)--------------			from Chap 2)
	1.939 "			

Interesting. In a way, we are only subtracting the measured end dimensions 2.218" and 0.279" from each other since these above figures were both subtracted form LIR or √7. We might look at this 1.939" as a real measured dimension. By the same token, the little offset 0.0035" could be called a measured dimension, being the absolute difference to the centerline between 0.143" and 0.136". Being an offset, it seems usable in this context. So we could try subtracting this tiny 0.0035" measurement:

(t)	1.939"	
(t)	0.0035"	
	(-)-------	
	1.9355"	(1.936", rounded to three places.)

With minimal manipulation of the measured figures, we find a resulting number *half a thousandth of an inch from the numerical "1936,"* more than close enough by my Rules 1 through 3, being as close to what would be a real date as one part in 1600. This is another kind of peculiar connection between BSC and 13L, and a fairly fundamental one at that--a simple *subtraction*. Notice that in a geometric sense, we could say this subtraction is another "sliding" of shapes or lengths.

Illus. N21   What happens if the distance Between Seal Centers (BSC) and 0.0035" Offset are subtracted from the Thirteenth Line? (13L)

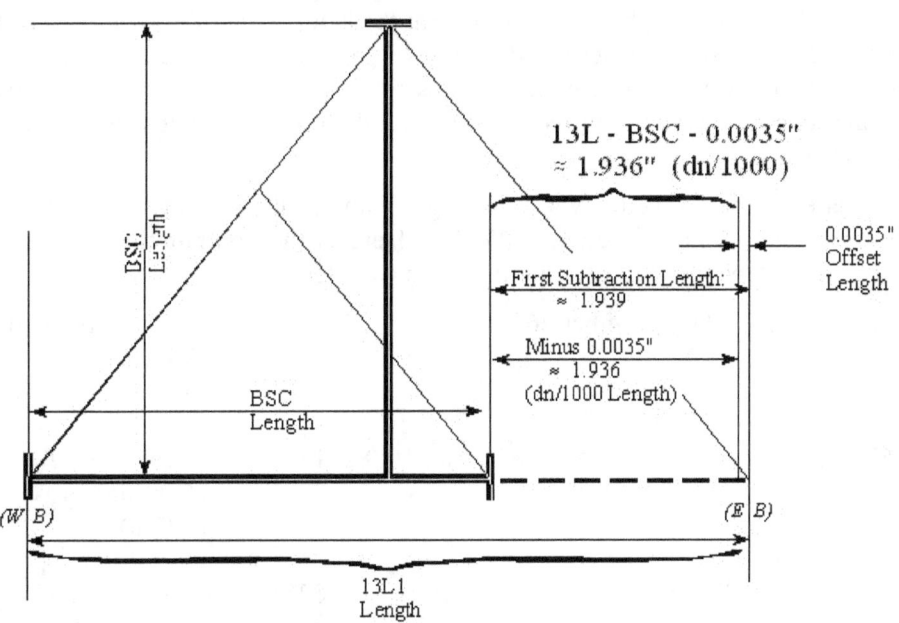

Another appearance of the "Date" September 16th, 1936:

The above discovery called for some comparisons. How different is this BSC5 length from the previously successful BSC2 of Chapter 3?

(t)	3.395402694"	BSC5	
(t)	3.394485080"	BSC2	
	(-)---------------		
	0.000916614"	or about one thousandth of an inch, in round numbers.

Ken McGrath

If instead, we mathematically "backed in" the computation for the BSC2 number from the other direction, starting instead as the height--what kind of 13L-type "base side" would it make?

(t) $\qquad 2 \cdot BSC2 \cdot \pi / 4 = 5.332046266 \qquad = 13L2$

This would have to be pretty close, or about a thousandth shy from 13L1. Now, if this 13L2 is subtracted from BSC5 *another interesting result appears*:

(t)	5.332046266"	13L2
(t)	3.395402694"	BSC5
	(-)-----------------	
	1.936643573"	13L2-BSC5

This is really close to the decimal form of a time late in the year 1936, nearly Davidson's special date. We have seen this trick before in Chapter 2, with subtractions between two line lengths where the lines already have a complex relationship of another kind. And I still don't think we are looking at the exact lengths that the Designer intended. Yet it seems amazing that the two tightly circumscribed pyramid forms--already connected by another route in the math and also to LOR--could be so crafted as to give a difference near a desired date. But if he can do *this,* I think we are only looking at an approximation of the real intended length. We will return to this line of thought shortly.

****

The Precise Date Number for September 16, 1936:

Let's now back up a bit and have a look at what the actual date of September 16, 1936 would be in its reduced form. This is a simple fraction. It appears in the dollar's design through at least three different means--two of which we have already seen. This route of investigation will eventually lead to another Thirteenth Line length, a new length which will provide a numerical answer for some of the Designer's complex tricks.

The true "date number" is actually a little bigger than the above number of 13L2 minus BSC5, and we will see forms of this number elsewhere ("dn"). Here is the computation of the "date," which is not unlike the original Independence Date calculation in Chapter 2:

(t)	September 16 is Day Number 260			(Sept. 16 being day number 260 in 1936. The year is divisible by four, so it is a leap year having 366 days)
then,				
(t)	260 / 366	=	0.710382513	(the fraction for the point 1936 Sept. 16, lying between 1936.0 and 1937.0)
then,				
(t)	1936 + 0.710382513	=	1936.710382513 (This is "**dn**")	("**dn**"= year+fractional "date" being the decimal form of Sept. 16, 1936)
then,				
(t)	1936.710382513 / 1000	=	1.936710383	(Divided by 1000)
		=	dn / 1000	(or Date Number divided by 1000)

This number differs from the second subtraction above by only one in the ten-thousandth's place and it is mathematically similar at one part in 29,000.

Hiding the Date Number through the Square Root:

We have *already* run into this 1936-number before in a mathematically encrypted form. Our clue from relatively simple math is the square of the simple number "44":

130

$$44^2 = 1936$$

This number is approximately the square root of the special date, and the exact root for that year. Forms of this turn out to be hiding in other decimal ranges such as the distance of the top line of LIR to the top of the outer ellipse of the Right Face of the Great Seal:

$$0.440^2 = 0.1936$$

(So, what happens with the other dimension 0.447" to LIR from the outer ellipse at the *Left* Face of the Great Seal? $0.447^2 = 0.199809$. Hmm. That's interesting. That's around *this* period of "time" if we multiplied by 10,000. See Illus. G5 pg 55, Chapter 3). Also, only a tenth as large is the "top of the top line of text" of the "UNITED STATES OF AMERICA" to the top line of the LIR rectangle = 0.044":

$$0.044^2 = 0.001936$$

This represents a clear pattern of the J7-type and a multiple reference to Davidson's date. This simple square root is not, however, the precise form of the Date Number as computed above. This is the square root of September 16, 1936 at ten floating decimals:

$$\sqrt{1936.710383} = 44.00807179 \ (= \text{The square root of dn})$$

Maybe we should be looking at this number in its *more precise form.* As one can see this would not be very different in appearance from the two smaller scale forms (0.44 and 0.044) and wouldn't be physically measurable--the "8" appearing at the very least at the 5th decimal place. Yet it might show up in mathematics by implication in a puzzle (R2) if a real, graphic demonstration could be found.

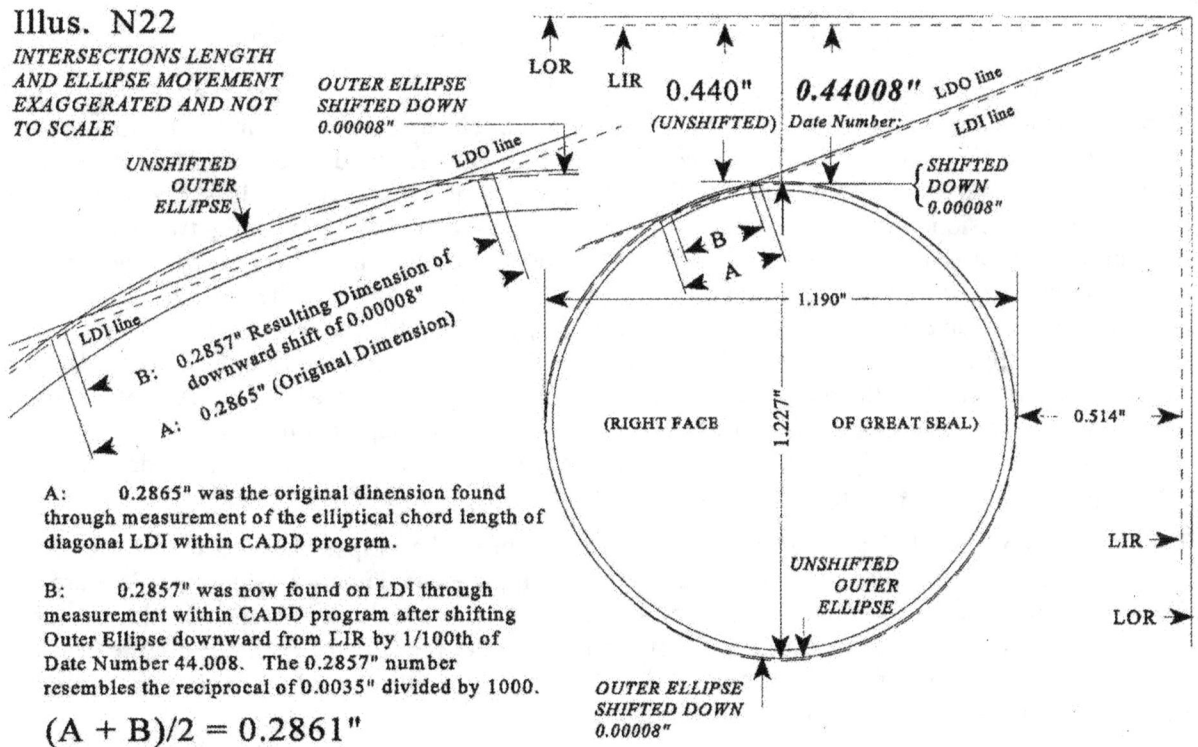

Illus. N22

INTERSECTIONS LENGTH AND ELLIPSE MOVEMENT EXAGGERATED AND NOT TO SCALE

A: 0.2865" was the original dimension found through measurement of the elliptical chord length of diagonal LDI within CADD program.

B: 0.2857" was now found on LDI through measurement within CADD program after shifting Outer Ellipse downward from LIR by 1/100th of Date Number 44.008. The 0.2857" number resembles the reciprocal of 0.0035" divided by 1000.

(A + B)/2 = 0.2861"

Once Again: *Another* **Close Look at the Elliptical Chords of the Previous Chapter:**

With the above observations in mind, while we are looking at 44-number related ideas, we should have another look at the Left Face of the Great Seal--the "nested problem" proposed on pg 113. In the re-examination of the grazing lines of the diagonals of LOR and LIR to the Great Seal faces, there was the number 0.2865", a length that appeared as a result of an elliptical intersection from the Large Diagonal of the LIR rectangle (LDI). This looked *strangely large* from the ideal D/1000 by about 4 ten-thousandths of an inch. My proposed hunch was that this was clue to a "nested problem." Even though the sum of elliptical lengths gave an exact and satisfying answer of $z^2/10$ (1.2217"), this one length seemed to be clearly set aside for closer scrutiny by it's slightly large size. For instance: if the sum of chords created such a precise answer of $z^2/10$ to the fourth decimal place, you would surely think the Designer could have made this particular chord length exactly 0.2861" if he had wanted to. This calls to mind the question: what happens if we use the corrected number 0.44008" instead of 0.440" from the top line of LIR? If we used this new date-number value it will have to move the Outer Ellipse down to produce a new, *smaller intersection length* inside the ellipse, possibly the ideal length.

My first intuitive guess was that the resulting elliptical chord would turn out to be exactly D/1000 after the tiny sliding motion. But fooled again, the Designer was much more subtle than that. In new measurements (scaled within a CADD program) I found that when I repositioned the ellipse slightly lower, using 0.44008"

(or dn/100, rounded to five places), the resulting length between new intersections is 0.2857". We should note that this number resembles the already observed *reciprocal* of the 0.0035" figure were it to be divided by 1000. Amazingly, when this new length is added to the original 0.2865" and averaged, the result is exactly our special number *0.2861"* :

the average of the two chords: (0.2865" + 0.2857") / 2 = 0.2861"

The above subtle, and overall arrangement of diagonals and ellipses is truly a marvelous tour de force of technical mathematics and clever clues, one of the more interesting arrangements to be found on the dollar. (See Illus N22.) It has the neat effect revealing itself to you, as you dig deeper. This conception is audacious, intricate and a masterpiece of organization. Beyond that it is *elegant*. It is a polished, single motion in two-dimensional space against four symmetrical sets of diagonals using two very slightly different ellipses, to arrive at a special average sum. Then, once the tiny discrepancy is noted in the largest chord, the very tiny shift is required as an experiment--and we find that the Designer has put even a deeper puzzle there--an *average*. And of course, this is way outside the measurement range of any tool--these differences are truly microscopic in a physical sense. These numbers can only be reproduced by following the above set of mathematical assumptions, as the computer program does. As noted earlier--few people capable of setting up this kind of problem. This was an huge computation in his time. We will see this technique repeatedly: the use of *two similar numbers* to identify one or more ideas, and *then using both of them again* as a basis to form another average to produce a third number.

This puzzle problem of translational or sliding movement of the ellipse shows greater potential for expansion in the mathematical sense, and this is not a complete mathematical solution. We are only looking at an approximation to create interest and I will not pursue the full problem here. It is enough for us to know now that there is a much greater depth to this puzzle, for which adventuresome readers may want to tackle. Another use of the "Date Number" or the 44-value will appear in a larger form in the next chapter, where it plays an important role.

****

Yet another route to the "dn-number" and returning to the Thirteenth Line Question:

Deeper than the above 13L2-BSC5 subtraction of lengths, is finding that an even closer form of the "date number" can be found by another formula using *only 13L2 itself*. The following is a variation of a

J3-type formula used in several places by the Designer. He switches to the reciprocal, (1/x: e.g., 5 to 0.2) then, a ten-fold decimal place switch (e.g., 10 is 0.1 or 0.5 is 5), and then a number taken to a power or a root (x^2 or $\sqrt{x}$). (See Chapter 3, J3 pg 41). Here is "the square root of the reciprocal of half of 13L2 times 10"

$$\sqrt{1/(13L2/2 \cdot 10)}$$ (Given: 5.332046269" = 13L2)

(t)	5.332046269 / 2	=	2.666023135	then,
(t)	2.666023135 / x	=	0.375090518	then,
(t)	0.375090518 · 10	=	3.750905186	then,
(t)	$\sqrt{3.750905186}$	=	1.936725377	

Which would be about the 265th Day Number of that "year," beyond the ideal Day 260. If we wanted to stop right here, this far more than satisfies Rule 2 for precision for an acceptable mathematical finding. But I think this is only an approximate result. We have seen this too many times and too closely associated to ignore the possibility of a common, exact source.

My intuition sensed trail that might be followed further. From the alternate route above to produce the dn-number *and* the geometric subtraction, it looks like the Designer had intended *two completely separate methods to make this 13L-length, simultaneously.* This seemed to indicate a higher level of complexity. What if the Designer intended *both* ideas above, within his idea of the Thirteenth Line length? *The subtraction together with this mathematical relationship.* Considered together, these open up another line of thought. Perhaps this 13L-line length could be composed from only two pure constants such as the Date Number, and the fundamental number π.

What happens if we try to compute this Thirteenth Line length *completely backwards from both ideas,* down the whole string of equations, using just the pure 1/1000th Date Number? (i.e., "dn/1000" or 1.936710383 as the *ideal, controlling factor* in the equation rather that a result of approximately measured lines.) In other words, using the previous equation above, but also tied to the above subtraction of reciprocal pyramids? How about:

(1)   1.936710383 squared, divided by 10, then the reciprocal, multiplied by 2, *minus the same 1.936710383* (this will equal 3.395418447, very much a "BSC-type number") and,

(2)   then after this, carry it a step further, since the above result is in the interesting range of BSC-type numbers--multiply by 2, multiply by π, then divide by 4? This would have to create a 13L-type number, which would allow *the exact Date Number subtraction from BSC and 13L lines:*

Maybe this satisfies *both* graphic ideas seen earlier. This can be said with a more compact expression:

(Given the ideal numbers: π and dn/1000 = 1.936710383)

(t) $[(dn/1000^2 / 10)/x \cdot 2 - dn/1000] \cdot 2 \cdot \pi / 4$ =   5.333510825 (being a new number, 13L3)

(Notice that the first part of the equation in brackets, following argument (1) above, creates yet a new and plausible form of the Between the Seals length, here called BSC6 = 3.39541845. Close to BSC2 in appearance but *very* close to BSC5 of the first subtraction.) The above equation creates a number amazingly similar to 13L2. How close is it?

(t)	5.33351083"	=	13L3	(Thirteenth line #3)
(t)	5.33348608"	=	13L1	(Thirteenth line #1 the original result by subtraction between LIR and 0.279".)

(-)--------------

0.00002475"

or, a difference of about 2 one-hundred thousandths, with an error ratio at one in 216 thousand, or 4.6 parts per million. I think this looks like it has a chance of being the Designer's intended number between WB and EB since it answers the Date Number subtraction problem between 13L and BSC lines. Or, I should say, this is *a plausible length* that fits between the two end balls, and doesn't differ by as much as a gnat's eyelash from the original number of √31.5" minus 0.279". "2 one-hundred thousandths of an inch" is not appreciable from any measurements to these ball shapes on printed paper--even if they were to have had precisely marked centers.

Is this the solution? Is this *actually* what the Designer had in mind? We will get back to this number, with strong *competition*. The competition comes from another very curiously similar number, yet arrived at by a totally independant theoretical route, which we will see shortly.

We know *rectangles* are fundamental to the Designer's general theme. We are still left in the dark as to what the full idea was behind the *height difference* to WB and the bottom line of LIR, which he highlighted. These parts of a rectangle produce a diagonal of 2.861" with a 0.0035" offset near the centerline. This height clearly measures 1.036". This has possible meanings which we will look at shortly. Curiously, this investigation based on 1.036" lead to another 13L-type of line, very much the same size as 13L2.

****

Gift Wrapping:

Here we should briefly take time out from the trail we have been following, to look at an earlier noted group of mathematical symbols: In Chapter 3 we saw that there was a tiny angular difference between the diagonals of LOR and LIR, that were the result of the difference between the rectangular proportions of the √7 and φ^2. What was significance of that number?

(t)	$\tan^{-1}(\sqrt{7}) - \tan^{-1}(\varphi^2)$	$= 00°12'01.25"$	(angular difference)
(t)	$\tan(00°12'01.25")$	$= 0.003496718$	(rectangular relationship from the angle)
(t)	$1 / 0.003496718$	$= 285.9824207$	(A hit as close as 1:16268 to "286")

There is that "349" number *again*. A similar number--surprisingly close to D/x, a difference of about one part in 2400 for an error ratio to the ideal D. Notice also, how similar this little *tangent function* is numerically to the above 0.0035 tiny *offset*. One is *dimensional* and the other *angular* (This another kind of J3-switch). This small tangent value multiplied by the LIR diagonal of 6" is also interesting in yet *another way*, in a *dn-type of way*, if raised to the fourth power (which is to say, on your calculator, squared twice. You will have to multiply your answer by 1,000,000,000 to read these digits on a ten place calculator):

(t)	$0.003496718 \cdot 6"$	$=$	$0.02098031"$	=a (This is a tiny tangent line length subtending the angle of 00°12'01.25" at a distance of 6 inches.)
(t)	a^4	$=$	$0.000000193752649"$	

This could also be thought of as being within an error range of about one part in 2400, proportionally, to a form of "dn" or September 16, 1936, if suitably reduced to the same decimal place, ten zeros to the right. These tiny details reveal a form of the Davidson's Displacement number reciprocal *as well as* a numerical derivation of "Entrance to the King's Chamber," all from the tiny tangent to its 6" diagonal. And refer to the tiny offset. These numbers could be ignored as coincidental, but this is astonishing.

Again we find incredible subtlety, made through very small distinctions of angle and distance. This arrangement points to a really good, *plausible reason* why the Designer chose this particular √7 ratio for this inner rectangle LIR and a diagonal of 6.0". Certainly there are endless groups of ratios that the Designer could have chosen for the dollar instead of this pair--$\tan^{-1}(\sqrt{7})$ and $\tan^{-1}(\varphi^2)$--but these values

134

were an exceptionally good choice, easily discovered by an investigator as a basis. After determining the suitability of the rectangular relationship LIR with the primary LOR, perhaps then he developed the rest of the immensely complicated and interlocking interior design involving the little Pyramid, rectangles, diagonals, ellipses and whatnot. The Designer makes really good, carefully thought-out and fortuitous choices for rectangles.

But what does this relationship mean? Is it merely ornamental? I have often wondered if this was merely the mathematical motif of Davidson's number pulled from the dollar's plan. Perhaps it is like an a rough design, repeated like gift-wrapping paper patterns, as though for an abstractly decorated gift box. Otherwise, it doesn't appear to lead anywhere in the scheme of the dollar. Yet this is most likely *another puzzle*. In several earlier cases of *near misses to an ideal*, a small adjustment or alternative starting point forces the precise form reveal itself. Sometimes the adjustment itself carries the message. If there is a solution in this case, I haven't found it. Perhaps some reader may uncover it. Other forms of this Date Number can be found in conspicuous places in the design, which we will look into in the next chapter.

****

Intersecting Arcs, and the Difference between Diagonals, derived from the Thirteenth Line:

Returning to the previous trail of the 2.861" arc diagonals from the Western Ball (WB) and Eastern Ball (EB) trail: What *exactly* happens at the bottom line of LIR *at the centerline of the dollar* where the arcs should have come together? They are offset by 0.0035" to the left and this we know is important having been the highlighted puzzle at the center of the dollar. When the circles are repositioned into balance, both converge at the lower line of LIR at a point 0.0035" left of center. The rectangles formed from the Thirteenth Line Partial Lengths and the adjusted height difference to the centers of the balls should be informative. Suppose we considered both balls as having the *same height* from the bottom of LIR, or about 1.036" (as measured) which was the result of the circle shift -idea discussed earlier. What are the lengths of the diagonals of these rectangles?

First Diagonal PLD1: From the Western Ball to the Centerline at the base line of LIR:

(t)	$x =$	2.67024304"	$=$	1PL13	(1st Partial length of the 13th Line, from above)
(m)	$y =$	1.036"		H13L1 $=$	(Height of the WB ball, at 13th Line-- as measured)
(t)	$\sqrt{(1PL13^2 + H13L^2)}$		$=$	2.864174208"	(PLD1: Partial Length's Diagonal #1)
(t)	$\tan^{-1}(1PL13 / H13L)$		$=$	68°47'41.13"	$= \theta\,1$ (Characteristic angle 1)

Illus. N23

Computing the diagonal lengths of rectangles below the Thirteenth Line from the Western and Eastern Balls, to the Centerline at the outer edge of LIR. The ratio for these diagonals to the ideal 2.861" is plus 1.0011 (West) or minus 1.0011 (East). This resulting ratio is 1.0011 or, the same as the ratio between British and Pyramid inches.

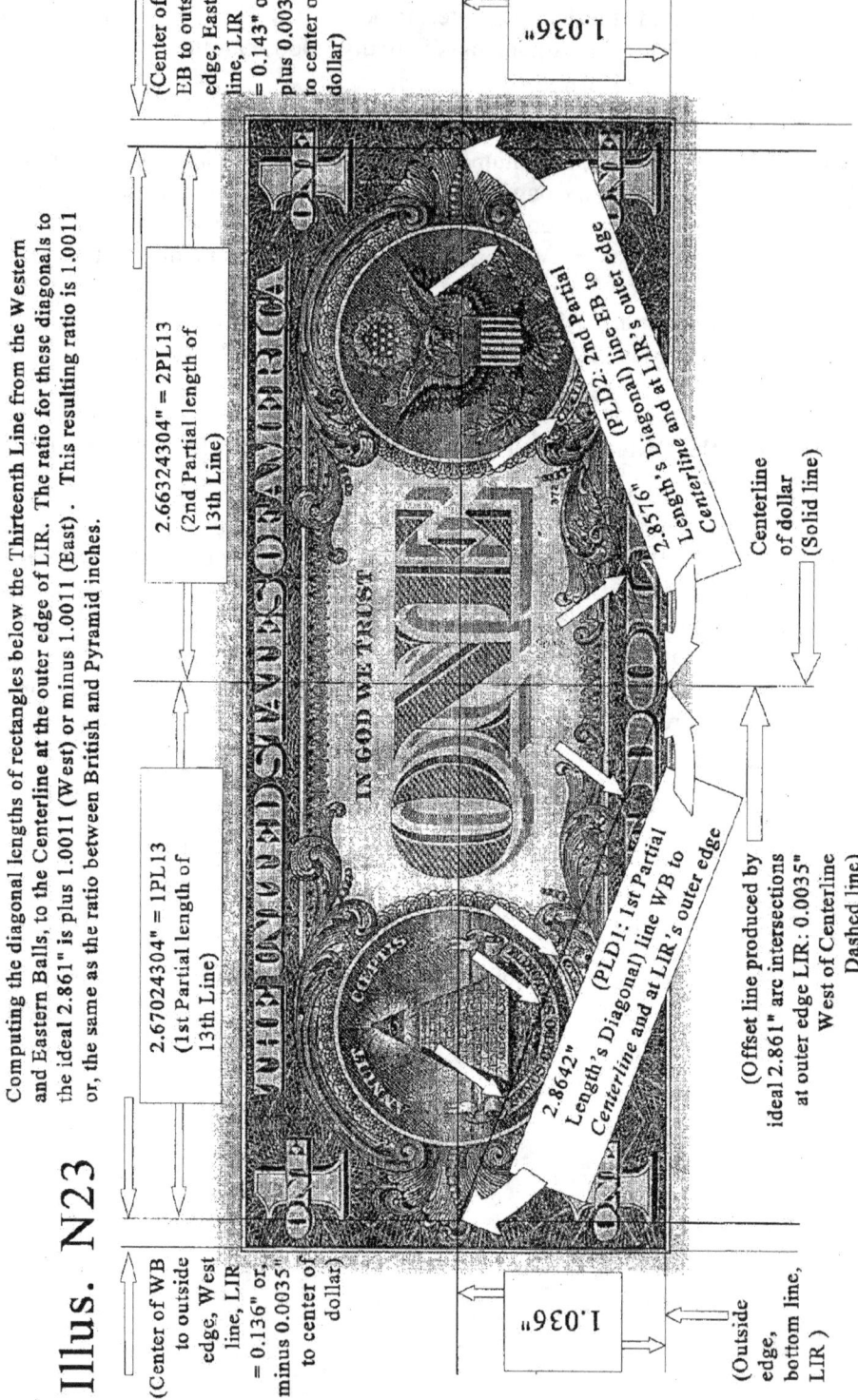

(Center of EB to outside edge, East line, LIR = 0.143" or, plus 0.0035" to center of dollar)

1.036"

2.66324304" = 2PL13 (2nd Partial length of 13th Line)

(PLD2: 2nd Partial Diagonal) line EB to LIR's outer edge

2.8576" Length's Diagonal at Centerline and at LIR's Centerline

Centerline of dollar (Solid line)

2.67024304" = 1PL13 (1st Partial length of 13th Line)

2.8642" Length's Diagonal) line WB to Centerline and at LIR's outer edge

(PLD1: 1st Partial

(Offset line produced by ideal 2.861" arc intersections at outer edge LIR: 0.0035" West of Centerline Dashed line)

(Center of WB to outside edge, West line, LIR = 0.136" or, minus 0.0035" to center of dollar)

(Outside edge, bottom line, LIR)

1.036"

This is about 0.00315" longer than 2.861022156", and this turns out to be very significant, in that as a divisible ratio it is *1.0011 longer proportionally* than the numerical 2.861022156 of the ideal D/1000. This is an unusual and telltale ratio, an unmistakable one we have seen earlier in the Pyramid lore. This seems to indicate an effort to communicate the awareness of the Pyramid Inch/British Inch ratio.

(t)   2.864174208 / 2.861022156   =   1.001101722

The Designer knew that we would likely know of this relationship from the popular lore, and this rectangular frame of the corrected Eastern Ball, Thirteenth Line, centerline and lower edge of LIR allow us to be certain that this Pyramid Inch value is what he intended.

Second Diagonal PLD2: From the Eastern Ball to the Centerline, at the base line of LIR:

(t)   x =   2.66324304"   =   2PL13   (2nd Partial Length from above)
(m)   y =   1.036"   =   H13L1   (Height of the 13L)
        $\sqrt{(2PL13^2+H13L^2)}$ =   2.857649294"   (PLD2: Partial Length's Diagonal #2)
        $\tan^{-1}(1PL13 / H13L)$ =   68°44'38.38"   = θ 2 (Characteristic angle 2)

Neither of the above angles (θ 1 and θ 2) looked like anything recognizable, nor did their tangent values as ratios. The second diagonal is about 0.00337" *shorter* than 2.861022156" and about the same inversely proportional relationship, 0.99882 (1/x = 1.00118) as seen above.   There is, of course, an arithmetic symmetry between these diagonals since both are offset the same way. And this can only mean that the *average* of these must be close to D/100 as we know from the circles' radii. Note here, that the *reciprocal* of the previously discussed 0.0035" offset bears a certain resemblance to 1/100th of PLD2, not to mention the "nested puzzle" discovery. But quite beyond these, we will bump *into another very similar number* to PLD2 much later, from surprising and unlikely sources (See pg 181 and 182).

Notice above, along both diagonals the Designer arranged to have his trademark alignments touching at *serifs* and *edges of round things* (See Illus. N23 at arrows). Lest we have any doubt that he intended us to concentrate on these diagonals: from left to right, the tip of the upper-left serif on the N of "NOVUS"; the exact upper-left serif on the E of "SECLORUM"; a grazing tangent at the right edge of the curved cartouche containing "THE GREAT SEAL"; the precise upper-left serif on the D in "DOLLAR."   Then on the right, likewise, the exact upper-right serif on the first L in "DOLLAR"; the true center of the O in the right-hand curved cartouche, in the word OF in "OF THE UNITED STATES"; (notice also, the E then O alignment thing again on opposite sides.) Then, the tip of the *last* prominent arrow in the bunch of arrows gripped in the talons of the Eagle's left foot.   Since we know that there are a total of thirteen arrows there, and if it was the *last* one, it could clearly be said to be the *Thirteenth Arrow*.

What is the average of these two numbers?

(t)   2.864174208"   PLD1

(t)   2.857649294"   PLD2

        (+)----------------   (PLD1 + PLD2)/ 2   =   2.860911751"= average)
(t)   5.721823502

Okay, so it appears a little smaller than ideal D/100, but not by much--and, we really don't know the exact y height *yet*, to any greater exactitude than the estimated thousandth with a 10 power glass. Nor can we truly say that we know the x length any better. (And, the average *should* be close to an ideal D/100 anyway, since the centerline *varies inversely* to the intersection from two arcs of 2.861.) But following as we did, the previously successful trick of using the theoretical difference between the two end dimensions and accepting the measured y height to the center of the Western Ball to the round thousandth, we have calculated a number from the graphic evidence that is plus or minus a ten-thousandth of an inch to the length of 1/100th of Davidson's special number:

(t)	2.861022156"	D / 100
(t)	2.860911751"	PLD3 (the average above)
	(-)-----------------	
	0.0001"	or about a ten-thousandth smaller

This is a clear case of the intent to accurately display the Displacement Factor number 2.861", *and* to illustrate the Pyramid Inch size and ratio at the same time. And we are only using estimated thousandths of an inch at WB and EB. But why the tweaked diagonals, why are the sources of WB and EB offset?

Having seen the previous elliptical chord-length puzzle, we should notice the enormous similarity between these two diagonals 2.8641" (PLD1) and 2.8576" (PLD2) to the elliptical chord lengths 0.2865" and 0.2857". Both are the result of (1) a *slightly offset reference position* (the centerline as opposed to the height of the ellipse). Where there was (2) an initial *offset clue that was also 0.0035"* (first west of the centerline, and then downwards for the Right Face of the Great Seal). Then, (3) both shifts are *rectangular* (east to west and up and down); and that (3) the *averaged sum of both pairs of numbers makes "2861"*. (Okay, the chords make 0.2861" and the diagonals appear to make 2.8609" by calculation but we should get the idea at one part in 26,000.) The numerals *look* about the same--yet there is sufficient precision in both relationships to be fairly sure that they are in reality, slightly different. The method is the *same*, but the numbers are a little *different*, yet the two relationships ultimately accomplish the *same 2861 message*. Maybe one message that we are getting out of this is the Designer may never allow a true proof of Davidson's number without a mathematical"decoding" through the use of *at least two* number elements. Notice that would also be true for *all exact* mathematical findings of this number (i.e., the length of LOR in the DFT theory, or the combined effect in $(1000 \cdot \varphi)/D$ or 5.655440261; the total of the elliptical chord sums.)

Shrinkage used to Create an Ideal Average, and in a slow state of "Motion"

We should take note here, of a really fascinating arrangement--an idea to truly stretch the imagination. The Designer appears to have developed a display of the shrinkage of the paper dollar, over the time of some months, with a odd use of the difference between Pyramid inches and British inches. This display makes use of the fact that the *difference between the two kinds of inches, is the same factor as the paper dollar's shrinkage from large to small: 1:1.0011*. The total shrinkage necessary would be within the expected life of a one dollar bill--said by the Treasury Department be about eighteen months. As we saw earlier, the geometry of the 13L line was set up to have a westerly offset of 0.0035", slightly skewing these diagonals to the centerline intersection. The dollar appears to have been designed slightly large in the beginning, from the point of view of British inches, so that it would *originally be scaled in Pyramid inches*. (This shrinkage would be about six thousandths of an inch, or 0.006".) When finished shrinking, if there was no distortion, all of the same dimensions in Pyramid inches would *later have become British inch dimensions*. British inches are smaller by about the same ratio as this shrinking. This arrangement produces some curious mechanical consequences for the skewed diagonals. When read *in the original, un-shrunken in British inch units*, or when the opposite diagonal is read *after shrinking in Pyramid inch units*, the numbers appear to reverse positions. We will look at the case in reversed time order, more or less as I first began to reason this out:

(I) *Measuring* from the Western Ball on the **Westerly diagonal PLD1** on the surface of an *old, fully shrunken dollar* (such as I originally had in my possession) to this centerline point on the lower line at LIR, will actually give 2.861 in **Pyramid inches**--not our larger computed figure of 2.8641 PI, since it is *now 1.0011 times too small, having shrunk slightly:*

2.8641 PI / 1.0011 = 2.861 Pyramid Inches

(II) Yet on the other hand, eighteen months earlier, *measuring on a freshly minted bill*, the **opposite Easterly diagonal PLD2** would be correspondingly *larger*. This means that one would actually read the number 2.861 *here too*, but now found in **British Inches**, since PLD2 with the smaller computed figure of 2.8576 PI, would then be *in the original, Pyramid inches, which are larger*. As it happens, this exactly cancels the "smallness" of this figure by precisely the same ratio of 1 to 1.0011:

2.8576" · 1.0011 = 2.861 British Inches   (See Illus N24)

This is dizzying stuff. Why did the Designer do this? It appears he intended that during the paper bill's process of shrinkage, one would *always be able to find an average of "2.861" numerically*. This *would have to be found* between the exact centerline point at the bottom of LIR and the diagonals to both WB and EB balls, *no matter what stage of shrinkage* the bill would be in. And this is (curiously) irrespective of units of measure--if both units are known. If you started your investigation on a new bill, the Eastern diagonal PLD2, would automatically measure "2.861" in British inches. If you started on an older bill, "2.861" would now only appear as an average of both diagonals PLD1 and PLD2 in British inches. So, this change runs from the East to become a general average of both sides. Oddly enough, if you measured the Western diagonal PLD1 in Pyramid inches, on an old bill--and then, measured the average of both sides of a new bill, *exactly the opposite will occur, with the same numbers in opposite units*. The first change runs from the East to a general average; and the second, or inverse form, runs from the West to a general average between diagonals. (This is definitely "J3". See Illus. N23 and N24) But strictly speaking, since the dollar is shrinking, the simplest relationship that would be noticed after uncovering the original use of Pyramid inches on the dollar, is *an East to West conversion* of "2.861" as read in *opposite units, moving to opposing sides.*

But stranger still, from the point of view of one trying to measure the dollar over a period of time, this average must then be *moving*. Since any dollar would always be in some state of shrinkage between two ultimate beginning and ending points, a correct average of "2.861" will *always be possible to find at some exact proportional spot between them, in either unit of measure.* Since the bill is continuously shrinking, this abstract place of *the average* could be said to be very slowly moving, "floating" as an abstraction, if you will, across the dollar in mathematical space, in one direction until it stops shrinking. A proportional equation might be created, for a super-average both kinds of units, to track this snail-like, average "motion." Notice the two-number-coded *average* idea once again, where two factors are needed to find an average, which is the ideal D/1000. In this case, the Ideal is a slowly moving target-number, *independently of units or points in time between time limits.* You might say that this framework and paper accomplishes this transfer while in neither unit. If the firs dollar was printed in, Spring 1935, the process at eighteen months should end in Autumn 1936. Isn't that amazing?

Its fair to say that this may take some study of the diagrams and re-reading to have a clear idea of what this is all about. If you find this setup confusing, I must commiserate—even after glimpsing this idea, I always have to go back to square one whenever I come back to it.

Illus. N24

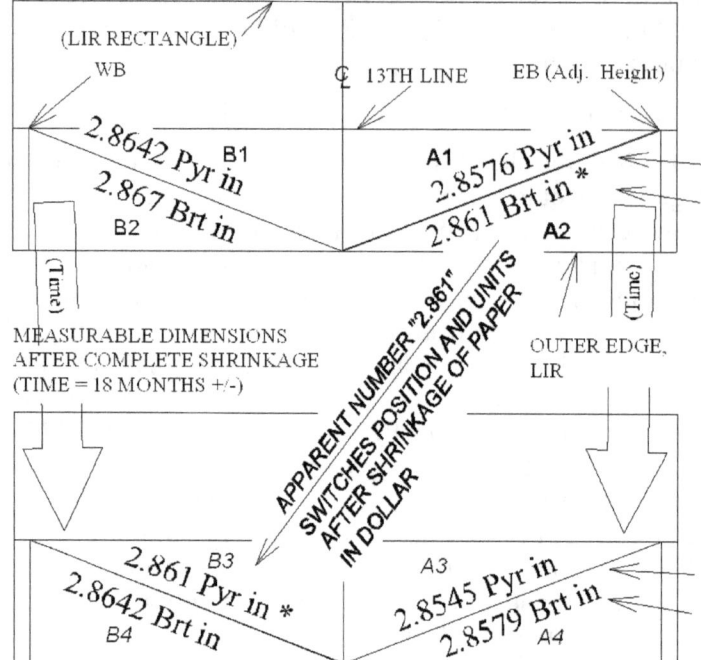

THE BEFORE CONDITION:

(A1 + B1)/2 = 2.8610 *
PYRAMID INCH
AVERAGE
BEFORE SHRINKING:
DAVIDSON'S IDEAL*

(A2 + B2)/2 = 2.8642
BRITISH INCH
AVERAGE
BEFORE SHRINKING:
LARGE

THE AFTER CONDITION:

*(A3 + B3)/2 = 2.8579
PYRAMID INCH
AVERAGE
AFTER SHRINKAGE:
SMALL*

*(A4 + B4)/2 = 2.8610 *
BRITISH INCH
AVERAGE
AFTER SHRINKAGE:
DAVIDSON'S IDEAL*

MEASURABLE DIMENSIONS
JUST AFTER PRINTING
(TIME = 0)

The "2861 number" could be said
to move from East to West over
time.

(The numerical values used
in this diagram are
theoretical and based on the
ideal 2.861022156, and the
ratio 1:1.0011. This will
differ slightly from above
calculations made from
measurement.)

ORIGINAL
PYRAMID INCHES

ORIGINAL
BRITISH INCHES

(Also note: reciprocal
exchange of numbers
from
B1 to B4; and from
A1 to A4 into
opposite positions
and opposite units.
Another result of
shrinking in this
format.)

SHRUNKEN
PYRAMID INCHES
SHRUNKEN
BRITISH INCHES

The Designer's intended investigator is always provided with a theoretical mathematical average of (t) 2.861", or an empirically measurable proof (m) 2.861" of this average. Here, our Designer has used *the defect* in the character of the paper material as an engine to drive another display of the art of his sacred mathematical scheme. This scheme simultaneously ties together (at least) these elements:

(1) the parallel of British Inch to Pyramid Inch as a *relationship of Units,*

(2) *Exact fidelity in information transmission of information* by inversely variable geometry, for the purpose of demonstrating the *numerical* message of "2.861" apart from units.

(3) The abstract, moving numerical average of 2.861", *in a state of "motion"* across the bill

(4) the specific message of 2.861 (D/100) *in both* units.

(5) *PLD2* a now familiar number, similar to (1) 0.0035/x; (2) the shifted elliptical chord length 0.2857". (*There is also similar number* which we will see later, from elements within the Great Seal, from the time of the American Revolution. See Chap 5, pg 181.)

(6) The *actual length* of D/100 in Pyramid inches in the after condition on the Westerly diagonal, or the before condition. In the initial condition it is the *average* of both diagonals, an arrangement that could be said to be automatically inversely-averaging into two diagonals composed of British inches over time.

Earlier in this chapter, we saw the Designer reach into the third dimension using the dimension z. *Here,* we see the Designer touching on the *fourth dimension,* by the slight, but actual use of the time/motion axis. In this case he uses 1/ z · 10, or D/100. (J7) One would have to look far and wide to find a comparably subtle contrivance as this arrangement above. With other things we will soon see about element (5) above, this is a truly amazing bit of work for tying many symbolic things together.

****

A *Deeper Look* at the Average and Motion Effect: The Passage of the Sun:

Again, is that all there is to this Mystery puzzle? Over time, I came to see that there is *even more* to this arrangement at another conceptual level and within another kind of symbolism. This a rebus, and its symbolism is outside of the bounds of Rules 1 through 3. The moving mathematics and geometry appear to have been set up in the form a symbolic tableau. Lets look at the two ideas of the riddle, taken together, but seen as clues of another sort:

(I) Over time, the numerical message moves *"from East to West..."* What would we say, moves from East to West? (obviously, the Sun) and,

(II) Standing back and looking at the whole picture of the dollar geometry, why does the favored intersection occur *at the bottom center point* of the design, at the bottom line of the fundamental LIR rectangle, rather than the *top*? The corresponding top area at LIR line seems to be ignored or *empty* with respect to geometric design importance.

This puzzled me for quite some time. At some point, the light bulb came on, and I began to wonder if the *empty top* wasn't the dark "North" in the arcane or symbolic sense as used in Freemasonry.

Illus. N25   The Symbolic Layout of a Masonic Lodge

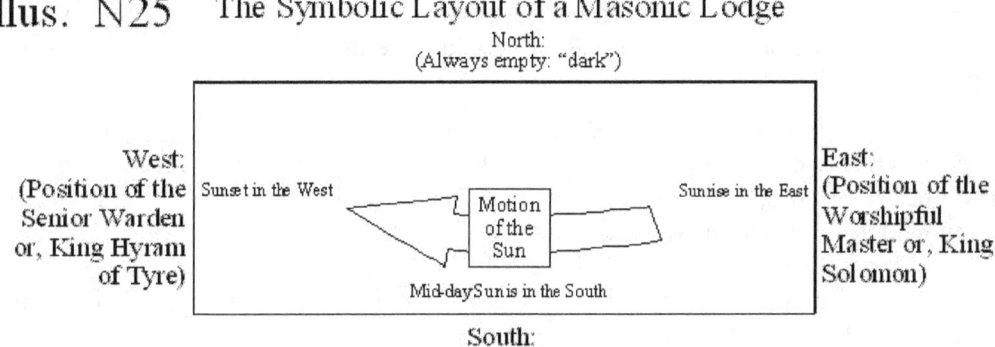

North:
(Always empty: "dark")

West:
(Position of the Senior Warden or, King Hyram of Tyre)

Sunset in the West

Motion of the Sun

Sunrise in the East

East:
(Position of the Worshipful Master or, King Solomon)

Mid-day Sun is in the South

South:
(Position of the Junior Warden or, Hyram Abiff, the Architect of the Temple)

If the special intersection area at the bottom line of LIR rectangle is intended to be the *South* in this design, then a recognizable whisper of a symbolic scene unfolds, well known and fully familiar to students of Masonic Ritual. The temple is also rectangular in form, running east to west. The South symbolically alludes to the place of the legendary *martyred Architect* of the Masonic lore, known in Masonic rites as the place of the Chair of the Junior Warden, and a specific gate of the Temple of Solomon. (See Illus. N25.) This is the place of "the Sun at meridian height, being the beauty and glory of the day..." The *movement of the Sun* is a principle element of lodge rites. Many will be able to automatically extrapolate other possible Masonic identities from the graphic and mathematical elements shown above, such as the diagonals, that might be said to communicate the special number from East to West, from the Eastern and Western Balls.

Paralleling Davidson's beliefs about the sacred number, the Designer appears to have equated the Displacement Factor number 286.1 with the Sun in symbolic passage across the sky from the point of view of the Temple. He appears to be supplying a mathematical substitute for the "Lost Word" of Masonry. He has arranged the geometry of these lines and the scale change of the paper dollar to make the fractional form of 286.1 "move" mathematically from East to West. The implied parallel to the Sun symbol appear to be a reference to the Biblical symbol of Divinity, probably from the Transfiguration of the New Testament.

This is a splendid example of the truly "minimalist" approach to symbolic design, favored by the Designer. A surprisingly large amount of things can be said to be generated around very little material evidence. For those who are now fully persuaded of some or all of the above interpretations, we should stop to look back and recall that *all of this* is based on two very small circles at the end of some leaves, and a rectangular shape. There are no "lines," really, *nor even ends* of lines. Any lines outside of the LIR rectangle are imaginary or are lines that *we have made*, such as the added line of hatching above the Chief on the Shield. We got here by estimated measurement and some mathematical guesses. This whole design presupposes an understanding of mathematics, the Bible, Pyramid lore, US History and a knowledge of the Ritual of Freemasonry. Yet one would be hard pressed to prove any of this to the skeptic, nor will these symbols be very convincing to anyone who has not partaken of the Mysteries.

But that is the nature of our Designer's art. Some people will see these things; others won't; and still others yet who will learn of them--but in time will forget the whole thing. Think of all the complex things the Designer has strung together here. And, as we will see later--I am fairly sure that we don't know the half of it. Later, we will see the elements of the *original Great Seal* hinting to part of the above diagonals, PLD2.

****

The 1/256th Power; The Pyramid and the Height of the Eastern Ball:

There is yet another relationship relating to the Thirteenth Line that may be worth investigating. The height above the bottom line of LIR, (about 1.036") could have the effect of tying some of these number patterns together.

The Designer evidently *did all of his original calculations from the dollar in Pyramid Inches,* and these are the same as Davidson's units. As a reminder, we should note that the numerical lengths are computed from theory based upon measurements made from fully shrunken bills, and--since shrunken--are therefore completely in measurable British Inches. A dollar in this condition *will be numerically the same as the original units* when measured in British Inches. This is mentioned since there will be some comparisons made with figures drawn from Pyramid Inch-based theory of Davidson, which might seem at first like comparing apples and oranges.

At some point in 1980--perhaps I punched the square and square root button too many times, and found a relationship between a side of the Great Pyramid in Pyramid Inches and the length of LOR, thus producing *yet another* competing source for the fundamental "5.655" number for the LOR rectangle length.

According to Davidson, the Great Pyramid's year length was the average made from several Processions of the Equinox, was 365.242465 [3]. Therefore, his length of the Total Base Course of the Pyramid was 36,524.2465 PI. Davidson's idealized side (one fourth) of the Great Pyramid base is 9131.061625 PI. The 16th root of this number (four button pushes at the $\sqrt{x}$ key) is almost exactly the same as *ten times the reciprocal of the DFT version* of the top length of the dollar, or, the length of LOR.[4] Early on I noticed that the reciprocal of 5.655 was about 0.1768, so I was already primed to latch on to anything resembling this number. I had seen the Designer's J3-switch of "the square root of 10 times the reciprocal" equation a few times before, so the connection was made fairly quickly for once:

(t)	$9131.061625^{1/16}$	= 1.768204859	then,
(t)	1 / 1.768204859	= 0.176820486	then,
(t)	1 / 0.176820486	= 5.655453298	(PST, Pyramid Side Theory)

This is *very* close to the DFT length, so close, that it is truly marvelous:

(t)	5.655453298	(PST, Pyramid Side Theory)
(t)	5.655440261	(DFT, Displacement Factor Theory)
	(-)----------------	
	0.000013037	or, about one *one-hundred thousandth* of an inch.

This is a precision difference of about one part in 433.7 thousand, or about 2.3 parts per million. One might almost suspect, inversely, that the DFT formula could have instead been used as the source of the Great Pyramid's sides. Turning these equations around going backward, starting with DFT, one will find an alternative pyramid Base Course of 36525.59365, closely approximating the 100x Sidereal Year length, one of Davidson's fundamental three base course derived lengths. This length said to be due to the special hollowing creases in the sides of the Pyramid. (See Illus M6, pg 94.)

But what does this have to do with the Thirteenth Line on the Dollar? Several years later, I was looking into the Thirteenth Line end points. Again poking the square button (x^2) a few extra times, I started with a form of 1.036 (this was 1.03626. See Illus N18 pg 125). This time I noticed the "1.768" number now appeared *up the ladder of squarings,* and this immediately caused me wonder if this lowly height 1.036" wasn't in reality intended by the Designer as *one of the roots of the Great Pyramid's sides.*

Now, turning this all around, this time using a tenth of the reciprocal of 5.655440261, or the top length of LOR using the Displacement Factor Theory, there is a similar "1.036" number to the 16th root from one side of the Great Pyramid:

(given: DFT or, 5.655440261")

(t)	5.655440261 / 10	=	0.565544026	then,
(t)	0.565544026 / x	=	1.768208935	then,
(t)	$1.768208935^{1/16}$	=	1.036265045	(HDFT or, Height by Displacement)

Interesting. The Great Pyramid's Side is also 4 powers of 2 in the *heigher direction* from 1.768208935. Looking back at the Little Pyramid on the dollar, I became suspicious that those "eight paired steps" were intended to be symbolic of the whole string of square roots from one side of the Great Pyramid. These are the eight Steps are the pairs of pairs of step widths, beginning at the top of the first step. "Pairs of pairs" with respect to "eight" units, could symbolically mean *"eight doublings, or the 256th power."* As if a confirmation, we can see the vertical line of the Whirling Square (See pgs 68 and 69, Table 1) splits Step Six and Step Seven, at a point that would mark the *mathematical position* analogous to the positional place that 1.768208935 (or its reciprocal 5.655440261) which would occupy in the middle of hypothetical a ladder of squarings and roots from either end. *And this same Thirteenth Line also ends just at this height of "1.036" above the LIR line.* (It might, then, be expected to split courses at the top of Step 1/bottom of Step 2, which is food for thought.)

Here is the 256th root of one side of the Pyramid, (or, more precisely, 1/4 of Davidson's Total Base Course of the Pyramid). Starting with 9131.061625, make eight pushes on the √x button:

(t)	$(9131.061625 \text{ PI})^{1/256}$	=	1.036264895 PI	(HPST or Height by Pyramid Side Theory.)

From the top of the ladder scheme, we find HPST as the final, 256th root of the number 9131.061625. Powers are not additive but are multiplied--16 roots in one direction (down) with 16 squares in the other direction (up) are 16·16 or 256. (Note again the J3-arrangement here.) Also, as seen earlier, the height of the Thirteenth Line that *cued this relationship in the first place* hovers just a fuzz above the top of the center of the pattern of the eight regular steps at Step Six. (This is a puzzle pointer perhaps, since it is a slightly "defective" alignment.)

How does this new HPST form affect the other patterns that we have seen? A new 13L length can be implied. This becomes possible since we *now know two theoretically exact variables* for the Eastern and Western rectangles under the Thirteenth Line: Davidson's Number (D/100) is a given and "1.036" is now given as 1.036264895. How would a 13L number appear if generated from this data? To find a new mathematical route leading from this new number for the height, where the only other given is the assumed ideal diagonal of 2.861022156, we must use this with the Cosine and then Tangent functions.

What happens if we start from this *new height*, using instead the ideal *diagonal* of 2.861022156, as a given, to get a new half-length of 13L?

(t) $9131.061625^{1/256}$ PI   =   1.036264895 PI   (=HPST) (the given new height)

Then, using D/100 as the diagonal and HPST as the height of a right triangle, an angle is determined:

(t) Cos^{-1}(HPST / (D/100))   =   θ 3
(t) Cos^{-1}(1.036264895 / 2.861022156)   =   θ 3
(t) Cos^{-1}(0.362200933)   =   68° 45'52.47" = θ 3

Pretty close to the earlier angle. Then, the Tangent of this θ 3 angle is determined, to then be multiplied by HPST to find the length of the new 13L half-length:

(t)   Tan(θ 3) · HPST   =   (or half of a new 13L)
(t)   (2.573433566) · 1.036264895 PI   =   2.666758865   (the new computed half length of a 13L line) then, multiplied by 2 for a full 13L length:

(t)   2.666758865 · 2   =   5.33351773 (13L3)

This 5.33351773 number is startlingly similar to the original 13L1, and also very similar to 13L2. Let's have a look:

(t)   5.33351773"   =   13L3   (Thirteenth line #3)
(t)   5.33351083"   =   13L2   (Thirteenth line #2)
        (-)----------------
        0.00000690"   about seven millionths of an inch, with an error value of one part in 773 thousand, or as small as *1.3 parts per million*. Nearly the same number to five places.

How about between the original 13L1 and 13L3?

(t)   5.33351773"   =   13L3   (Thirteenth line #3)
(t)   5.33348608"   =   13L1   (Thirteenth line #1)
        (-)---------------
        0.00003165"   about three one-hundred thousandths.

The precision/error ratio between these two is about one part in 168.5 thousand, or about 6 parts per million. What happens (once more, reproduced from above) between the original 13L3 and 13L2?

(t)   5.33351083"   =   13L2   (Thirteenth line #2)
(t)   5.33348608"   =   13L1   (Thirteenth line #1)
        (-)---------------
        0.00002475"   or about two one-hundred thousandths of an inch, with a precision ratio of one in 216 thousand or, 4.6 parts per million. All three of these are *very* similar. Lets look back to where these numbers came from:

13L1 was the result of the first, simple subtraction of the apparent offsets from the ideal length of LIR:

$\sqrt{31.5} - 0.279 \quad = \quad 5.33348608 \quad$ 13L1 $\quad$ pg 126

13L2 came out of the recursive Sept 16th 1936 (dn) square/square root and dn subtraction formula:

$[(dn/1000^2 / 10)/x \cdot 2 - dn/1000] \cdot 2 \cdot \pi / 4 \quad = \quad 5.333510825 \quad$ 13L2 $\quad$ pg 130

13L3 is derived from another recursive formula using the 1/256th root of the Pyramid's side and D/100 as the given diagonal:

$\text{Tan} [\text{Cos}^{-1}(HPST / (D/100))] \cdot HPST \cdot 2 \quad = \quad 5.33351773 \quad$ 13L3 $\quad$ pg 133

To get us at least within the numerical neighborhood, the Designer provided us with relatively simple evidence that can have the result of a line length of 13L1 through subtraction of the end distances 0.136" and 0.143" from $\sqrt{31.5}$. Maybe this was intended as a starting approximation of the 13L length. This is a close approximation of both 13L2 and 13L3. Yet all along it seems that the Designer must have intended a number manifestly similar to 13L2 or 13L3 for a whole raft of reasons seen above. Although these are from *completely different sources,* and can have no causal connection, they both are very close, and paradoxically, the both share some evidence from the design of being intentional.

What happens if we test these two 13L2 or 13L3 with "Occam's Razor?" We might say 13L3 (based on one ideal side of the Pyramid) is a better candidate as the intended number, since it requires a simpler formula, and a slightly more plausible *source*--the Pyramid's side, per David Davidson. But 13L2 uses the distinctly Davidsonian date "September 16th 1936" together with a π-based Pyramid formula, for which we have substantial graphic evidence from the dollar. Although it is a little more round-about, both formulas have a recursive element, and both might be said to have about the same number of terms.

I prefer the 13L3 explanation, since it seems a more straight forward. And on top of that, using it's half-length, this number can be run backward through the 13L2 formula, whose result closely approximates the Date Number 260 of 1936, to within one theoretical "day," (day number 259) since both numbers are so close. But until we learn something else about these formulas, one looks as good as the other as a tentative explanation for the logical source of the Thirteenth Line. Its close to being a single formula.

****

Before we move on, here are some tidbits we missed earlier, on the dimensions to the WB and EB balls from the ends at LIR. We saw that the WB ball was 0.136" east of the outer edge of LIR, and that the EB ball was 0.143" west of the outer edge of LIR. A total of 0.279". As seen earlier, doubling the right hand number makes the number 0.286". The difference between 0.143 and 0.136 creates a 0.007", which told us that the center of a line drawn between them would be half, or 0.0035" west of the centerline of the dollar. What happens when we instead *add* the difference 0.007" to 0.279"? we get 0.286" *again.* Neat. 286 displayed *two ways.* (Looks like magic, but this will happen with any two numbers like this.)

Now let us consider the fact that 0.279" is a sum of lengths on a line extended from the Thirteenth Line and the Between the Seal Center (BSC) line. Since both of these were bases of π-based pyramids: what would be the height of a tiny pyramid with a side of 0.279? Four times 0.279 divided by π, divided by two:

(m, t) $\quad\quad$ $(0.279 \cdot 4 / \pi) / 2 \quad = \quad 0.177616916$ $\quad\quad$ (About one 10000 th of 1776.)

A very small π-based pyramid with a height symbolic of the year of Independence. A persistent math researcher will also find "2861" also hiding in this little number. Isn't that curious.

****

145

Discovering Pyramid Inches:

When I reviewed my early successes in my investigation of the Pyramid relationship to the dollar, I was often puzzled by the fact that I didn't detect the use of the Pyramid inch unit anywhere in the design of the dollar. It seemed obvious that a designer of the exacting technical knowledge as he evidently was, would be fully aware of the Polar inch. And this designer did not seem to be one to do anything by halves. Was it a choice? Still, why would he use to inches--wouldn't Pyramid units be more in keeping with the traditional values? But regular "inches" appeared to be working pretty well, and I hadn't given any thought about the nature of *paper,* which was the answer to the question. I was using "shrunken" paper, and hence smaller units. It is entirely coincidental that the paper shrinks exactly right for this.

My original bills had become a little worn through use, and it became time to get new ones. I was mostly ignorant of paper and it's characteristics. Yet I had often noticed in survey work how old paper maps stored in drawers had shrunk with respect to scale, in comparison to Mylar[5] maps that do not seem to shrink much at all. Over time, paper blueprints made from the original Mylar drawings will not match up to originals. I had read somewhere that the expensive rag paper that the dollar was made out of was "quite dimensionally stable." With the idea that they couldn't change much, left it at that. I might have figured out that this was not completely correct in time, but it finally dawned on me after buying some new bills at a bank, and doing some measuring.

Strictly speaking, the last group of discoveries above showing the Pyramid inch relationships came long after I had come to terms with the shrinking dollar, and so this is out of order in the sequence of discovery. But I think the reader is better served by an effort to make logical sense the facts, than to get tangled-up in the order of logic as I did in my original bumbling. At the beginning of my investigation, I was inclined to think of the dollar in terms of my survey experience. This was to see all good technical layout work in terms of strictly regular 90° rectangles and familiar units. And I had been very successful with regular inches, which are British inches. I concluded that the Designer must have used inches, and in a regular way. Yet among the first oddities I noticed, was that the lower edge of the dollar was strangely a trifle longer than the top length. This annoying irregularity was a puzzle I avoided for quite some time. Soon, I had reason to be much more annoyed. All of my measurements were wrong.

After happily bringing home my little sheaf of untouched, crisp, sharp new bills, all in sequential serial numbers--none of the numbers appeared to work anymore. (I think this was 1989 or thereabouts.) The dollars had the correct proportions alright, but they were *too big*. I thought there was a chance that they had picked up a lot of moisture in transit to the bank from the Treasury, or that there was a quality control problem in the printing for this batch. Other than the ones that I had originally used, I hadn't really looked at any other bills. When I finally divided the lengths taken from the new bills into the ideal length (DFT 5.655440261") to see how much difference there was, I discovered that they were too large by a *familiar fraction* ranging from 1.0011 to 1.00104, an average I later found to be somewhat more than 1 : 1.00107. The ideal ratio of Polar inch/British inch is about in this range. (Davidson published "1.00106" but probably used 1.00108, see Appendix A, pg 183.) Soon I realized much to my chagrin, that these dollars must have been originally etched and printed in Pyramid inches. But as we saw above, Pyramid inches left to decay into familiar British inches in time.

Eventually the fog of surprise lifted when I sorted things out and came to understand the facts. I was coming to live with the occasional feeling of disillusioned foolishness. Then came the awestruck realization that the measurements must be true in both units of inches, which seemed altogether suspicious. Strictly speaking, this could only be correct from only two perspectives: first, at the very instant after printing, (Pyramid units) and second, at the very end of shrinking, (British units) whenever that came to pass. I had read the "eighteen months life span of the dollar" in more than one place. This meant that the dollar's units would be slightly "incorrect" for 80 or 90 percent of the life of the dollar, only being usable in British inches toward the end. Yet the *proportions* would be correct all the way along. This vaguely set the stage in my mind for the discovery of the changing diagonals. If wasn't for the dumb luck of starting to work on old bills, much of this might not have been discovered.

Now I felt better: the dollar design was taking on the appearance of an idea as a logical, seamless whole. The Designer's brilliant plan must have started from the realization that the paper shrinkage ratio was about the same as the Pyramid inch/British inch ratio. As a printer, he would have been well aware of this effect--all paper shrinks. Next, *even if he had wanted to use only Pyramid inches,* he was stuck with eventual shrinkage into what become British inch -sized units. His genius was to make use of this unchangeable fact, and create a remarkable framework of lines that would work for both units, and preserve both units as time progressed.

****

Ten Dollar bills as Measured in 1996:

The following is a statistical table showing one reasonably new collection of one dollar bills, measured on June 5, 1996. These are close enough to the typical run of experience for illustration. These values will then be compared to the ideal lengths of the DFT theory and Davidson's Primitive Inch factor:

Ten One Dollar Notes, Horiz., Vert., Diagonal Measurements. Serial numbers, beginning L20697053F and ending L20697062F. Measured June 5, 1996

Series 1995:

No. Bill	Top.	Bottom.	R. Side	L. Side	UR-LL	UL-LR
(1) -53F	5.657"	5.664"	2.166"	2.160"	6.064"	6.062"
(2) -54F	5.662"	5.663"	2.162"	2.161"	6.060"	6.060"
(3) -55F	5.658"	5.6635"	2.162"	2.161"	6.059"	6.063"
(4) -56F	5.659"	5.663"	2.160"	2.160"	6.060"	6.061"
(5) -57F	5.658"	5.663"	2.163"	2.161"	6.061"	6.061"
(6) -58F	5.660"	5.659"	2.162"	2.161"	6.054"	6.059"
(7) -59F	5.657"	5.663"	2.160"	2.160"	6.058"	6.060"
(8) -60F	5.662"	5.667"	2.161"	2.160"	6.061"	6.062"
(9) -61F	5.661"	5.666"	2.160"	2.160"	6.061"	6.061"
(10) -62F	5.659"	5.662"	2.1615"	2.160"	6.051"	6.060"
(+)------						
Sums:	56.593"	56.6325"	21.6175"	21.604"	60.599"	60.609"
Mean:	5.659"	5.663"	2.162"	2.160"	6.060"	6.061"
Combined m:	5.6613" = x		2.1611" = y		6.0604" = dia.	

(The occasional half-thousandth seen above is an estimate, which if combined with other similar half-step estimates can improve the total measurement. These could be said to round-up or -down to an even one thousandth of an inch. This wasn't necessary in this particular case. The fourth digit seen in the final averages, sometimes called the "doubtful digit" of one place estimated past the finest unit of the measuring device, is included here in the Combined Mean of top and bottom mean values.)

Using the values of the DFT theory, how do these ratios compare to the ideal values of Primitive Inch conversion?

(m,t)	5.6613 / 5.6554 = 1.00104	(DFT LOR Length =	5.655440261)
(m,t)	2.1611 / 2.1602 = 1.00046	(DFT LOR Height =	2.160185958)
(m,t)	6.0604 / 6.0540 = 1.00106	(DFT LOR Diag.=	6.053958037)

With the exception of the vertical average, these are very much what might be expected from the early stages of shrinkage. In time these all will closely match the ideal in British inches, where the m and t will approach 1:1 to the ideal, where the abstract "Sun" will set in the West. Even at this stage--were we to just multiply the Combined Mean values by Davidson's ideal conversion of 1.00106--all will fit the DFT theory fairly closely. The following is an experiment for comparison using Davidson's 1.00106:

(m,t)   5.6613 / 1.00106 = 5.6553   (t / m - 1)/x = error of 1 : 56553   (Length)
One ten-thousandth difference: should be 5.6554

(m,t)   2.1611 / 1.00106 = 2.1588   (t / m - 1)/x = error of 1 : 1542   (Height)
1.4 thousandths difference: should be 2.1602

(m,t)   6.0604 / 1.00106 = 6.0540   (t / m - 1)/x = error of 1 : (infinite)   (Diagonal)
(no difference to theory.)

The differences seen in the Means (before creating the Combined Mean) appear to be more or less consistent in several samplings. But there does appear to be an *intentional skew* of diagonals and top and bottom lines, which we will look at elsewhere. Three things are notable here:

(1) An average of the top and bottom length *result in a number* that is apparently the ideal IDT length of 5.6554. (One could also say, that the ideal DFT length only occurs if measured across the dollar at the exact midpoints at the ending vertical lines of LOR.)

(2) The bottom line is larger than the top by about 0.004", (see Means above), being a little bigger than a tenth of a *millimeter*--which although quite small, will have certain consequences in the true geometry of the LOR rectangle;

(3) The differing lengths of the diagonals together with the four rectangular mean lengths of LOR, will allow computation of a quadrangle to make a revised, or the true LOR shape.

What remains yet unclear, is the purpose of the skew built into these lines. There may well be a message here.

Notes:

---

[1] Pg 109, Secrets of the Great Pyramid. With Appendix by Livio Catullo Stecchini by Peter Tompkins, 1971

[2] Pyramidographia: or a Description of the Pyramids in Ægypt. By John Greaves, Professor of Astronomy in the University of Oxford. 1646

[3] The modern calculation for the Tropical Year is about 365.24219 solar days, but this difference has little effect overall, so I have stuck to Davidson's version for consistency. He said that the choice of this number (365.242465) was based on a "good average for several Processional ages." These numbers can be re-worked with Davidson's base geometry using modern astronomical values, and the results will not invalidate the precision and odd coincidence of astronomical proportions found in this geometric relationship.

[4] The fractional exponent here is another way of noting roots. $2^{1/2}$ for instance, is the same as $\sqrt{2}$, or the "square root of two," and so on.

[5] Mylar is a commonly used polymer for map sheets since it changes very little dimensionally over time. Acetate, a much older and more primitive plastic, changes radically in size with changing moisture and heat, and is prone to warp.

CHAPTER 5: UNCOVERING OLD TRADITIONS—

Section I

Between 1980 and 1985, my research split into several odd directions. This was not a continuously dutiful study of the dollar. But I kept a small notebook and calculator in my top pocket, and occasionally worked out ideas amid other things going on. Many unusual directions turned out to be productive, often quite distant from previous geometry and measurements. There were a few moments of success from new measurement, but the big surprises came from research. My wife provided me with a clue for a radically new idea, a development that we will look at under "Logarithms" later in this chapter. (Judy and I married in 1981, and early in 1985 we left Chicago and moved to Prescott, Arizona. The first three chapters of this book were written in Chicago and Prescott.)

Discovery of other Date Numbers: Some "dates" of the American Revolution,

with a *larger form* of September 16, 1936--the Date Number of Davidson's Theory:

In the early 1980's, I noticed that the distance between the Eastern cornerstone of the little Pyramid and the Eastern edge of LIR, appeared to measure about 4.215", plus or minus one or two hundredths. This is about the *square root* of 17.76", an already familiar number. And the general form was familiar: a decimal one-hundredth division of the Declaration of Independence date.

Illus. O1

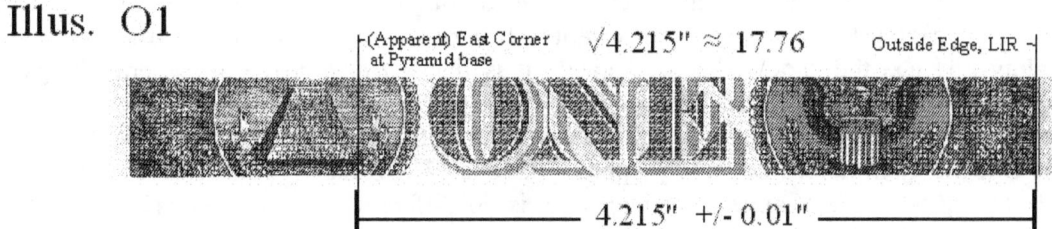

Although very exciting, the discovery got stashed away to work on later since I was involved in the ellipses of the Great Seal at the time. After a long stretch of procrastination I decided to spend some time on this idea. It was very likely another message technique, and almost certainly, several other important discoveries would be found along this line of investigation.

But right of the bat, I was factually *wrong,* if we measure precisely. At first glance--from a single measurement--it seemed that it had to be the square root of a tenth of the "Declaration of Independence date", the ideal of about 17.765. That date is on the base in of the little pyramid in Roman numerals-- what else could it could be? As usual, it wasn't simple, and wishful thinking was at the helm once again. It wasn't really this length: the correct dimension to the actual cornerstone point was more than one and a half hundredths of an inch longer (0.015"). Yet the true length turns out to imply another date: one every bit as interesting as the Declaration of Independence as we will see below. Due to the *obscured character* this corner point, the Designer might be said to have arranged to allow for *both dates.* The *actual point* of the corner of this little pyramid, noted earlier, is annoyingly blocked by an artfully placed plant or bush-like feature making direct measurement very difficult. This is the same annoying bush with the little "flower" we saw earlier in Chapter 3, in Illus. K2.

This was another ruse, evidently intended to force the examination of many mathematical theories and trial dimensions. The Designer wants you to read his mind, to guess what the intended plan was. This device allows multiple schemes to be displayed. It is a mechanism to force the suppliant investigator to look through all possible cases for significant meaning. The Designer strategically controlled the *width of the variables by graphic parameters.* He provided for various *kinds of possible answers* which are implied by inference, by the finding of mutually supporting references and by traditional clues.

This tiny muddled bush-area gives the Designer much latitude for "mathematical poetic license." A pair of mathematical ideas can be made to look like one another by size and shape. The *obscurity of where they end or begin graphically* can be a "rhyme" like a pun. The ideas could be like the LOR and LIR shapes, having *the appearance of being concentric.* Yet as soon as these rectangles are examined, fundamental differences appear in their designs. A large variety of measured dimensions to the corner under the little "bush" are likely from anybody's measuring tools. Maddeningly, the actual corner must be estimated. On top of this, there is a desire to "make something work" out of one's measurements--a source of error as well as legitimate inspiration. Since it appeared that since the Designer provided no "target corner point" or easily measured end point, it seems as though he may have wanted to be free to say more than one thing. Here he makes the investigator laboriously analyze and solve for an involved riddle having an exact answer. The real "poetry" of his work is where he makes the solutions both decidable and un-decidable *at the same time.*

There is for example, no dead certainty for the either the Independence Date Theory (IDT) or the Displacement Factor Theory (DFT) ideas. *Complete certainty* is veiled because these ideas are defined by immeasurably small differences. You are "invited" into tiny regions far beyond the measurable or the plausible. All we can say is that the *data will fit both theories, nicely, to an unmeasurably small level of precision.* But not incalculably small. Which is part of the problem: calculation can go far beyond what could prudently be inferred from measurement. And yet *we now know,* to a large extent, what numerical symbols were preferred, since some paths are traceable with substantial evidence. There is quite a lot for the Davidsonian DFT theory that appears irrefutable. Although we seem to have left the Declaration of Independence theory (IDT) in the dust with Chapter 4, *here* it will reassert itself in several places with great certainty. Hence the "poetic licence" idea: it is like a *freedom* to say several things within *a rigorous convention of similar measurements.* This like a rhyming poem, but calling for interpretation in some cases and *small amount estimation* to create exact identities. We must remember: the *observer* is creating part of the math here, from likely approximations. How many "puns" can be put into a verbal statement? The question, as it appears in language, is no different than in a numerical identity. Under some conditions a very large number of puns can be jammed into a carefully worded sentence. A geometric shape or an equation is like a sentence in mathematics.

In the measurements and associated mathematics surrounding the base of the little pyramid and to the ending LIR length, there are several apparently "legitimate" arrangements of *dimensions appearing in distinct groups.* These are *alternative groupings having a sum of the same total distance* (the LIR length, $\sqrt{31.5}$), each legitimate as seen from rules we have already discovered. These might be said to be like rhyming sentences, with the same "meter". And there are interesting symbolic distinctions as before-- right at the edge of measurability. One might observe several, slightly differing positions for possible end points from to compute, at the little pyramid's corner from the East line of LIR. Each could be a potentially important symbolic number like "date" or some other identity computed from a measurable length. As it turns out, there are a number of possible symbolic meanings can be proposed from various lengths ending inside this area under the tiny bush--many consistent with previous findings.

Probably, the Designer wanted only a small list of these identities to be seriously considered from bits of evidence, before one finds the ultimate answer to the Designer's riddle. Yet, the very *first dimension* anyone would like to know, or *need to know,* to start to solve this puzzle--was made a puzzle in of itself. This the base length of the little pyramid.

The base *length* is the key to the little Pyramid problem. This is analogous to the historical theory on the base length for interpretation of the Great Pyramid. Clues for an exact length must be looked at with regard to a careful estimate of the height of the apex: there is almost certainly an important ratio between these two. (This is a J5-argument: special or familiar proportions.) As a technical problem, this point was made as a *invisible apex;* it is a *virtual* intersection of the edge lines of the lower portion of the little Pyramid that must be measured first. This apex is about one thousandth of an inch above the apex of the Pyramidon and about a thousandth or so to the East of it (See Illus. O2).

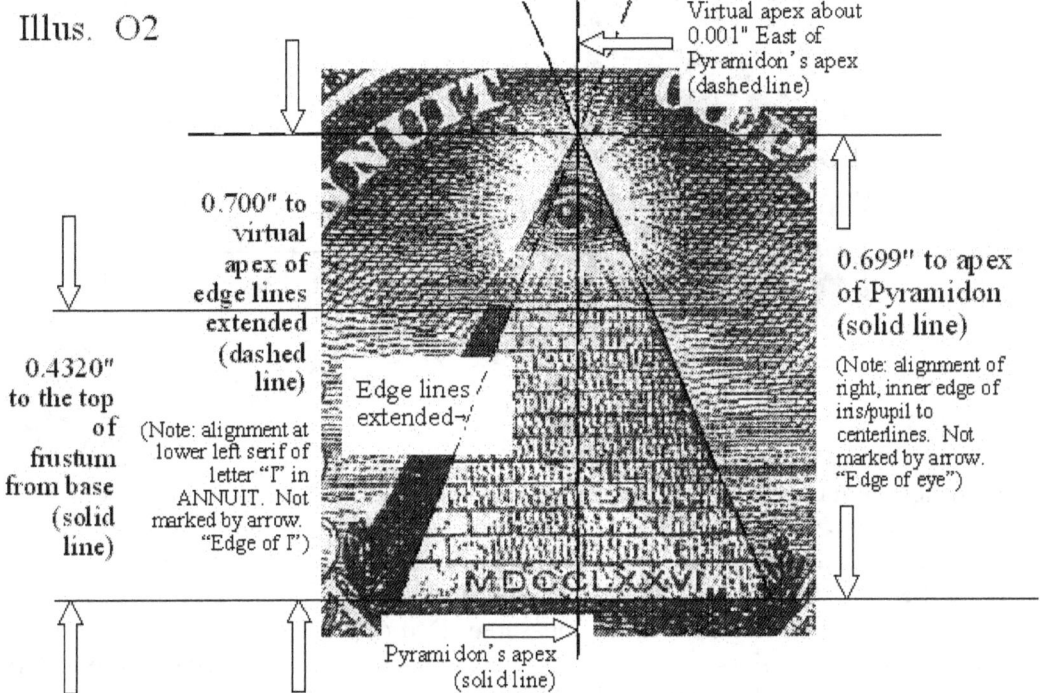

Illus. O2

Virtual apex about 0.001" East of Pyramidon's apex (dashed line)

0.700" to virtual apex of edge lines extended (dashed line)

(Note: alignment at lower left serif of letter "I" in ANNUIT. Not marked by arrow. "Edge of I")

0.4320" to the top of frustum from base (solid line)

Edge lines extended—

0.699" to apex of Pyramidon (solid line)

(Note: alignment of right, inner edge of iris/pupil to centerlines. Not marked by arrow. "Edge of eye")

Pyramidon's apex (solid line)

Franklin D. Roosevelt [1] ordered a *reversed* the position for the original Great Seal faces. In the *original* dollar, before the modified layout, this right-hand corner would have been a clean and sharp point on the *east side of the dollar, facing inward.* It would have been a much easier point to measure directly to as we will soon see. Since the little pyramid appears to be symmetrical in the Seal Faces *at the base*, this is a resolvable problem, provided the base length is *exactly known.* But this does not change the nature of the problem of the *unknown length of the base for either design,* since the *one corner* is always obscured in both designs and the problem for arithmetic is the same.

In terms of measuring, any base length would depend on a start from the sharp and clearly illustrated West corner stone point to the Westerly Edge of LIR. In true measurement, the overall length of this base might be too small to make an accurate numerical fix in distance, even if *both* sides were sharp and well defined in printing. This is because of small size, and uncertainties in the Eastern *edge line* of the little Pyramid and corner area, even if they were made fully visible. (See Illus. O3).

The Designer's plan clearly calls for one end (the West Corner of the little Pyramid: WC) to be fairly sharp and measurable, with the other, (or East Corner: EC) mostly blocked from view (See Illus. O2 and O3). But this EC corner is not without some graphic clues. It would seem to be a fairly simple matter to project the base and edge lines together to find the hidden end for a base length. Yet this problem has several small, significant difficulties. The Eastern edge line of the little Pyramid is sufficiently rough that *several* plausible, alternate base intersections can be found by a straight-edge or electronic graphics at the +/- 0.0005 of an inch range. The area is blurry at any magnification. Although any such intersection must eventually be solved by mathematical methods, the graphic clues from which we base measurements for calculations are unusually shaky. These do not provide for much better than a analog solution--a scaled measurement; an approximation. This is partly due to the physical smallness of these shapes with respect to the wideness of the lines they are made from. This gives a large built-in error of uncertainty. (And this uncertainty for the Eastern edge line is not only large at the bottom at the base line, but will be the same problem at the apex, also.)

Illus. O3

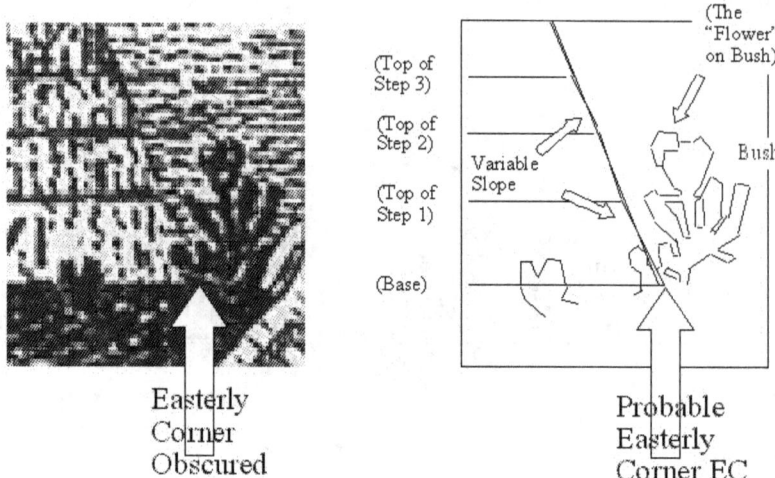

Easterly
Corner
Obscured

(Top of
Step 3)

(Top of
Step 2)

(Top of
Step 1)

(Base)

Variable
Slope

(The
"Flower"
on Bush)

Bush

Probable
Easterly
Corner EC

If these puzzles are looked at just from information theory, the measured dimensions are a lot like a telephone number or a spelled-out name. At only three digits for a decimal fraction, including another digit in front of the decimal point, many good guesses can be made to clues hidden by the Designer. The two main LOR theories (IDT and DFT) that are made from the length of LOR, begin with numbers having a theoretical precision *just beyond* the measured third decimal place. That's near our limit. Yet, guesses from as few as four floating digits may lead us directly to a special theoretical place, close enough to find meaning. And as observed earlier, one suspects the Designer mostly intended these dimensions to be taken to the third place. Then, with other subtleties, juxtapositions or arithmetic, his *exact ideas may appear* from research and calculation, showing some ideal or intended refinement that may extend into the fourth or fifth places past the decimal.

Strangely enough, the geometry problem of the slope/base intersection on the Seal Faces is not greatly improved by enlargement of the area. The problem of line width to line length error ratio doesn't go away with photographic expansion. The Designer undoubtedly began with a large scale plan for illustrations, perhaps blackboard sized in pencil, in which he had carefully crafted crucial obscurities and intriguing clues for target ideas to be discovered by measurement and calculation. These would then be scaled down and given to an engraver so that graphics could be etched in the original steel printing die plate for the dollar. My guess is this data would be no less puzzling if the original plan was seen. (It would be interesting to see his original group of source equations, however.) As before, the Designer would have found several good numerical symbols, all more or less the same general length, which he would have arranged to make for crucial points or lines *just blurred enough* to contain all potential possibilities. Ultimately, I believe only one of the ideas, or a coherent group concept is favored by the Designer.   In each case, a multiple references will emerge from careful measurement and calculation showing his intent.

****

152

The Problem Variable Structures:

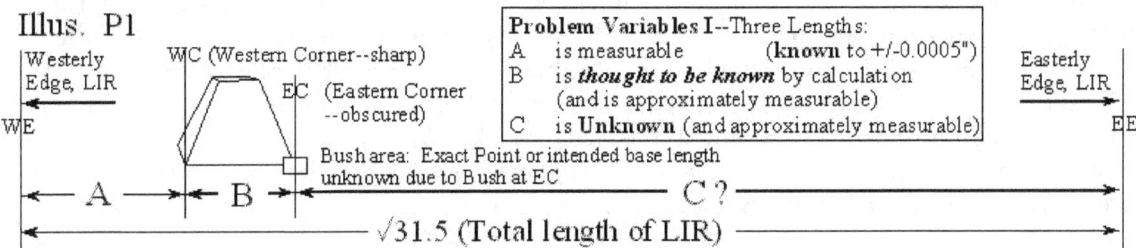

The design appears to force certain abstract logical forms: (I) *If you do know* what this pyramid base length really is (variable B), then you can measure from the clear corner on the Eastern side of the little Pyramid base to the East edge of LIR, A; then add this to the base length, A+B; and then subtract both from the length of LIR: √31.5 - (A+B). This ought to reveal the exact, hidden dimension C, from the invisible corner to the East line of LIR. (Could a lucky hit--like something similar to the C-length, such as a recognizable "date number"--demonstrate the correctness of a theoretical B-length?)

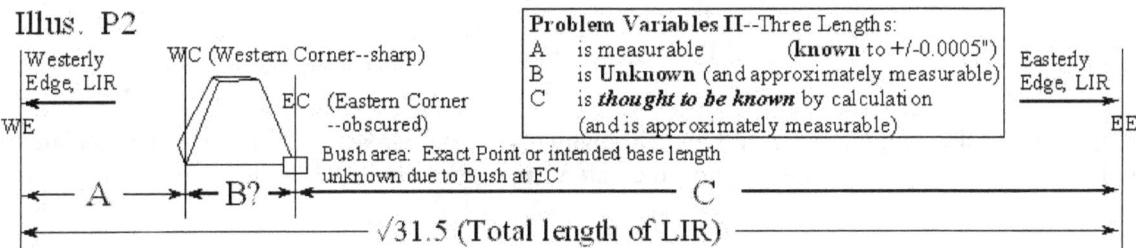

But alternately, the other, *opposite* case (II) could appear from similar problem conditions. For instance, suppose you *think that you know* what the long distance C was, from the hidden corner to the East edge of LIR. This could be added to the clear tie A to the West edge of LIR, (A+C), and then subtract this total from the LIR length √31.5 - (A+C) to reveal a length for the base, B. This might have meaning also. (Will your C-length idea show some validity of a theory for pyramid base B?)

Notice that there are generally *two* undetermined variables, (B *or* C plus A) with respect to *one known* variable: √31.5. This places the logical form in *an indeterminable situation* for the investigator, where the *two possible sets of answers* are always possible. A real answer, must answer *two sides* at once. Both variables *hinge* on the hidden point in the obscured area, like a arithmetical teeter-totter. No matter what we try, the answers must always bear *a reciprocating relation* to one another--a longish base: makes a shorter sum distance, a longish sum distance: a shorter base (Note a J3-arrangement). Any real solution, if there is any, must ultimately uncover *both unknowns*. As we know from experience, various interesting things can be hidden under incomplete clues, and that the Designer likes to hint at several types of solutions. (So far, we have ignored A as only a measurement, with no special symbolic identity. Soon, we will look at a third solution to this problem requiring a theoretical investigation of measured variable A, here accepted as "known.") A real solution would be "a best fit" of both possible answers for graphics with no more than a few ten-thousandths of an inch of error, based on mathematical results. In other words, one or both numbers B or C have to be guessed, where their sum plus A will always equal the total √31.5. (This suspiciously has the appearance of a quadratic problem, as have some of the other problems. But work done on these problems use *Newton's Method*, sometimes called "Cut and Try.") Our numerical solution must occur together with a philosophically convincing, traditionally persuasive, or otherwise connecting theme already known from the Designer's special interests.

In this stage of investigation, the very smallness of the shapes becomes an important problem because of the limits of measurement. More uncertain guesswork and mathematics are required due to ambiguous meanings that are possible in small shapes. The following steps show reasoning based on number forms we have already looked at in earlier chapters. Intelligent examination of the features within the Great Seal areas are probably not possible without having already solved a number the earlier riddles in this design of specific length and proportion. Without these discoveries, it is doubtful that much of the smaller picture can be convincingly explained. Having graduated to the next step, there is somewhat more difficulty and a greater reliance on educated guesswork. We are *now* somewhat more "educated" than before, and we now know some of the wily Designer's favorite numbers and some of his methods. These are powerful tools, much more useful now than before.

****

A Number Trail for a Connection to the Little Pyramid:

After several unsuccessful trial ideas, a more or less convincing trail appeared: an idea based on existing exterior proportions and data from the dollar. Eventually, manipulation of this information revealed a believable and reliable length for the base of the little Pyramid. This theoretical pattern starts by arriving at a height from the base to the frustum from familiar theoretical reasons: using the Davidsonian Displacement Factor Theory (DFT) length of LOR, I divided by φ^2 to obtain the end height of LOR:

(t)　　　$5.655440261 \cdot \varphi^2$　　$=$　　2.160185958　　(height of LOR per DFT)

Then, doubling this length, makes a dimension which we have already seen is already associated with LOR and two of the Designer's other themes to create virtual pyramid forms in the last two chapters.

(t)　　　$2.160185958 \cdot 2$　　$=$　　4.320371916

We have used this number form before. This is a number, when multiplied by π and divided by 4 that gives a variation of the BSC length, seen earlier.

If *Divided by 10*, however, it bears a very close resemblance to the measured, "unfinished height" of the little Pyramid, or frustum, which we saw in Chapter 3 measures vertically about 0.4320". This little Pyramid obviously does not conform to the Great Pyramid of Giza shape; it is altogether too tall and narrow. Its projected edge-lines appeared to intersect at a measured height of 0.700", and the tip of the Pyramidon was slightly West of and below this invisible intersection (See Illus O2 & P4).

I think the Designer ultimately used π to make the base anyway, but in a much different way than we might expect from the Great Pyramid. Here are some preliminary ideas leading up to the length finally accepted, the theoretical Little Pyramid Base length LPB3:

Earlier, we saw that the height was apparently related to the height by φ:

(m)　　　$0.699" / 0.432"$　　$=$　　1.61805　　($\varphi = 1.618033989$)

If instead, we then tried a *tenth* of twice the theoretical DFT end length of LOR, multiplied by φ, what would that tell us?

(t)　　　$0.432037192 \cdot \varphi$　　$=$　　0.699050862　　(little Pyr. Height #1, or LPH1)

This is the J5-argument: this certainly doesn't disagree with the apparent measurement to the apex. And, this is already a familiar number in one way or another; it is a *tenth of the cubic diagonal of the DFT box* (Chapter 4 pgs 108 and 110; Illus. N1 & N2), or twice the reciprocal of D/1000:

(t)　　　$0.699050862" / 2$　　$=$　　$0.34952543"$　　$(1/x = 2.861...)$

If this fractional form is adopted as a regular *height* of a Giza-type pyramid, we would multiply this height by 2, then by π, and then divide by 4. Yet this would make a base-side of about 1.098", which is too big, and it really doesn't appear to match anything graphically in the picture. (It will match something, but we don't see it just yet.[2]) If instead, we tried using the above number as a pseudo-height, where we skip the first multiplier of 2, we find a number that looks close and seems nearly the size of the base, possibly the odd proportions of the little Pyramid:

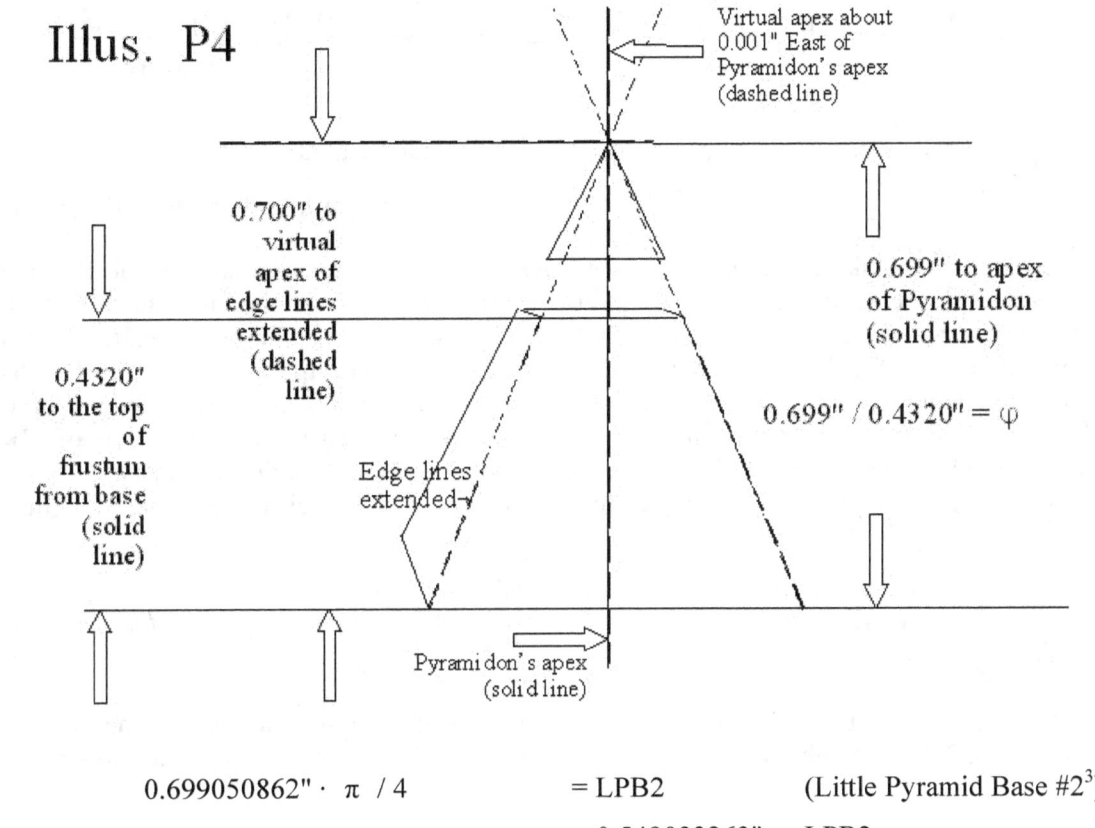

Illus. P4

Virtual apex about 0.001" East of Pyramidon's apex (dashed line)

0.700" to virtual apex of edge lines extended (dashed line)

0.699" to apex of Pyramidon (solid line)

0.4320" to the top of frustum from base (solid line)

Edge lines extended

0.699" / 0.4320" = φ

Pyramidon's apex (solid line)

(t)	0.699050862" · π / 4	= LPB2	(Little Pyramid Base #2[3])
(t)		= 0.549033263"	LPB2

Although this is in the range to the estimated base length, it is *still* a little small, by an easily visible six thousandths. This can't be it. The estimated 0.555" seen way back in Chapter 2 still seems measurably closer in size, considering bush and all, with less room for guesswork. The base must still be bigger than this, *but smaller* than the result of subtracting the square root of 17.76 from the LIR length to get to this position. Suppose instead, that we used ten times the apparent height (the full cubic diagonal of the DFT box), and *divided* by π instead of multiplying (note the J3-argument--dividing vs. multiplying):

(t)	6.990508619" / π / 4	=	0.556287"	LPB3	(Little Pyramid Base #3)

This is about a thousandth of an inch bigger than the estimated measurement of 0.555" for which I can see no objection--well within limits for estimation of this base by graphics. This is also a very intriguing number. For instance: if we look just above to the top line of this base course line, there is a horizontal top edge at the next step measuring about 0.500", presumably an intended exact "half-inch." (See Chapter 2, pg 63 Illus J1.) Were we to *subtract* an exact half inch such as this from LPB3 there is an interesting result appearing from the difference (see Illus P5):

(t,m)   0.556287 - 0.5   =   0.056287

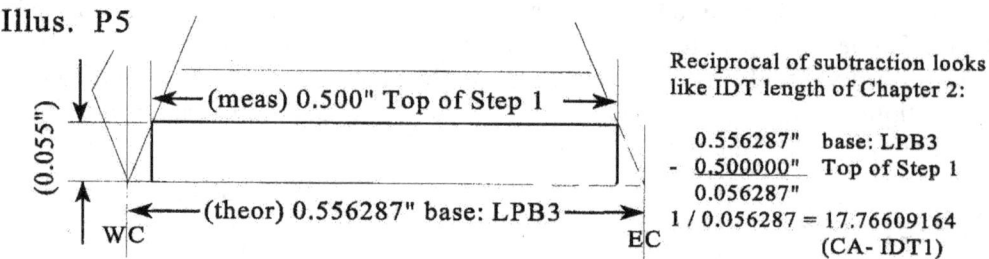

Illus. P5

(0.055")

(meas) 0.500" Top of Step 1

(theor) 0.556287" base: LPB3

W|C   E|C

Reciprocal of subtraction looks
like IDT length of Chapter 2:

```
    0.556287"   base: LPB3
 -  0.500000"   Top of Step 1
    0.056287"
1 / 0.056287 = 17.76609164
              (CA- IDT1)
```

Not so interesting a number in itself, apparently, but the reciprocal (J3) is:

(t)   0.056287 / x   =   17.76609164   (CA-IDT1 "Close Approximation to IDT #1")

Doesn't look much like coincidence, does it? Again we find a reappearance of the IDT idea *in yet another form.* Is this an explanation for why the Designer made the top of the First Step a half an inch or, 0.500" to cue a subtraction from the base? LPB3 has become a likeable number *already.* Was This is a Close Approximation of IDT #1 (CA-IDT1) all that the Designer was up to here? Not likely. There are lots of other intriguing things going on near by through the use of this number. Why *this* base LPB3? How does LPB3 as a length relate to the unknown distance measured from the East line of LIR? The number is well within the family of already familiar numbers at a tenth scale. For instance, ten times the reciprocal of this number proves interesting as the circumference of a circle (also see pg 113, Illus N4):

(t)   (0.556287 / x) · 10   =   17.97633236   then,
(t)   17.97633236 / π   =   5.722044308   (a diameter) then,
(t)   5.722044308 / 2   =   2.861022154   (a *radius or pyramid height.* The last
digit should be a "6".)

In other words, base LPB3 is a tenth of the reciprocal of the circumference of the circle having a radius of 2.861022156, a number often seen in Chapter 4. This does not prove this base theory, though.

****

Testing LPB3:

Measuring from the West Corner (WC) of the little Pyramid to the West edge-line of the LIR rectangle gives what appears to be 0.826". (This is "A" in Problem Variables I and II above.) If the above LPB3 was the actual length of the base, the horizontal location of the East Corner (EC) should be determinable by subtraction. (This is in the form of Problem Variables I plan.) The method for obtaining an unknown length is to use the shorter measurement to the near side, and subtraction from the theoretical LIR length, as before. But in this case as previously discussed, the long measurement is *blind,* or without a second check dimension due to the tiny bush. How far then, will this corner be from the West line of the LIR rectangle?

(m,t)   0.826" + 0.556287"   =   1.382287"   = A+B
                                                          then, subtracting from LIR ($\sqrt{31.5}$),
(t)   5.61248608" - 1.382287"   =   4.23019908"   = C

As measured by estimation from the East line of LIR, this number originally appeared to be about the square root of 17.76, or approximately 4.215" (plus 0.01" or more) somewhere in the bushes. But, this is a tiny bit larger number. What does it become if squared?

(t) 4.23019908^2 =   17.89458426

Here is *another* possible "date number." This could pass for one hundredth of the date of the *founding of the Republic* from the point of view of the date of the Adoption of the U.S. Constitution, on September 17, 1789. Numerically, the above result is close enough to make a historical date connection, with no further digging--considering the likely historical/political symbols. The square root of one hundredth of the correct date would be about 17.89712329, or "cn". (See "cn" the U.S. Constitution Number: computations in note [4]). From the precision values that I adopted, this is pretty close--and a measurement of 4.230" would be more than close enough at one part in 4200. Looking at the greenery-area of EC, it does seem close to a likely graphic solution for an intersection of the Easterly edge line and the base line, or the Easterly Corner (EC).

Illus. P6

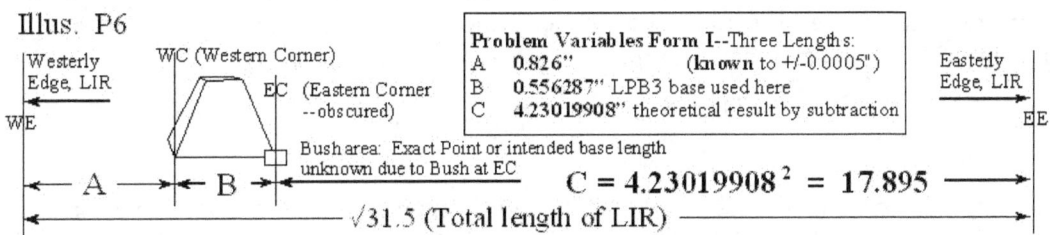

Now, were we to use the actual Constitution date (cn) instead, plus LPB3--backward mathematically, there is no appreciable difference in the resulting theoretical tie to the West line of LIR which computes to be a number (t) 0.825699902" [4]. This would be about *a third of a thousandth of an inch smaller* than the measured 0.826" or no measurable difference. If you think about it, this historical date *ought* be the base date of the little pyramid, if it is to represent *the actual founding* of the Republic. But with the Independence Date of 1776 in Roman Numerals on the little Pyramid as per the legal description of the Great Seal enacted into law (See pg 186 and 187), it makes sense that the Designer would resort to the "slight-of-hand" of shrubbery to hide what could just as well be both lengths at the EC corner. And this not to mention the above half-inch subtraction trick from the top of the Step that creates the reciprocal of 17.766 (CA-IDT1). There is *more* to the area ending under the bush, which we will look at later.

Discovery of a Special Number--A Digression:

Let's look at the square root of 17.76508197, or, the Independence Date number (IDT) subtracted from LIR, but without yet considering the next subtraction of the tie length (0.826") to get a base:

(t) $\sqrt{17.76508197}$ =   4.214864407 then,
(t) $\sqrt{31.5} - 4.214864407$ =   1.397621673   = a

This number, what would be a base plus a tie length, is *very interesting in its own right*, very close in size to another very interesting number, the double of the apex to base height or, LPH1:

(t) $2 \cdot 0.699050861$ =   1.398101724 (2h)   = b

Suppose instead we subtract this theoretical 2h number b from $\sqrt{31.5}$ and square it to find a "date":

(t) $(\sqrt{31.5} - 1.398101724)^2$ =   17.7610355

Which is still close, or at least within the "year" of the IDT date number. This horizontal "double the height" number (1.398) reappears later, but as a small, horizontal offset length, *also* at the base of a pyramid--this time *the actual Pyramid of Giza*. Curiously, this length is the same as the difference between Ludwig Borchardt's scribed line on the paved surface at the Pyramid, to the measured centerline, being 0.0355 m offset to the west: 1.398 Pyramid inches (See pgs 215, 216, 227, 229, 230 and 231 Appendix A; Fig 1-A, 1-B and Illus. 4).

What have we seen thus far? LPB3 was a derivation of the DFT form of the LOR exterior. We saw that subtracting LPB3 as a base for the little pyramid plus the measured 0.826" from the base to the west edge line of LIR, the square root of the date of the adoption of the constitution is found: 17.895. And there is no significant adjustment if we were to prefer 17.89712329, or "cn": 1/100th the 260th day of 1789--this rounds to the same number. And the top line of the first course (Step 1) of the base course of the little pyramid, subtracted from LPB3 gives a 17.76 number (CA-IDT1) as a reciprocal.

****

Evidence for a large "44-form" Date Number for September 16, 1936:

An important point similar to the base corner, is at the top at the frustum (EF: East Frustum corner). If the lower corner is important--certainly the top corner is also. This would be every bit as important design-wise for a "date" if the base corner was. When I measured to the East line of LIR it was found to be 4.401", and this works out to be the square root of 19.368801. Then with more care, and measuring from the more reliable, shorter direction route (under 2 inches) from the WE line, this was confirmed by subtraction from $\sqrt{31.5}$ (LIR) and then squaring:

(m,t)	1.2115" - 5.61248608	=	4.4009
(t)	4.4009^2	=	19.3679

The resulting number is very close to 100 · DN, having an error ratio of about one part in 24000 to the ideal 1/100th of the Date Number 19.36710383, as shown in the previous chapter. If we had any question before about the dn-numbers found earlier--this is a directly measurable, one step relationship through the square root function. The measured was about 4.4009", and the ideal should be 4.4008".

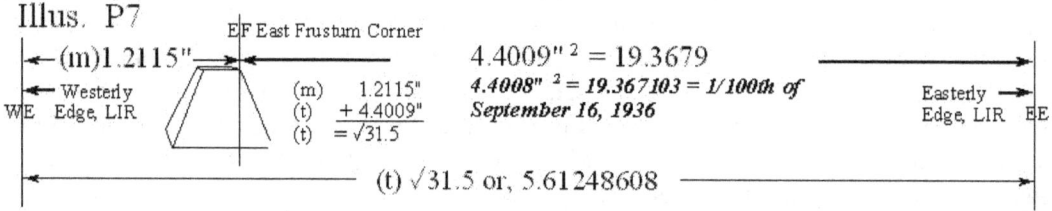

Illus. P7

So now, there are two good examples of known "dates" based on the Easterly edge of the little pyramid as measured to the Easterly Edge of LIR. This gives justification for wondering if the edges at the Step lines or the Thirteenth Line's position etc., have intended historical meanings.

****

1936 and the Bottom Line Length of LOR--A Digression:

While we are looking at Davidson's Number in this form, we should look at another connection on the dollar--as a side track in reasoning. Curiously, the one-hundredth of dn number form is related in yet *another way* to other graphics and formulas that we have looked at in chapter 4. Here is the *fourth root* of 1/100 · dn multiplied by φ:

(t)	$(19.36710383)^{1/4}$	=	2.0978100912, then,	
(t)	2.0978100912 · φ	=	3.39432803	(BSC7)

As a comparison, this new Between Seal Centers (BSC) number is numerically only 0.000158049" smaller than BSC2, which was a successful starting dimension we saw earlier. So, all of these number groups tie together to a large extent--at the very least to the fourth place, or around the ten-thousandth of an inch. Beyond this though, lies still *another* interesting relationship to look at.

The above BSC7 appears to be the start of a pattern, or a mathematical string, that may pertain to *the skewed LOR lengths* that we saw at the end of the last chapter. We know that the BSC-numbers (e.g.,

3.39) can be run through the π-based pyramid formula to make a number about twice the height of the end lengths of the LOR rectangle (4.32), a number easily related to the outside framework of LOR by φ^2.

Using the BSC7 number, a relationship can be found closely predicting the slightly long *bottom length* of LOR, the outermost rectangle of the dollar's design, as shown in the measurement Means in chapter 4. *Suppose* we multiplied BSC7 by 4 (a pyramid base), divided by π, and then divided by 2 to get a height:

(t) $\qquad$ $4 \cdot (BSC7) / \pi / 2$ $\quad = \quad$ 2.160896338 $\qquad\qquad$ (height LOR')

Then, create a new theoretical LOR length from this end length using the φ^2 proportion:

(t) $\qquad$ $h \cdot \varphi^2$ $\qquad = \qquad$ 5.657300059 $\qquad\qquad$ (length LOR')

The Mean of the measured dimension for *the bottom length* of the new 1995 bills (pg 147) is 5.6633". This, when divided by the pyramid inch factor 1.00106 gives 5.657303259 or 5.6573. Pretty much the same dimension. No appreciable difference as measured from the theoretical result just above. So this oddly long length as measured *could be said to represent the dn date, all by itself.* Subtle, but well within the odd parameters and known rules of engagement the Designer has already shown.

In the early examination of the top and bottom lengths of the dollar, there was only a consideration of an average of these lengths and this average was the accepted as the length of the LOR rectangle.

What about the top dimension which is shorter?

One of the conclusions that I came to earlier about the top and bottom lines as measured, was that *the average of the two lengths at top and bottom* gave the DFT length of 5.6554. The formula for the average is both lengths added together and divided by 2. Which *also* means, that the sum for the average before dividing by 2, *would the same* as 5.6554 multiplied by 2. If we subtracted this somewhat larger number (5.657300059 or length LOR') rather than the average from the average doubled, we would expect to get a somewhat *smaller number* than the average as a result. Since the Designer has used average-constrained tactics before, this could be of interest,. Suppose we subtract the above theoretical height LOR' multiplied by φ^2 from $2 \cdot$ DFT length:

(t) $\qquad$ $(2 \cdot 5.655440261) - h \cdot \varphi^2$ $\quad = $
(t) $\qquad$ 11.31088052 - 5.657300059 $\quad = \quad$ 5.653580463 $\qquad\qquad$ (a)

How does this compare with the somewhat short top dimension as measured? From the measured Means of the last chapter, (pg 147) the top is 5.6593". If we divide by Davidson's Pyramid inch:

(m, t) $\qquad$ 5.6593 / 1.00106 $\qquad = \qquad$ 5.6533 $\qquad\qquad$ (b)

Then, subtracting the measured length from the theoretical:

(m, t) $\qquad$ a - b $\qquad\qquad = \qquad$ 0.0003 or, about 3 ten-thousandths of an inch.

We will then be able to add the two lengths to obtain the average DFT 5.6554". Pretty close fit to this theory: ABL, or, Alternate Bottom Length theory. The difference of theory to measurement is about one part in 28,000.

Perhaps the Designer also intended to leave evidence of his devotion to this special date in the outer framework of LOR. Although this is a more complex relationship than the square of a measurement to the frustum corner on the top of the little pyramid, it is more smoke from the same fire.

We should note the gathering of evidence of a repeatedly used puzzle format, a specialized form of the J1- and J3- formats. This form is:

(1) a doubling of a graphic linear value,

(2) setting the whole slightly it off center, or giving either of the "doubles" an tiny, extra length,

(3) causing the calculations with significant answers to use the doubled average in some way,

(4) making the tiny difference at the center (or edge) crucially important to the meaning of the calculation,

(5) and often hiding a larger value in a small distance or tiny difference, sometimes as a reciprocal of some important number.

This last form is interesting here, since the top and bottom lines of LOR have the same bilateral symmetry in a vertical, rather than horizontal form. Here we see the Designer's puzzle method unfolding, and a display of one of the many ways that he hides numerical messages.

****

What we have seen in this segment, is the clearest example of "September 16, 1936" as a simple square of a directly measured, LIR-based length. And related to this, a J3-connection to the LOR length of the Displacement Factor Theory (DFT) by way of a golden ratio squared calculation. (The last segment was inserted here rather than the last Chapter, since it required some prior discussion of Polar Inch units.) In the next segment we return to the previous trail to test possible Problem Variables from another position, an alternative solution.

****

Now, a swing in another direction on the mathematical teeter-totter, whose fulcrum is the uncertainty of the bush-area. The result of this favors the Declaration of Independence Date number-form: Testing LPB4:

Lets have another look a the "17.765 date", the initial argument at the opening of the chapter, a point that hides under the bush and not at the corner EC. This problem is now considered under the "Problem Variables II" form. Here we will try $\sqrt{17.76508197}$, the IDT number, as variable C. What happens if we try to guess the base-length B, from just the "knowns" A+C subtracted from LIR?

(t, m, t)	$\sqrt{31.5}-$	(A+C)		= B
(t, m, t)	$\sqrt{31.5}-$	$(0.826+\sqrt{17.76508197})$		= B
(t, m, t, t)	5.61248609 -	(0.826+4.214864407)		= 0.571621673

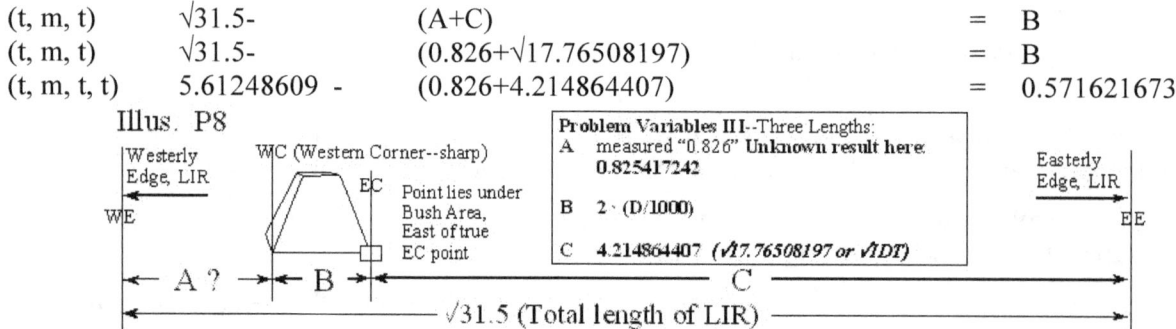

Illus. P8

It looks a hundredth and a half *too long* for the base as measured, yet--as we saw earlier--it is about in the right place, under the bush. But, what makes this really interesting is that this number B is really very close to *twice 0.2861*. This is a familiar number, a number seen just above. *Very suspicious.* What if something is being "said" by a variation which is *near in size* but clearly some distance by measurement? What should now be tried, is to accept the exact form of *twice 0.2861* as B, together with C as above to examine the dimension *A.* Now what we need is a *Problem Variables III* condition to interrogate dimension A:

What happens to our idea of the *measured dimension* A, were we to accept B (as $2 \cdot D/1000$, hereinafter **"LPB4"**) added to C (the IDT-form or 4.214864407 above) and subtracted from the $\sqrt{31.5}$ length of the LIR rectangle?

| (t, t, t) | $\sqrt{31.5}-$ | (B+C) | | = | A' |
| (t, t, t, t) | 5.61248609 | (0.57220431+4.214864407) | | = | 0.825417242 |

What we found is the measured tie-length is well within the error range to be expected. Subtracting the original A (0.826") from A' above gives about 0.0006". A' is very similar to variations of this length considered above.

What is especially interesting is that A' is similar to *the reciprocal* of the subtraction of the square root of a one-hundredth of dn as seen just above. (5.61248608 - 4.400807179 = 1.211678901, then consider:

1 / 1.211678901 = 0.825301157, called "a"). This suggests a longer mathematical string of dimensions, like a longer equation statement.

What happens to number C *if all three exact, theoretical numbers are used?* This should make a final subtraction making a C-type number very close to the square root of 17.765 or an IDT-number:

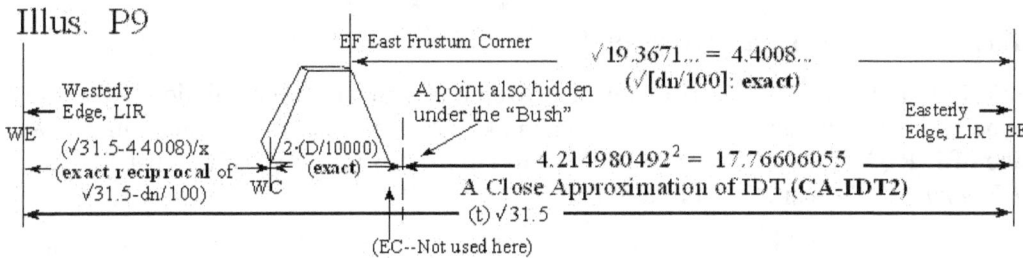

Illus. P9

An interesting result of this trial, is that the three theoretical numbers: (1) the Date Number; (2) a pure reciprocal of the exact 1/100th of the Date Number; and (3) twice D/10000 (LPB4), can be used to create a very close version of the IDT length if squared. (This would be Day Number 221, or, August 8, 1776.) But notice the astonishing similarity to the "top of the Step 1 subtraction and reciprocal -trick" or:

1/(LPB3 - 0.500"), or 0.056287 / x = 17.76609177. (This would be Day Number 222: August 9, 1776).

This is a numerical coincidence on the order of one part 570,000--suggesting that these were a careful arrangement and not happenstance. One must wonder at the nearly exact identities of these numbers in both setups. (Is the date meaningful?) These suggest the Designer might have used a common algorithm to layout these numbers. The presence of the "Close Approximation to IDT" theoretical Length (CA-IDT2), and the nearly identical number resulting from the reciprocal of the subtraction of LPB3 from 0.500" (CA-IDT1), it would not be unreasonable to think *both* the LPB3 base theory with Constitution Date theory (cn) were intended together. Both the CA-IDT1 and CA-IDT2 displays with the LPB3 or the alternative LPB4 base appear to be a single, complicated arrangement. From a comparison of these two CA-IDT numbers, it could almost be said that the Designer was giving a kind of proof or approbation for both schemes--since *both appear to implicate each other*. Reciprocating displays:

From Page 152 to 163, we have seen evidence of a two-sided scheme hinging on the EC/bush area that:

(I) Allow the Designer to refer to the Adoption of the U.S. Constitution in 1789, by the square of a number C measured from the use of the base LPB3 by subtraction with respect to the east edge of LIR. (LPB3 can easily be derived from the Displacement Factor Theory, DFT) Then, we can subtract the half-inch top line of Step 1 (0.500") from the base LPB3, B, take its reciprocal and find 17.76609164 or close approximation to the Independence Date theory: CA-IDT1.

(II) Alternatively, by use of twice one-hundredth of the displacement or 2 · (D/1000) as another little pyramid base LPB4, (B) plus the *reciprocal* of the square root of one-hundredth of the Date Number for September 16, 1936 (1/100 dn), (A) this sum A+B subtracted from of $\sqrt{31.5}$ (LIR) gives a number that is the square root 17.76606055 or CA-IDT2. Difference error = 1 : 569,097.

Both sides of this mathematical teeter-totter use parts of each other; both use elements we have already seen elsewhere in the investigation; and both are strictly logical constructions. We could even say Day Number 260 has been implied again, once for year 1936 and also for 1789. This is an enormously clever and very difficult puzzle arrangement.

****

Reciprocal Parts of the Little Pyramid:

This little Pyramid is full of surprises, most of which appear out of calculation. One the J3-oddities that struck me as interesting and possibly fundamental, is the fact that measured from the WC corner to the West line of LIR (0.826"), turns out to be close to being *the reciprocal* of the distance measured from the *upper EF corner* to the same West line of LIR:

(m)   1 / 0.826   =   1.2107"   (EF to West line edge of LIR)

Then, if subtracted from the LIR length $\sqrt{31.5}$ and squared for comparison:

(t) $(\sqrt{31.5} - 1.2107)^2$ =   19.376

Illus. P10

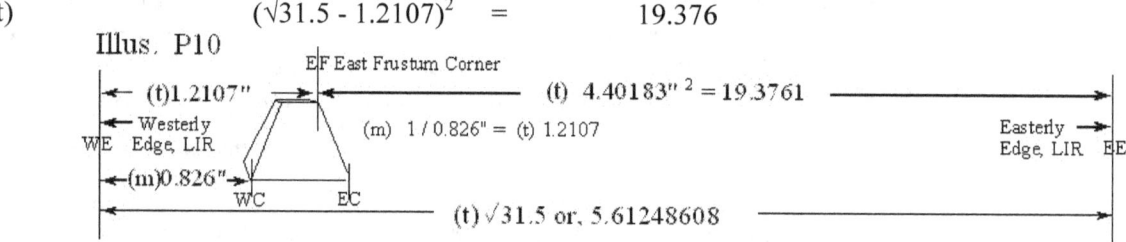

What happens from the reverse direction in the mathematics, using the ideal Date Number (dn) as a start? What does this theoretical "tie-length" to the West line look like?

(dn/100 = 19.36710383)

(t) $\sqrt{31.5} - \sqrt{(dn/100)}$ =

(t)   5.61248608 - 4.400807179   =   1.211678901   then,

(t)   1 / 1.211678901   =   0.825301157   = a

Not very far from the measured dimension. We should also note, that this doesn't disagree much with the computed check dimension from note 4, Page 157. What happens to "a" using the dn/100 from the east, then subtracted from $\sqrt{31.5}$, then the reciprocal?

(t)   b   0.825699902"   Check dimension, note 4 based on 17.897

(t)   a   0.825301157"   Last reciprocal, above

                       (-)------------------

                       0.00039"   or, about 4 ten-thousandths of an inch difference.

The elements a and b here, represent two terminal limits of the reciprocal ends of the above equations. As it turns out, the *average* of the two above numbers a and b, 0.825500529, has the effect of being able to produce 17.89881022 from one direction at the bottom, mathematically, and 19.36967961 from the other at the top, through the same processes used above.

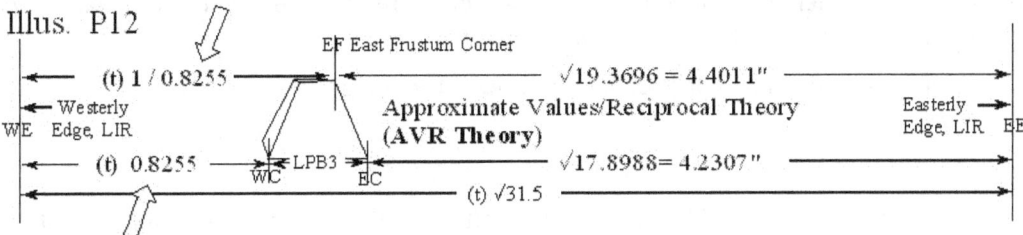

Illus. P12

EF East Frustum Corner

(t) 1 / 0.8255 — √19.3696 = 4.4011"

Westerly Edge, LIR — WE — Approximate Values/Reciprocal Theory **(AVR Theory)** — Easterly Edge, LIR EE

(t) 0.8255 — LPB3 — WC — EC — √17.8988 = 4.2307"

(t) √31.5

Which is to say, this looks suspiciously as though the Designer might have planned the graphical dimension to be exactly 0.8255"--(at the half thousandth of an inch)--looking suspiciously like he meant to create wide flexibility that would allow open conditions for this family of related puzzles. If this was the intent, from the WC to the West line of LIR, he could:

(1) tie in *both* special "dates" *to both* of his Pyramid corners (EC and EF) within a rough "date" region to an accuracy of a few ten-thousandths of an inch at one part in 7,000 (J7, J6),

(2) *and* use the special, internally related LPB3 base length with complete precision (J5),

(3) *and*, that all of this would be uncovered through the use of their *reciprocal* values using the *bottom left* and *top right* corners of the little pyramid frustum (J3, J5).

But beyond this--this arrangement will allow for rational speculation that is concordant with all above theories. Notice the AVR argument *is about reciprocals--not dates*. This is the theory's simplest form.

There are many more possible subtleties within this idea not fully developed here for lack of space. The approximate values and reciprocals theory (AVR) is a scheme that seems very attractive as a possible solution to some of the little Pyramid's graphic geometry. This is based on what we know of the Designer's signature J3-reciprocals and previously discovered number forms. The AVR scheme seems to provide a variety of useful lead-ins of interesting clues that may have been intended to initiate interest an investigator familiar with outer aspects of the design. From *reciprocals* the investigator might in time find that much of the message can be guessed from these clues, or otherwise backed-in mathematically as was done above. But there is a lot going on here, of enormous subtlety and craft.

Some Tentative Conclusions about AVR and the CA-IDT schemes:

The previous puzzle problems have variables small enough to be *bracketed within* the graphic approximation of the AVR scheme. The Designer, fully aware of the two principle solutions that an investigator could derive from these clues, needed a slightly ambiguous framework to string them out on. He settles on one that he liked that fit the J3-format of reciprocals.

What was necessary then, was to add a shadowy bush area, to visually close off the one and half to two hundredths of an inch *east* of EC so that the 2 · (D/1000) width could be thought of as an equally possible mathematical argument for a base length. And *then*, he made the top of the first Step exactly half an inch, 0.500", so that the subtraction/reciprocal -thing performed with LPB3 would confirm both number results, (CA-IDT1 and CA-IDT2) of either of problem condition.

Possibly there was an unknown, ideal CA-IDT number underneath CA-IDT1 and CA-IDT2. At some point early on he would have noticed this subtraction/reciprocal relationship in LPB3 related math (maybe even before the design of the dollar, in the planning stage), and then chose to build in the 0.500" number as a graphic feature as the top length of the first Step, where it would be near the base--where it might be used in a subtraction. The so-called teeter-totter of double solutions might be unresolvable if it wasn't for the placement of this graphic 0.500" length.

The subtraction of 1/2" or 0.500" subtraction/reciprocal relationship provides an inescapable connection to the LPB4 alternative base subtraction, also making a connection to 0.556287 or, LPB3. The bare placing of the 0.500" length at the top of the first Step allows the Designer to *lock both solutions* into giving a sort of validation of the other.

These Problem Variables are complex and subtle "structures" which have yet to reveal all of its secrets.

****

There are other ideas to be found here, and surely others will locate other relationships that I have not seen. When I look up from this forest of relationships, maybe the most interesting thing that occurs to me about the general whole, is the rigorous logic and attention to detail. The Designer sticks to his plan and modus operandi through thick and thin, and had faith in the power of later investigators. I'm fairly sure that I am missing many things here, but I am certain that those more gifted in math and investigation will have even more success than myself. (We will later see hints of a similar problem rotated 90°.)

Still, this has mostly been a technical exercise, and something is being said if we could only discern it. What does this all mean? What can be safely extrapolated from all of these symbols and numbers? In some places the Designer appears to almost wax lyrical with geometric praise to the Deity, and can certainly be said to have religious ideas. There are significant alignments to the All Seeing Eye, the letters HE, and so on, as noted earlier. The 2861 number has now been seen in perhaps as many as 15 different forms, both direct and quite subtle. For the Designer, this number must be a sacred abstraction, blending the lore of Masonic, Pyramid and Christian themes. Does he have a historical, prophetic or political scheme?

****

Some Interpretation of the Mathematical and Graphic Symbols:

What do the numbers mean? It seems evident, that the Designer considered of the beginning of the United States as being the adoption of the U.S. Constitution in September 17, 1789 by Congress, even though he makes you work to come to this conclusion. Additionally, the relationships in his setup, using the length of the LIR rectangle, the bush, the little Pyramid base length and a larger pseudo-length (2·D/1000) allowed him to include the 1776 number in some interesting mathematics. These numbers 1789 to 1936 leave little doubt the ascending courses of the Pyramid are time-related as graphic symbols of the United States' progress toward Divinity, or *The New Order Of the Ages,* represented by the All Seeing Eye. Curiously, both dates occur on the same Day Number 260, (September 17, 1789 and September 16, 1936). This is hard to ignore--perhaps this is fundamental to the Designer's message.

The triangle or Rock above the little pyramid now appears to be an allusion to an Ideal Society, perhaps a perfected republic, by extension of the lower frustum of the pyramid which can now be connected to historical development of the United States. But the New Order form will be distinct from the old, the geometry of the Pyramidon is very different from the unfinished pyramid. The graphic design seems to say that the Old Order, *made by Old World construction,* ends with the unfinished top of the pyramid. And after some sort of interim, a *Divinely formed* good order (the ideal Pyramidon) replaces the old in upward development. (The Pyramidon idea appears to be related to the element of the Latin motto "NOVUS ORDO..." or "new order". Later on we will see a possible mathematical relationship taken *directly* from these Latin words, apparently followed as a mathematical symbol. This is seen echoed in the little pyramid *alternate base* LPB4 and the *reciprocal of the apex height* of the Pyramidon.)

These symbols remind us the Book of Daniel, where there is the well known vision of the Great Image, an allegory of the Empire of Babylon. His vision is a scheme of the world's long range development. In many interpretations, this Image is a gradual corruption of religious and political order that will dominate some point in the future. In this vision there is a small rock *"cut out of the mountain, not made by human hands"* that somehow destroys the Great Image, and later becomes a great kingdom, "...that shall stand forever." (Daniel 5, 44 et seq.) The Great Image is said to be made of Gold, Silver, Brass, and Iron mixed with Clay. These four elements of the Image are allusions to a time progression of historical empires into corruption, and the final "Iron mixed with Clay" is said to be our times. The *small rock* then breaks-up the feet of Iron mixed with Clay. The vision appears to follow a prophetic form similar to the primal David and Goliath story: David collects "five smooth stones" from a brook and defeats the Giant Philistine. (i.e., the small rock that takes down the Giant was "...not made by human hands.") The Giant, like the Great Image is a symbol of the towering profane evil challenging God's people. This is the first act, the first move at the source of the prophetic, never-ending kingdom of David. Is the tiny Pyramidon intended to be a symbol of the small rock, not made by human hands?

Beginning in September 17, 1789, the little pyramid Step courses come to an unfinished point that is pretty clearly intended to represent September 16, 1936, the date supplied by Davidson's writings, making a period of exactly 147 years.[5] But, the United States did not come to a stop there, nor can we say that was there any noticeable change at that time. Although the U.S. was in indeed *unfinished* at just that same symbolic tier, exactly at that place in time, just as the dollar was being made: it still looks as though the top edge of the frustum is supposed to signal the *end of a specific period*, as seen from the Pyramid lore. But, the graphic pattern seems to end there at the interim. Due to the radically different character and altogether different angle of the Pyramidon, which does not follow the same angular pattern or center of alignment, nothing beyond the rock base ending at 1936 can be extrapolated with any certainty. Except perhaps for the topmost intersection point of the invisible apex/centerline of edge lines, the Pyramidon appears to signal Divine construction.

The top of the important Sixth Step gave about 4.318" from the East line of LIR, which when squared would be 18.645: this would translate to a date near the turning point of the Civil War, something like June of 1864. We cannot really know if this exact date is meant, there being far too much measurement error in these small dimensions. But I think it is safe to say, that this could easily point to some historical message or commentary, what with the important events of that time. As for generating a historic date here, unfortunately we do not seem to have any celebrated (and solidly technical) check-in points, as we have had for the Declaration of Independence date, the Adoption of the Constitution, or The Entrance to the King's Chamber. And we do not know what the Designer would have thought of as significant for this point in time. Direct measurement is probably not the answer, there being too little precision or repeatability at this small scale of measurement. Of course, there may be no intended "date" meaning here at all. But, if the apparent 0.432037192" is the intended height of the frustum, this length could be pro-rated between the two dates, and *if* a rigorous pattern might be found for the various Steps--meaning each Step could have an exactly computable "date."

But this is pay-dirt in a way; because *if there are* exact mathematical messages marked along the edge, we can now use the dn- and cn-numbers as lengths, as a way to mathematically anchor the little Pyramid's slope angle against the height to the frustum. (And, as we saw earlier, there may be some alternate frustum heights, or other possible angles.) I have no doubt the Steps could turn out to have some sort of intended meaning in a historical sense, if the exact mathematical positions of these steps can be determined. What seems to be required is a convincing solution to the three separate series of Step heights together as a whole--and if possible, a solution that clearly ties into other known elements and mechanisms of the Designer's schemes. (But again, a note of caution here--keep all historic or prophetic extrapolation ideas "at arms length," that is: as theories, only. All of these puzzles were obviously intended as solvable mathematical puzzles problems first and foremost. After which, the riddles may finally yield up interesting kinds of information--very possibly both political commentary and prophecy.)

****

The Pyramidon:

Like the little Pyramid, the Pyramidon presents a real measurement problem, since the line width to shape size make for a much larger error ratio of uncertainty than before. And in this case, the problem may be larger for another reason. This must be the crucial display of the Designer's art, and though we would like to know what it's all about, it may prove very difficult to figure out. At this time it is unsolved. Even though the tiny triangle is a clean shape, and not obscured in any of its corners, the width of the graphic lines in a shape this size act almost as the same type of obstacle we saw in the "bush." At the level of precision that we are seeking, the corner points of the Pyramidon appear rounded and not sharp. But, following already established practice, we measure as best we can, compute and analyze and see if we tell what the Designer had in mind.

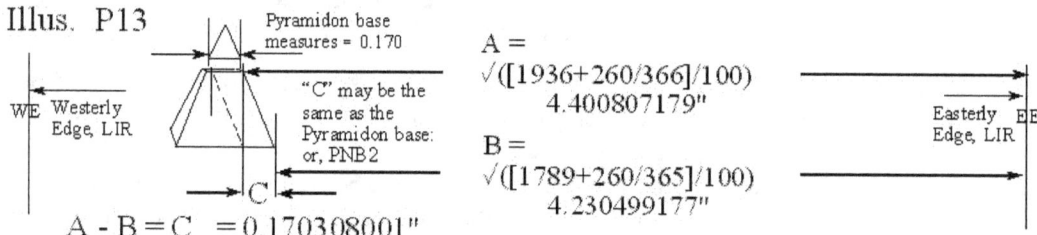

Illus. P13

Pyramidon base measures = 0.170

"C" may be the same as the Pyramidon base: or, PNB2

WE Westerly Edge, LIR

$A = \sqrt{([1936+260/366]/100)}$
4.400807179"

$B = \sqrt{([1789+260/365]/100)}$
4.230499177"

Easterly Edge, LIR EE

$A - B = C = 0.170308001"$

The base of the Pyramidon measures about 0.170" or, PNB1 (Pyramidon Base # 1). The difference between the two lengths of the cn date $\sqrt{([1789+260/365]/100)} = 4.230499177"$ and the dn date length $\sqrt{([1936+260/366]/100)} = 4.400807179"$ is 0.170308001 (C), fairly close to the length of the Pyramidon base. The PNB-base may have been intended to be 0.1703" but there is no real way to be sure. We'll call this base PNB2. There are, however, some possible indications in favor of several possible lengths, for PNB--all of which are smaller than the above C, but are equally well supported by the apparent size of the base, and by other parts of the mathematical pattern.

Measuring from the Western corner of the Pyramidon (WP) to the West line of LIR gives an average of about 1.0136". Measuring from the Eastern corner of the Pyramidon (EP) to the West line of LIR gave an average of about 1.1825".

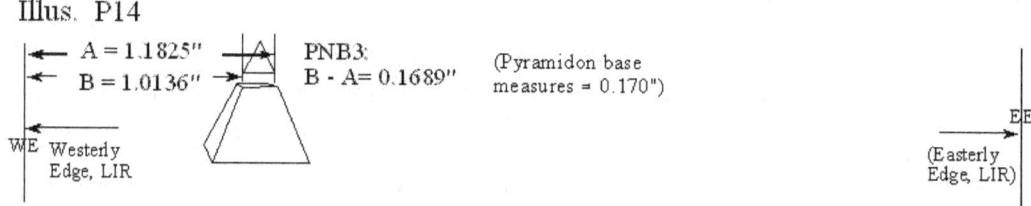

Illus. P14

A = 1.1825"
B = 1.0136"

PNB3:
B - A = 0.1689"

(Pyramidon base measures = 0.170")

WE Westerly Edge, LIR

EE

(Easterly Edge, LIR)

Subtracting to find the difference gives an apparent base of 0.1689" or PNB3. This is 0.0011" smaller than 0.170" a difference of about 0.00055" at either corner, about a half a thousandth of an inch. (The above figure is so close to a tenth of the square root of D/100, or 0.1691 that it should not be ignored, and will be assigned "PNB4", to set it apart.) The difference seems to be the eye's choice of places on the points to measure to: the outer edges of the figure, or from the center of the lines that the Pyramidon triangle is made up of, when taking longer measurements. It is hard to say in a small figure like this just where the ideal measurement should be taken from.

Two interesting things stand out here: the first is that the dimension "a" to EP is very nearly half of the reciprocal of the square root of 2.861:

(t) $(\sqrt{(D/100)})/2 \cdot 1/x$ = 1.182413515

166

Close enough to suspect that the measured distance to this corner as seen above, may really be this number. We can't argue with a ten-thousandth of an inch. But *this* is an important number, I think.

And, there is *yet another* suspicious similarity here--a textual one--to an observation made in Chapter 3 pg 74, where the New Testament is quoted to show the "Corner Stone is to become the Capstone" -idea. Is this a numerical representation of *the chapter and verse number* of Psalms 118:2 divided by 100? In a certain way of looking at it, *this is the most important number found here--it is the corner point of the Pyramidon.* If it is a textual reference as a message, then it is doubtful that a "date" is being signaled. Also note this number is the square root of 1.39810172, a distance from the West edge line of LIR to EC/bush area. (See pgs 157, 158.) Curiously, these numbers are already present in a crucial spot in the mathematics of the Great Pyramid, as shown in Appendix A, Entasis (See pgs 215 and 227).

The 1.1824" dimension terminates at the crucial corner point in the design (EP). Also the diagonal of the Small Inner Rectangle (SIR) passes through this point. This place is where the "shifted" or eye-aligned diagonal (Illus. E of Chapter 2, pg 47) intersects the right corner point of the Pyramidon.

A second issue that stands out, is the *average* of a and b, at a point that would be *the center point* of the Pyramidon base PNB:

(m)   (a + b) / 2   =   1.09805

Illus. P15

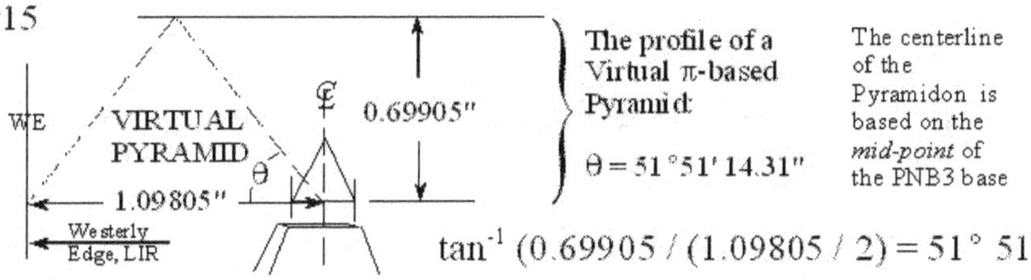

The profile of a Virtual π-based Pyramid:

$\theta = 51°51'14.31''$

The centerline of the Pyramidon is based on the *mid-point* of the PNB3 base

$\tan^{-1}(0.69905 / (1.09805 / 2)) = 51°51'$

We saw this number earlier as the possible base of a pyramid 0.699 tall (LPH1). In other words, measuring from the West line of LIR to the center point of this base, *makes one side of the theoretical pyramid* that should have resulted from the 0.699 height to the invisible apex--the thing we couldn't see just then.

This is however, *totally unmarked* with the exception of the *average distance* and to the West line of LIR implied height, here again an "averaging tactic" to make a hidden device open as a symbolic display.

****

The Kings Chamber Proportion Theory of the Pyramidon:

Through a combination of scaling of photographic enlargements and computing the internal angles of the triangle of the Pyramidon, I eventually came to two conclusions:

(1) The height of this triangle appeared to be 0.1748" which lead me to the conclusion that what was intended was (t) 0.174762715 ("h") is the *reciprocal* of 2 · (D/100), or 5.722044316, a number 10 times the theoretical form of LPB4 above. (i.e., a J3-reciprocal switch). And,

(2) With respect to the base, the rectangular position of the Pyramidon's apex appeared to be offset by *a ratio of* 0.53335 or, numerically very nearly *a tenth of the length* of the 13L line.

This lead to a search for a satisfactory base length for the Pyramidon that fit the observed measurements and the above two apparent theoretical constraints.

In the section beginning the Between the Seals Center discussion in Chapter 3, Page 51, one of the proposed widths was "The King's Chamber Proportion" which posited a length 3.400298666 based on the height of the SIR rectangle (1.520659792) multiplied by the square root of five, which although never found, seemed to be an attractive idea. (Here we are using the DFT theory as a of source the LOR length, to make a basis for an SIR rectangle.) Half of this, 1.700149333, if divided by 10--appeared to answer this search for a base of the Pyramidon. This is PNB4.

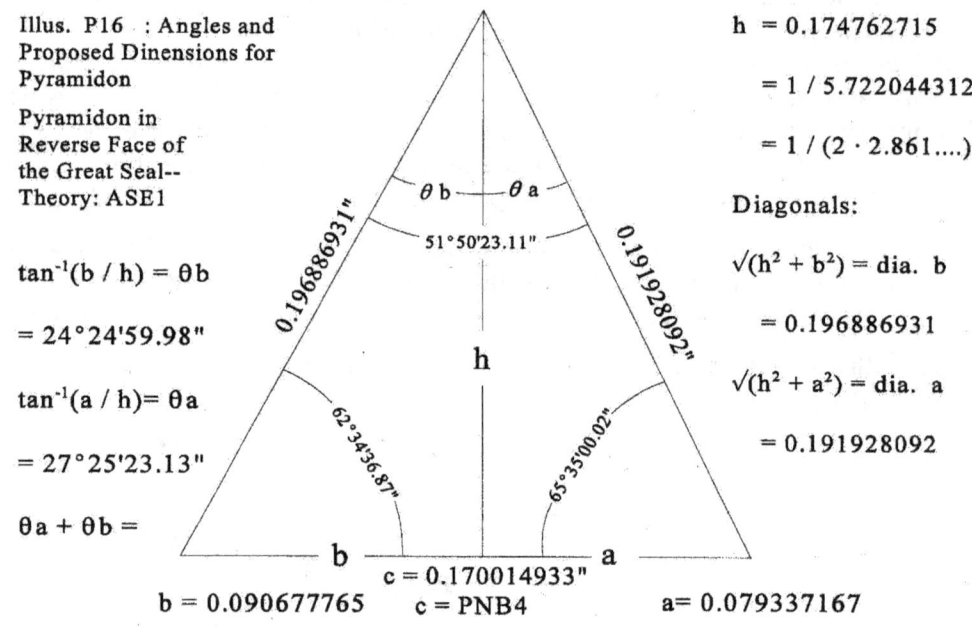

Illus. P16 : Angles and Proposed Dinensions for Pyramidon

Pyramidon in Reverse Face of the Great Seal-- Theory: ASE1

$\tan^{-1}(b / h) = \theta b$

$= 24°24'59.98"$

$\tan^{-1}(a / h) = \theta a$

$= 27°25'23.13"$

$\theta a + \theta b =$

θb —— θa

$51°50'23.11"$

h

$62°34'36.87"$ $65°35'00.02"$

0.196886931"   0.191928092"

b   a

c = 0.170014933"

b = 0.090677765   c = PNB4   a= 0.079337167

h = 0.174762715

= 1 / 5.722044312

= 1 / (2 · 2.861....)

Diagonals:

$\sqrt{(h^2 + b^2)} =$ dia. b

= 0.196886931

$\sqrt{(h^2 + a^2)} =$ dia. a

= 0.191928092

Adopting 13L3 divided by 10 as the actual ratio at the Pyramidon base for a vertical to the apex, a triangle could be produced, from the two resulting right-triangles. First I multiplied 0.170014933 by 0.533351773 (13L3/10: see pg 133) and obtained 0.090677765 "a" the semi-base to the East of the vertical. Subtracting this number from 0.170014933 gave 0.079337167 "b" for the semi-base to the West. It is then a simple matter of solving two right-triangles, attached as one oblique triangle:

(t) $\tan^{-1}(a / h)$ =   27°25'23.13"   or, θa the angle on the East side of the apex, then,

(t) $\tan^{-1}(b / h)$ =   24°24'59.98"   or, θb the angle West of the apex, then adding these to see the apex angle: a + b = 51°50'23.11", or within seconds of the angle computed for this Pyramid profile BSC3 in Chapter 3. (See Illus. L1, pg 78)

But this is only one of several theories. There is good reason to suspect that there are other possible implied solutions. Briefly: The *sum* of the above sides comes to 0.558829956, which seems close to the chosen base for the little pyramid of 0.556287. And scaling back to this length 0.556287 / 0.558829956 = 0.995449488 will give us a multiplier that makes an intriguing base for the Pyramidon:

0.995449488 · 0.170014933 = 0.16924128

or close to the square root of D /10000 or, PNB4 seen above. PNB4 is the value on the right side:

$0.16924128^2 = 0.02864261$ vs. $0.169145561^2 = 0.028610221$

But if the ideal value were true, the height of the Pyramidon would be *too small* to be the precise reciprocal of 2 · D/100 or 1 / 5.722044312 = 0.174762715. But (2 · D/100)/x seems like the intended number, considering all other clues available. Perhaps then, my calculation of the apex angle is somewhat off, or the diagonals are not quite right, or something else. (I am sorry to say I have never found a satisfying solution to this crucial triangle.) Beyond this, the height of the Apex appears to use 0.699 or the LPH1 number-form, with respect to the height of LIR:

Bottom edge of LIR to the base of the Pyramidon measures:	(m)	1.309"
the base of the Pyramidon to apex about:	(m)	0.1748"
		(+)---------
	sum	1.4838

(m) sum / (t) hight of LIR =
1.4838 / √4.5 = 0.699470028 vs.0.69905086 = LPH1 = (t) 2 · (1 / 2.861...)

I suspect that the Designer arranged the vertical axis of the little pyramid, Pyramidon and the LOR height to be a puzzle in the same fashion as the Problem Variables forms. (Mentally, turn the little pyramid, Pyramidon and total *height* of LIR 90° to the right and consider it as an analogous form to the Problem Variables apparatus discussed above.) I have not solved it, but we could look at one of the clearer of the numbers. The bottom edge of LIR to the base of Pyramidon measures about 1.309":

√1.309 = 1.1441154

compare this to Davidson's Capstone idea (see Plate XXIV, Illus. M8, pg 97) which would have a total base course of 4 · 286.1022156 = 1144.408862: 1000x = 1144.408862

Practically the same number if divided by 1000. Notice that this is pertinent in a symbolic way to Davidson's idea *as the base of the Capstone of the Great Pyramid* (See pg 97 and 98. J4, J5 and J7). Suppose we used the exact form of (4 · 286.1...) / 1000 and squared it, we will get 1.309671639. There is an interesting relationship within this number: if it were a circumference of a π-based pyramid--what is it's hight? First we divide (t) 1.309671639 into π, then divide the result by 2, giving 0.208440716. The square root of this number is *exactly an 1/8th of the side of the Great Pyramid* if multiplied by 10,000:

√0.208440716 = 0.456553081, then 0.456553081 · 8 = 3.65242465 · 10,000 = 365.242465

this theoretical number is the precise form of Davidson's 100 · base-course length of 36,524.2465 having re-emerged from the math from the use of Davidson's capstone base number-form. Here is a symbolic connection based on a hint provided from a scant four digit number 1.309 as the clue. This would doubtless be one of the *vertical* Problem Variables, perhaps analogous to the measured dimension "A" seen in the various forms above. Variable "B" sideways might be like the height of the Pyramidon, where a similar difficulty will arise at the Apex of the Pyramidon as was found at the EC point, such as PNB4 seen above. This point is also ambiguous in terms of measurement, even though openly visible, since it is small, rounded when closely observed and being the point of an oblique angle. The LIR height would again be the constraint: in this case now the length is √4.5 or 2.121320344. Although this is as yet unsolved, I have little doubt that a similar types of puzzle problems await the careful investigator.

****

The Symbolism of Top Three Steps of the little pyramid:

In Chapter three a claim is made to have found a solution to the arrangement of the top three steps of the little pyramid. This isn't completely true, or I should say more accurately, I am not completely satisfied with the solution. But (at least) the source of the arrangement of the top courses of the little pyramid can be seen below. Around 1990 in the midst of doing something else unrelated to the dollar, I noticed that some divisions of the height of the Great Pyramid could be made to form the Step/SIR fractions seen on the dollar. Here is the original table in the IDT theory from Chapter 3, showing the internal relations of these widths with respect to the SIR rectangle length. The "39.37 1/16ths of an inch"

(2.460625") is the length of SIR given here. Nowadays I would use the DFT-based number 2.460479229, although it won't make much difference in this case. Here the thickness of the Steps decrease going to the top:

Steps 11, 12 and 13 (measured); Divided into the (theoretical); SIR Length of 39.37 1/16ths of an inch:

Step 11 (m/t) 0.0225" / 2.460625" = 1/109.36
If the intent was a 1/110th: (t) *0.022369318"(0.0224")* = *(1/110)*
a difference of: 0.00013" being 0.001" in rounding at the ten-thousandth of an inch.

Step 12 (m/t) 0.0205" / 2.460625" = 1/120.03
If the intent was a 1/120th: (t) *0.020505208"(0.0205")* = *(1/120)*
a difference of: 0.000005" there being *no difference* in rounding at the ten-thousandth of an inch.

Step 13 (m/t) 0.019" / 2.460625" = 1/129.50
If the intent was a 1/130th: (t) *0.018927884" (0.0190")* = *(1/130)*
a difference of: 0.000072" being 0.0001" in rounding at the ten-thousandth of an inch.

What happens if we divided these Step-fractions into the Great Pyramid's height? This height is from Davidson: 5813.014373 Pyramid inches (36,524.2465 / π / 2 = 5813.014373). Notice here, the *opposite direction* of fraction identities going from small to large. The hinge-point in *this* teeter-totter is the middle Step 12. In the three numbers above and below the middle Step 12, two are about the same and the *top and bottom values are reversed* (J3) Compare *value order* in this table with the table just above.

110th of Great Pyramid's height (t/t) 110 / 5813.014373" = 0.018923056
(See 0.019" or *0.0189* above at Step 13)

120th of Great Pyramid's height (t/t) 120 / 5813.014373" = 0.02064334
(See 0.0205" or *0.02050* above at Step 12)

130th of Great Pyramid's height (t/t) 130 / 5813.014373" = 0.022363612
(See 0.0225" or *0.022369* above at Step 11)

Evidently, this is the source of these fractions, and there is an implication that yet another virtual pyramid is present somewhere in the dollar's graphics. If we started with 2.460479229 as a length of SIR, but used it as a height of a π-based pyramid, we will get a circumference of 15.45964694.

The *circumference of the LIR rectangle* is close: 2 · ($\sqrt{4.5}$ + $\sqrt{31.5}$) = 15.46761285 a difference of 0.007965907 at one part in 1,940. We should note the height of an LIR-perimeter π-pyramid numerically resembles the square root of the measured LOR diagonal (t) 6.0602 PI, (See pg 147). But these solutions do not lend themselves to any easy explanation as to why all of this was done in the dollar's graphics.

****

This is the end of Section I, Chapter 5, and the end of the measurement based work on the dollar in this book. I hope it has been interesting and fun. It should be evident in the last segments that there is at least as much work to be done in the vertical aspect of the Left Face of the Great Seal graphics as was found in the horizontal. Out of generosity or laziness I have left the east side dollar's mostly untouched. The intrepid investigator is to be assured that the Eagle's eye; hooked beak-tip; feather tips; the thirteen stars; arrow points; bottom point of the shield; the various and sundry Acanthus curlicues; points; balls and so on, all showed signs of being incorporated in the Designer's geometry. This is not to say that every thing has been accounted for on the west side of the dollar, or the multitude of unseen geometric relationships that cross the center. They are not shown here, there isn't room for them all.

In the next sections the traditional origin of several mathematical motifs are explored. Strangely, these are very old ideas, some stretching back to Federal Law adopted in the 1830's and at the time of the American Revolution. We will see in Appendix A, ideas found on the dollar appearing the Great Pyramid mathematics, some so unique that they are almost certainly from the Designer's knowledge of the details of the Cole Survey of the Great Pyramid.

****

Section II—The Sources of the IDT and Pyramid Tradition:

An Old Friend Reappears: IDT

The U.S. Declaration of Independence Date, this time as *a discovered ratio*, mysteriously built into U.S. Coinage out of the 1830's using Pyramid Numbers

Summer of 1980, I acquired a copy of Piazzi Smyth's book, *Our Inheritance in the Great Pyramid,* which was then being reprinted from the original 1880 edition by the Bell Publishing Company. It is now called The Great Pyramid a hundred years later. Due to the great interest generated by Peter Tompkins' book, Secrets of the Great Pyramid printed some years before, the book world had begun locating and publishing all sorts of books on the Pyramid, both old classics and the new. Tompkins' book was the first really important, popular book on the Pyramid in recent times. With it came a public awareness the pyramidology concept, or "the Great Pyramid as something mysterious and special, and more than just a great big tomb" -idea. The book, Our Inheritance in the Great Pyramid was even more "British" and old-fashioned than Davidson's book and it provided a good insight into the tradition that preceded Davidson's ideas. Smyth's book, like Davidson's, is strongly religious in character. It is very heavily flavored by Victorian thought of the 1860's, as well as the moral view of English Protestantism, expressed in almost every breath.

Smyth's book mainly covers interior dimensions and proportions of the Pyramid, for which he did a substantial amount of original survey work. The Professor and Mrs. Smyth sailed for Egypt in 1864 with a large amount of expensive and especially designed measuring equipment and surveying instruments. There, on the Giza site, they spent four months living in nearby caves with a retinue of native Arab servants, measuring and collecting survey data. Ultimately, he produced an immense three volume work called Life and Work at the Great Pyramid of Jeezeh during the Months of January, February, March and April, A.D. 1865. The later volume that I owned, Our Inheritance in the Great Pyramid, appears to be the popular summary of his ideas and arguments for his theories.

Smyth wrote on the history and theory of the Pyramid, and made much speculation on the weights and measures aspect of his Pyramid theory. Much of the beginnings of "British-Israelism" find their source in Smyth's ideas, including an early form of the prophetic chronology theory, a development from Robert Menzies theory. From Smyth's work, it is evident some of the theory of the Pyramidologists' Pyramid Cubit can be coincidentally tied into the metrology research of Sir Isaac Newton. He quotes Newton as proposing a length of a "sacred cubit" of the ancient Hebrews as being estimated to be 24.88 British inches. This is cited in support of the idea that the ancient British and Hebrew peoples may have a shared metrology of the Pyramid Inch and 25 Pyramid Inch Cubit, provided somehow by of the Biblical character Seth. The sacred cubit was supposed to be in ancient competition with another cubit of 20.68 British inches, which he called the "profane cubit." (This unit is well known to Egyptian archeology, being about 0.525 meters.) Smyth also speculated on alignments from ancient astronomy, the Pyramid's mathematical form and the Earth-commensurate character of its dimensions. (He was, as noted earlier, the Astronomer Royal for Scotland.)

The main value of this book for me, was that it allowed me to notice smaller and deeper subtleties that were built into the Pyramid, in contrast to the otherwise more popular, large-scale mathematical features. But the next segment is about *lore*. I had no reason to think there would be anything in his book that could affect my research on the dollar. But there was more in it than little features of ancient stone work. It will become evident the dollar's Designer probably read this book also.

U.S. Currency, according to Piazzi Smyth's source Dr. Watson F. Quinby:

Deep within Smyth's many details was a fascinating and peculiar side note about *American money and the Pyramid's mathematics* in Part III Chapter XV., Heat, & Other, Mensurations. This little article immediately caught my attention from the U.S. currency connection to the Pyramid. At the outset it didn't seem like a probable connection, seeming too much like a coincidence. But it shed some light on Pyramid geometry and the paper dollar.

Throughout Our Inheritance in the Great Pyramid Prof. Smyth labors to show that the Great Pyramid has enclosed a great many measures and natural constants of the Earth in its form, and refers to all of this as a *naturally occurring system* of weights and measures. In the section called Money, he responds to inquiries from the public about how *money* might figure in to the Pyramid-based system of ideal weights and measures. He writes:

"... Wherefore many inquirers have demanded, "What about money on the *Pyramid* system also?"

"I can only answer them, that I have not been able to find out anything about that subject in the Great Pyramid.

"And is that to be wondered at? Only look at any piece of money whatever: whose image and superscription does it bear? That of some earthly Caesar or other. Therefore is money of vain human inventions, and of things swiftly passing away. But all of the Great Pyramid measures hitherto investigated, being evenly commensurable in every case, either the deep things of the planet world, or the high things of heaven above, are to be considered as virtually impressed rather with a typical effigy of some attributes of the creation of God; in praise moreover and honour of God alone.

"Far be it from me, however, to circumscribe by my small knowledge, the bearings of any part of the Great Pyramid system. And, just as several other portions thereof have yielded fuller returns and more definite answers on taking into account the experiences of the Anglo-Saxon population of the United States of America, --so here, there is something further to be learned.

"The coinage of the grand family of Republics is the only example of the money of a truly great people without the effigy of Caesar: and also is the only one known which bears certain internal numerical relations to the King's Chamber of the Great Pyramid, when the dimensions of that chamber are expressed in British inches.

"This astonishing coincidence has but recently been recognized and been given to the world, by the acute Dr. Watson F. Quinby, of Wilmington, Delaware, U.S., and has been expressed by him nearly in these words:--

"'Our (U.S.) silver coinage corresponds in grains to the measures of the King's Chamber in the Great Pyramid, in British inches. So that the length of that chamber being 412.5 of those inches, the standard weight of the "Dollar of the Fathers" is 412.5 grains; the half-dollar, weighing 206.2 grains represents the breadth of the same chamber = 206.25 British inches; and the quarter-dollar of 103.1 grains represents in inches the half-width of the same chamber, or the "touch-stone" length as it has been called of so many of the Great Pyramid's measurements.

"'At the same time the grander golden coin, the American Eagle, contains 232.5 grains of pure gold, or the number of Pyramid Cubits in the vertical height of the Great Pyramid; and the 'half-eagle' contains 116.25 of the same gold in grains, equal almost exactly to the length of the Antechamber of the King's Chamber in the same Pyramid expressed in Pyramid inches.'"

Can this be true? I wondered. These are indeed "Pyramid Numbers" apparently from Pyramid inch measurements of King's Chamber. And, if true, does this have any bearing on the paper dollar? In this passage, a recurring anti-idolatry theme is touched upon, and the Great Pyramid's design, which has a total lack of ornamentation, all seen as concordant and parallel to the righteous teachings of the Prophets of the Old Testament. He doesn't speculate on the source of the numbers. This odd idea on grain-weights and Pyramid numbers and U.S. coinage from the good Dr. Quinby is provided without any further commentary by Smyth, who proceeds onto his next subject. Perhaps he thought this was due to the workings of Providence on behalf of the Anglo-Saxon Race, but he makes no further comment.

After thinking about this off and on for quite a while, the next obvious step was for me to find out what the actual coin weights were of that period. I finally went to a local library and went through the coin book section. Several books spelled out the original values of U.S. coins from the Act of April 2, 1792, which showed various original coin weights. The coins marked below with an asterisk are the coins Dr. Quinby should have been talking about, and these are the original weights as established about two and a half years after the adoption of the U.S. Constitution:

(I) 1792:

Metal	Denomination	Value	Weight in Grains		Fineness
			Fine	Standard	
Gold	Eagle	$10	247-4/8	270	.916 2/3
	Half Eagle	$5	123-6/8	135	.916 2/3
	Quarter Eagle	$2 ½	61-7/8	67-4/8	.916 2/3
Silver	Dollar or Unit	$1	371-4/16	416	.892 2/5
	Half Dollar	50¢	185-10/16	208	.892 2/5
	Quarter Dollar	25¢	92-13/16	104	.892 2/5
	Disme ("Dime")	10¢	37 2/16	41 3/5	.892 2/5
	Half-Disme	5¢	18 9/16	20 4/5	.892 2/5
Copper	Cent	1¢	264	264	1.000
	Half-Cent	1/2¢	132	132	1.000

As shown here, at the time of the Revolution, no such Pyramid-dimensioned grain-weight numbers appear to be present. The above silver Unit would be the actual "the Dollar of The Fathers," incorrectly referred to by Dr. Quinby in his letter. The above table of weights is the *original* bi-metal standard of U.S. currency: gold and silver being established at a *fifteen to one ratio* (15:1) between the metals. The dollars (or "Unit") of this period were to be based on the *"Spanish milled dollar as the same is now current..."* (Section 9, Act of April 2, 1792.) Here is the exact ratio of Gold to Silver shown in the grain-weights as stamped at the mint for that period from the above table:

1792--1837:

$1 gold / $1 silver = 15 : 1 ratio

(247.5 gr./10) / 371.125 gr. = 14.99494949 : 1 (result of division, about 15:1.)

But strictly speaking, *this is not the coinage* that Dr. Quinby would have had at hand, unless he had been a collector. The weights of the coins were changed later, due to the fact that the 15:1 ratio was no longer true economically and had become completely unworkable. Their *face values* had dropped below the *value of the weight* of metal in the coins. This made the actual metal by weight worth more than the value stamped on the coin. Why would you keep $100 of coin, when it could be made into $105 of jewelry? The profitable difference in the actual to statutory ratio of silver and gold were to eventually cause the coins to be "melted up for bullion." The difference caused the coinage *to disappear.* The last coins of that period were minted in 1803. Dr. Quinby would have been writing to Piazzi Smyth some years before the 1880 publication of <u>Our Inheritance in the Great Pyramid,</u> (i.e., "...has but recently been recognized" above) so *these coins* deserved a look. Dr. Quinby must have been looking at something.

After 45 years of following the original standards of 1792, the U.S. was now practically without coins of *specie*, or of precious metal, and new official coinage was desired. Beginning with the Act of January 18, 1837, a new series of gold and silver coins were authorized, with a slightly *lesser amount* of gold and silver, and at a *larger relative ratio* of metallic values. To discourage people from melting down the coins, the metal ratio spread had to be made much larger. The coins were also made with less precious metal than their statutory value, for the same reason.

These coins do indeed have numbers that look as though they may have been derived from the Great Pyramid King's Chamber dimensions. Here are the U.S. coins that the acute Dr. Quinby may have handled,[6] all the way up to, and beyond the late 1880's:

Metal	Denomination	Value	Weight in Grains		Fineness
			Fine		Standard
Gold	Eagle	$10	232.2	258	.900
	Half Eagle	$5	116.1	129	.900
Silver	Dollar or Unit	$1	412.5	453.3	.900
	Half Dollar	50¢	206.25	229.2	.900
	Quarter Dollar	25¢	103.13	114.6	.900

"412.5" is similar to the length of the King's Chamber: 412.132 PI (Davidson); "206.25" is similar to its width: 206.066 PI (Davidson). (Oddly enough, if you were to multiply "103.13" by the square root of 5, you would get a passable height of the chamber: "230.61". Compare: 230.389 PI per Davidson.) The above series of coin weights can (all) be reduced to two, simple and general correlations--one for gold and another for silver. Gold: 232.2 grains / 10 per $1 and Silver: 412.5 grains per $1. All the coins are simple divisions by two of these weights. (Dr. Quinby got the gold value slightly wrong by 0.3 grains.)

Gold: In Pyramid Cubits, the ideal height given by Smyth for the Pyramid, is about 232.52. Half of this number--due to the nature and even divisiblity fraction of 25 Pyramid inches per Pyramid cubit--is also *twice* the nominal height of the Pyramid in Pyramid Inches, divided by 100: or about 116.26. How close are these, numerically, to the above grain weights for the gold coinage? From the first number 232.2, the second number 116.1 is merely a halving of the first with the same fineness. Since, presumably, the series is aliquot, or in evenly divisible parts, the ratio of this calculation will apply for all gold coins:

(Grains /Gold @ $10)   /   (Pyr. Cubits @ Pyr. height)   =

(232.2 / 232.52 - 1)   /   =1:727   error from Smyth's ideal number.

Silver: The ideal length of the King's Chamber in British inches in British inches from Smyth is 412.132 pyramid inches, (about 412.58 British inches.) How close is this, numerically, to above grain weights for the silver coinage? The above statutory grain weight is 412.5, and here again, the following numbers (206.25 and 103.13) are only halvings of the fundamental $1 weight in silver with the same fineness, so the following calculation will apply for all silver coins that Dr. Quinby had:

(Grains / Silver @ $1)   /   (Kings Chamber length in   =
                                          PI
412.5   /   412.132-1/ x   =1:1120   error from Smyth's ideal number

The first case above of 1837 U.S. gold coinage, is the weaker correlation of the two, but not negligible. The silver correlation, however, is higher. (Considering both together as an average, there is a difference from real to the ideal number values on the whole of about one part in 923. Which is fairly convincing--this doesn't look like a chance relationship at all. Were it in British inches: 1 : 3000.) Piazzi Smyth's book, of course, was not available at that period in the 1830's, having not yet been written. This raises an interesting and enigmatic question: *where did* the lawmakers of that time get these numbers from?

The puzzling thing here, is that of the published data on the Great Pyramid that seems to have been available in 1837, *no* account appears to have given the correct height of the Pyramid. Nor, does it seem that would they have listed any such height in Pyramid Cubits. The 25-inch unit as applied to the Pyramid *appears* to have been Smyth's idea, probably not appearing earlier than the 1860's. Sir Isaac Newton's Sacred Cubit would have been around in the 1600's, but it does not seem to have been applied to the Pyramid's height. (This unit was 24.88 British inches, and is similar to Smyth's Sacred Cubit of 25 Pyramid inches and the theoretically identical Collet Unit, one ten-millionth of the Polar Radius. The ideal number would be 232.52, if were to have been provided by Smyth.) Even if the Pyramid's height was more or less correctly known, and even if by a leap of imagination, someone had applied Isaac Newton's sacred cubit to it, one *would still not* have got the number "232.2" from existing information. Using the above 24.88 British inch Newtonian Sacred Cubit, with a correct, ideal 5819 British inches for Pyramid height, the best one could have got would be about 233.9.

Curiously, John Greaves, the first careful examiner of the Great Pyramid, in his book Pyramidographia written in 1646, gives what would be about 5772 British inches for a height of the Pyramid: "...if we measure the perpendicular, it is foure hundred eighty one feet;" [7] And, if we then used *this* distance *and* Isaac Newton's Cubit we will get:

$$(12 \cdot 481)/24.88 \quad = \quad 231.99 \text{ Sacred Cubits of Isaac Newton} \quad \text{(about "232.0")}$$

That's pretty close. To get "232.2" with this unit you would have to start with a height of about 5777 British inches, five more inches than the height given by Greaves in the 1640's. So it is at least possible that some small variation of (1) height or (2) Cubit values could be the source of the 232.2 number.

After the French returned from the war in Egypt in the early 1800's, it is quite possible that the work of Napoleon Bonaparte's surveyors and the able Edme-François Jomard in Egypt at the Pyramid, may have become available. Jomard's work was technically quite good. He played a large part in the beginning conception of the idealized Pyramidology sentiment, especially on the subject of weights and measures, and purported ancient knowledge of mathematical proportions and geodesy. If something near to a true measured height of the Great Pyramid from French military surveyors was divided by the Collet Unit, which is entirely possible, the "232.2" number would have emerged from division. (The Collet Unit was about 25 inches in length, or 0.6357 m. See Chap. 4, pg 91). The historical record should be closely scrutinized, since this relationship in the coinage is not a fluke.

But, finding this 232.2 number is still a big problem. I don't think we know the whole story. This number-form seems *much too close* to twice the decimally reduced diameter number 116.26 to be chance. It is possible that other published (or unpublished) works in that period may reveal something interesting from Pyramid theory. But these coin weights appear to be evidence a hidden school of thought in this country (and Britain) that seem to have included an idea of a 25 inch unit. The signal that *these numbers are special and were intentionally chosen* has additional evidence we will see below.

The pyramid-number correlation is closer in the 1837 *silver coinage*. Being so suspiciously close to ideal number-forms, it is difficult to ignore as chance. The source is almost certainly John Greaves. Few if any writers on the subject in English could have been available to the readership of the early American Republic. Also, we know his book *was* available in a Colonial library and an illustration in it may have been the source of the original unfinished pyramid for the Great Seal. See below.[8] In Pyramidographia, John Greaves gives his measurement of the length of the King's Chamber as "...thirty four English feet, and 300 and 80 parts of a foot divided into thousand (that is 34 feet and 380 of 1000 parts of a foot.)"[9]

$$34.380' \cdot 12" = 412.56" \quad \text{British inches.} \quad \text{(This equals 412.123 PI.)}$$

It seems likely that this statement by Greaves was the origin of U.S. legislators' numerical choice of the *silver* grain weight. (Smyth and Davidson's figure would have been 412.132, a difference of one-hundredth of an inch, with an error difference of about 1 : 46,000.) The implication is a clear Great Pyramid connection to gold and silver symbolism, even if not through John Greaves. [10]

Beyond these numerical symbols, there are two other important things that point to additional symbolic manipulation of the numbers. Each of which, if known separately and casually would not arouse undue attention. But within the context of the above thought and symbolism, and from what we now know about the design on the back of the paper dollar, these numbers have a greater meaning.

Fact number one: the change of *the fineness* in the metal alloy of the coins in 1837. Although it does not mean much in terms of value, the original fineness ratio of 1485/1664 (0.892427...) was changed to 9/10 or the ".900" as seen above. This is not an important change in the metallurgy of the coins. But to the eye of students of Biblical symbolism, if a symbol of "a monetary quantity of 9/10ths" were alluded to--there is one likely connection. This 9/10ths can be said to be the same as proportionally symbolic of *a tithed income,* perhaps symbolic of the righteous remainder of ones' income, *minus the tithe.* Which is to say, a tenth of ones' income was prescribed in Hebrew law, a ceremonial amount due the temple (See Micah, etc.). Perhaps nothing of factual import is being said about the coinage, but "a remaining nine tenths" is suggestive of the sacred ratio. But here, the Bishop of Occam might frown doubtfully and say that this "may have just been done for the sake of simplicity at the mint."

Fact number two: a far more interesting fact to us investigating the dollar, is the nature of the new gold/silver ratio. What was the actual gold/silver ratio in the coinage of 1837? The first time I solved this little division problem my eyes were as wide as saucers. "Ratio" seems to be the name of the game through the whole investigation:

Post 1837:

$1 gold	/	$1 silver		=	ratio	
(232.2gr. / 10)	/	412.5gr.		=	1 / 17.76485788	(result of division.)

Now we can see the source of the tradition of the IDT motif. Here it is, in a single equation, from solid historical fact. Most readers will now recognize the two decimal shift number-form from as far back as Chapter 2 and the discussion of the Independence Date Theory (IDT) as well as the Close Approximation to the Independence Date Theory (CA-IDT). Even the height of the tiny pyramid having a side of 0.279" may have been inspired by this, (see pg 145). The Bishop is grumpily silent. Something must be afoot.

Notice that we find pyramid information *and* July 1776 in one equation. This is the clearest and most solid source of the two principle mathematical themes found on the dollar: (1) Pyramid dimensions built into the currency (and then later for the dollar, Davidsonian refinements, Displacement Factor, etc.) and (2) the patriotic symbolism--the capturing of the date of the Declaration of Independence by means of *a mathematically derived process connecting both symbol groups.* Note that this connection device is the *entire essence* of the dual Problem Variables display shown earlier in this chapter. Was this crafting of carefully chosen grain weights formed in emulation of the Pyramid ideas in the Great Seal legislation? The same might now be asked of the dollar design. Yet the dollar of 1935 appears to have incorporated even earlier symbols than these. Both ideas (the date 1776 and Pyramid symbol) are originally found in the Reverse Face as described in the "Remarks and Explanation" text adopted by Congress in 1782.

Here in yet a another discovered form: again appearing as the approximate date of the Declaration of Independence *and* in the "divided by 100" form. But *this* is probably the source of the number-form. Notice this number more closely resembles IDT than the two CA-IDT forms. Certainly, this cannot have passed unnoticed in the last 163 years before now, and must appear in numismatic literature. It would appear this was an intentional symbol by the lawmakers of 1837. But perhaps it is more likely that some special person crafted this coinage relationship, who then quietly introduced this symbolic freight through a committee without debate, where it was passed through a vote by the typical crew of sleepy Senators just trying to stay awake.

Just how close is the "1/100th of Independence Date number" to the above gold/silver ratio of 1837?

(IDT Date Number / 412.5gr /(232.2gr./10) - 1)/x =

(17.76508197 / 17.76485788 - 1) / x = 79277 : 1

This could be said to compute to be the theoretical Day Number 177, or June 25, 1776--9 "days" shy of the Declaration of Independence Date number. There is a large correlation to at least one or another aspect of the earlier IDT theory. At the very least, we can say that something of a tradition is the source of the idea behind the length of the LOR rectangle on the dollar. The relationship reaches a long way into our past, and we must wonder how this idea got started. The finding adds to the hints of a possible hidden tradition. Perhaps this kind of "signature" should be looked for elsewhere, within other State related items and designs that could bear symbolic ratios or special numbers.

At this point in my research, (after the 1980's) I had lost my ardor for the patriotic IDT theory in favor of the Davidsonian DFT idea. Though this later discovery didn't shake my faith in the newer DFT theory, it remains a dizzying revelation that things may *still* not be as they seem. The earlier conclusion still held: the 1935 dollar's Designer must have found a fortunate happenstance in the numbers that allowed a close proximity between the IDT and DFT lengths, and that he had gone with the flow and woven the ideas together. (e.g., CA-IDTs and Problem Variables III trick). We can logically speculate that the equation ((232.2gr. / 10) / 412.5gr.) = 1 : 17.76485788 was his source material for getting started on his paper dollar design. Later he exploited the numerical or geometric proximity of other symbols. The above coinage data could have been provided as evidence of an American tradition, were the embarrassment of the curious nature if the dollar's design to have come up. (e.g., using π and the 1837 gold/silver ratio number, one could propose a dollar length of 5.65472989". This is not very different from 5.654801219 (IDT) which could well have been one of the Designer's original target numbers.)

The above findings permit some reasonable conclusions about the concepts behind the 1837 coin-ratio, even without researching the unknown author(s) of this law, enacted some 163 years ago:

(1) At or before 1837, some important or influential person(s) in the U.S. Government, was deeply interested and involved in fairly *accurate details* of the Great Pyramid.

(2) *Both* gold and silver weights were carefully chosen (by persons unknown) to represent the *height* of the Pyramid and the *length* of the King's Chamber.[5] Meanwhile, Congress simultaneously accomplishes a monetary solution to the dissolution of coinage--by a wise improvement of the precious metal ratio.

(3) Through the arithmetical relationship of the specific gold and silver grain weights, one-hundredth of approximate Independence Date was enshrined *between these Pyramid derived numbers*. That is to say, this ratio is an inescapable mathematical result of division, for anyone who cared to work out the arithmetic. But it appears that both of these numbers had to be intentionally shaved or modified slightly (and rounded) from some unknown original values to achieve the special 17.76485788 ratio ((232.2 / 10) / 412.5)/x. This must have been a compromise: If, for example, you try the *true* ideal proportions from the Pyramid: ((232.52 / 10) / 412.132)/x the number 17.72458283 results, which is unsuitable as the date of the Declaration of Independence. Therefore, the natural ratio must have been adjusted for date symbolism.

(4) A simple, sacred Biblical ratio of "9/10ths" appears to have been included perhaps as a final colophon, signaling a fundamental, "before all else" symbol of a tithe cut from income at the beginning. This is as if in symbolic adherence to Judeo-Christian scriptural law.

The fact that "1776" and a pyramid appear together in the original description of the Great Seal, must be source for this the rebus in U.S. currency tradition. I think that the early American Republic of 1837 still had many students of the Mysteries, or their equivalent acting in important places. (Chief on my list of suspects is John Quincy Adams, a former President, later a long time member of the U.S. House of Representatives. In his time, as a boy of 8, he watched the battle of Bunker Hill in 1775, and later he served in Congress with Lincoln. I have no proof, but he is "one of the usual suspects" --a Founding Father or, in this case, a founder's son. But I could be all wet here.) Whoever set this idea in motion *knew* with some precision the height of the Great Pyramid in Polar inches and Pyramid Cubits. This person was cleaver enough to create this math relationship, and had enough moxie to push it through Congress. The amazing thing is that this tradition is still preserved in the 1935 dollar of today.

Did the Designer discover this one day in Smyth's book, and then decide to craft all of these symbols in the dollar? Or is there separately a hidden doctrine, or informal coda of rules somewhere at the U.S. Treasury, a general collection of accreting traditions under the surface of our currency? In what boils down to *only two* fundamental numbers as hard evidence (the gold and silver grain weights), there are a *lot* of odd and unique similarities to be drawn here between this 1837 currency and the 1935 dollar. These can be derived as both separate forms as well as conjoined symbols. To recap various points:

(a.) A general pyramid motif, by numerical symbols

(b.) Specific identification to the Great Pyramid of Giza, (not just any ideal pyramid)

(c.) The unusual use of *numbers* as specific symbols

(d.) Using the *Declaration of Independence Date*

(e.) The Independence Date in *a 1/100th fractional form*

(f.) Use of a *mathematically hidden* form of the Independence Date

(g.) Using specific numerical quantities of a symbolic nature, creating a symbolic ratio that appears *between them* as a result of arithmetic

(h.) Use of a *numerical form* to link to Biblical lore (i.e., 9/10ths fineness)

(i.) Using currency as a *medium to bear mystical symbols*

One of the few dollar parallels missing from the above list of symbols in the 1837 coinage, is unusual knowledge of mathematical manipulation, or special knowledge of arcane ratios such as φ^2. Yet as we shall see later, there is even evidence of φ^2 from *an earlier period* than 1837, from the very earliest acts of our new government.

Smyth may have provided an important insight about symbolism. When Christ had someone hold up a coin and says "Whose image is on it?" *Caesar's* ownership became evident. Perhaps the creation of the dollar imagery and Great Seal symbolism is intended to assign a sacred authority to money, and government. (e.g., "In God We Trust.") If one had to compute a total probability of the likelihood that these numbers occurred by chance from the above error figures from the two cases of metals and lengths, it might be on the order of one part in a million, nominally. But, when an interconnected and specific *ratio* that lies between them is also considered, the chance element relationship is astronomically high against coincidence. Yet in terms of astronomical unlikelihood, what chance would there be of any group of legislators passing into law a thing of such subtlety?

****

Section III--Logarithmic Relationships:

The Designer was a user of logarithms. We will look on the dollar first, preparing for related facts that surface from about the time of the American Revolution. At the bottom of the dollar, just above the LIR line there are *two and only two vertical features in the filigree*. These centers measure 3.677" apart:

$$\log_{10} 3.677 = 0.56549 \quad \text{(or, 0.5655 rounded to four places.)}$$

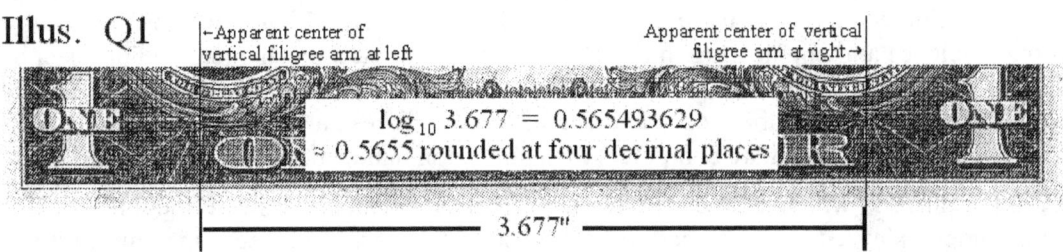

Illus. Q1

–Apparent center of vertical filigree arm at left

Apparent center of vertical filigree arm at right→

$$\log_{10} 3.677 = 0.565493629$$
$$\approx 0.5655 \text{ rounded at four decimal places}$$

3.677"

The log of the base of 10 of 3.677 is a number form we've seen before, about a tenth of the length of LOR. The next surprise that jumped out was when I tried "natural logs," logarithms based on function e. (On your calculator these logs are called "ln" for *log natural*. 2.71828... is the base "e" rather than 10. The inverse of ln is e raised to the x or the "e^X" button. The x here, is whatever is on your calculator screen before you execute this function.)   I tried $e^{5.655}$, and this gave 285.716487, which is pretty amazing--since we have seen this number-form several times before.

$$e^{5.655} = 285.716487 \qquad \textit{Note:} \qquad 1 / e^{5.655} = 0.003499973$$

Look at the astonishing similarity to the 0.0035" offset of the last chapter--one part in 130,000. Also, notice the number-form similarity to PLD2 = 2.857745339 (See pgs 137 and 139).   This raises the question, what does the Natural log of the familiar Davidsonian number D look like?

$$\ln 286.1022156 = \quad 5.656349144$$

This would be a passable approximation of the diagonal for a 4"x 4" square, as we saw in Chapter 3, (pg 47) to about one part in eleven thousand. Again, *two more* of many possible "competing sources" found for the theoretical origin of the LOR length number.

Where do logarithms come from?   John Napier (1550 - 1617), a mathematician and fanatic Scottish Presbyterian anti-Papist, invented logarithms. John Napier was at the cutting edge Seventeenth Century mathematics.   He was a "Scottish Laird", born The Baron Merchiston in the Castle Merchiston in Scotland.   In a biographic sketch of him, author John McLeish writes: "...had he been asked, he would certainly have said his major work for humanity was his campaign against Roman Catholic supremacy in Scotland and later in England."[11] But his great legacy to the world was *logarithms*. In his leisure time he developed a device called Napier's Bones, a calculating technique using marked rods. He also pioneered the use of the decimal place point or comma. Without getting into a long discussion on the different of bases of logs, we should credit Professor Briggs (Savillian Professor at Oxford), for the suggestion to Napier in 1614 that logs should be made on the base of ten ($\log_{10}$). The suggestion was so that they be tabled for greater ease of use. Napier's logarithm concept caught on like wild-fire in Europe.

> "Neper, lord of Markinston, hath set my head and hands a
> work with his new and admirable logarithms. I hope to see
> him this summer, if it please God, for I never saw book, which
> pleased me better, and made me more wonder."

> March 10, 1615, Henry Briggs, from G. Huxley

Napier passed away in 1617. Briggs' tables, which also introduced the decimal point to a large audience, revolutionized mathematics. A Swiss, Joost Bürgi, published accurate tables in 1620. Large tables were developed all over Europe within a very few years, some with an accuracy to 14 places. Napier is mostly known for his book *Mirifici Logarithmorum Canonis Descriptio,* (Description of the Marvelous Rule of Logarithms) that opened the subject. Logs gave rise to Newton's Calculus and modern science.

It is hard to appreciate at this point in time, just how valuable the use of the logarithm method once was. I was graduating to high school when the transition from log table books to the electronic calculation was beginning. Slide-rules were common then, and were based on graphic logarithms. (Slide-rule precision is limited to about three places past the decimal.) If you wanted to do any serious computation in science and engineering in the 1960's, you acquired a mechanical adding machine and an expensive ten-place table book of logarithms. Most textbooks for math, science and navigation had formulas that were arranged for the theory and use of logs, and these are almost incomprehensible today. There was some access to time-share computers then, and soon hand calculators were everywhere.

I knew a boy in 1968 who had an expensive HP scientific calculator his dad bought him. This space-age tool was widely admired. All you had to do was press buttons. The electronic chip in it that did the work, was the same used in the Apollo on-board computers. His calculator could create Natural logs, Logs to the base of 10, not to mention *exact* Sine, Cosine, and Tangent functions--quite awesome then. As I graduated high-school, British and Japanese electronics companies were selling 16-function, ten-place floating decimal calculators for less than a hundred dollars. In time, they were to be found in blister-packs on hooks in drugstores. Logarithms, no longer needed because of the simplicity of calculators, are mostly forgotten. What did logs do--what were they good for?

Maybe the simplest way to think about logarithms, is to look at the work and time they saved. From Napier's time to (almost) the present, if you became seriously interested in astronomy--such as computing celestial motion and so on, the hours *or days* of time spent in computation was enormous. Logs are almost like dehydrated numbers. The beauty of logs was that you could *look up* the logarithms of two numbers you wanted to multiply, *add* these together, then *look up* this sum in the Antilogarithm Table to find the answer. Simple as that. *Subtracting* logs from one another is the same as division. But this isn't the half of it: suppose you had to multiply and (or) divide 40 six place numbers together with a pencil and paper? This is a *lot* of work by hand. How much paper would you need? But with *logs,* this might all be done on a page in one column, in a fairly short time. So how does the table-book know the answer to these problems? It seemed like magic--the tables had somehow captured the essence of multiplication and division.

The "you don't get something for nothing"-sense of fairness, seems to be faintly abused here. What's lost in the process here? How did we get away from adding all those staggered columns of numbers, the normal business of multiplication? The genius of Napier was to see that *adding exponents* was the *same as multiplying.* Looked at from an information perspective: all the multiplication busy-work has been *done once and for all.* It was done through the original process of the work of tabling these values. Someone did a lot of numerical labor to create a table book of exponents (the logs) and their inverse relationships (the antilogs). The resulting tabled form of numbers *preserves* the previous labor of calculation in what is now re-usable form, and with any other logarithm. The massive and tedious work *is still made useful,* frozen in the character of the tables. Was anything lost? Yes, the technique is limited by the precision of the table--that is to say, the amount of decimal places to which it was tabulated. But we accept this limitation in electronic calculators, to the given extent of the display. Finer and larger calculations require a bigger and better table having more decimal places.

It is no surprise that the dollar's Designer was familiar with logarithms. His period of time was the "golden age" of logarithms, a time when the technique was reaching it's maximum use. Anyone as technically competent as Designer Edward Weeks was at the time of the 1930's would have known all about logs. Unless he was a mathematical *Savant,* good table books are about the only thing that would explain his prodigious calculation ability.

The above sketch of the logarithm concept is presented not so much in relation to the dollar and it's graphics, but in light of other questions about the Great Seal that emerge from the American Revolution. In Eighteenth Century Europe, many accurate tables were published to seven places (one part in a million) which would have allowed a freedom for calculation equivalent to a six-place display electronic calculator. As we will see shortly, someone at the time of the final formation of the legal description of the Great Seal may well have been a user of Briggsian logarithm tables.

****

Alphanumerically Encrypted Logarithmic Relationships:

Although my wife Judy is not a mathematical hobbyist, she provided a surprising idea that turned out to be immensely fruitful. After listening to me ramble on about my project on dollar, she asked whether it was possible that numerology-letter relationships existed in the dollar design. This hadn't crossed my mind, and I couldn't think of any reason not to look into the idea. At first I doubted that anything would be found, since the origin of the wording on the Great Seal dated from the time of the American Revolution. In 1980, I had no reason to think that anyone from that period could have had any hand in dollar-related symbols. But as seen above, evidence began to unfold that there was a hidden, pyramid related number agenda as little as 40 years after the time of the Revolution.

The first step was adding numbers mapped in alphabetical sequence as they appeared in the left face of the Great Seal. (At first, I ignored the Roman numerals at the base of the little pyramid.) Following this idea, I examined groups of numbers, sums, fractions and other relationships. For "ANNUIT COEPTIS NOVUS ORDO SECLORUM." The total came to "415,"(See table A):

Table A   For example: "THE": T=20 H=8 E=5 a sum of "33"

A	=	1	K	=	11	U	=	21
B	=	2	L	=	12	V	=	22
C	=	3	M	=	13	W	=	23
D	=	4	N	=	14	X	=	24
E	=	5	O	=	15	Y	=	25
F	=	6	P	=	16	Z	=	26
G	=	7	Q	=	17			
H	=	8	R	=	18			
I	=	9	S	=	19			
J	=	10	T	=	20			

From the above table, codes and numerical sums can be made of letters and words:

ANNUIT	=	1+14+14+21+9+20	=	79
COEPTIS	=	3+15+5+16+20+9+19	=	87
NOVUS	=	14+15+22+21+19	=	91
ORDO	=	15+18+4+15	=	52
SECLORUM	=	19+5+3+12+15+18+21+13	=	106
			(+)-------	
(30 letters, Cumulative Summing of digits = 1)			415	(Two mottos)

Certain surprising relationships appeared between the words and groups of words (see below). But, although some of these were interesting--they did not seem spectacular or beyond chance. However, when I tapped the "$\log_{10}$" button--I saw "2.618048097."

$$\log_{10} 415 = 2.618048097 \quad \text{Very close to:}$$
$$\varphi^2 = 2.618033989$$

There were some awestruck minutes, then I checked the translation of letters to numbers and addition. No doubt about it, the resulting number was just what I saw--very similar to φ^2. From a historical point of view, what this could mean? I vaguely remembered that Napierian and Briggsian logarithms had appeared in the 1600's, so it was conceivable that this contrivance was possible for someone at the time of the Revolution (Rule 2, extended into that earlier time period). The golden ratio was known in the Eighteenth Century, but where would it have appeared in the squared form? Although $\log_{10} 415$ is not exactly the square of the golden ratio, it is very close at 1:185,575 or about 5 parts per million. From what we know of contemporary log tables at six or seven places, this is well within the world of the possible. Another interesting fact here (looking back to the original equation in Chapter 2, pg 43) is that this doesn't significantly disagree with the original equation, made from the original measurements, if considered only to the *fifth decimal place:*

$$5.655 / 2160 = 2.61806 \text{ vs.} \qquad \log_{10} 415 = 2.61805 \qquad (351,000 : 1)$$

Could this number be the original source of inspiration for the Designer's idea of the φ^2 ratio?

After that, the next step was to add *all* of the original letters (Roman numerals and all) from both sides of the Great Seal. The log of the sum has got to be interesting, I guessed:

E	=	5	= 5	(motto No.3)
PLURIBUS	=	16+12+21+18+9+2+21+19	= 118	"
UNUM	=	21+14+21+13	= 69	"
MDCCLXXVI	=	13+4+3+3+12+24+24+22+9	= 114	(Roman Numeral Date "1776")
			(+)-------	
(22 letters, Cumulative Summed digits 9)	=		306	(remaining sum)
			415	(sum from above)
			(+)-------	
(Total from above: 52; Summed digits 1)	=		721	(total)

$$\log_{10} 721 = 2.857935265$$

Another familiar surprise. But my perspective at that point in time was different from the present reader, who has now seen this number-form all throughout this book. When I first looked at this result in 1980, I wasn't convinced it was anything important, other than a partial coincidence with respect to the number 2.861022156. It seemed as though the sum needed another letter--perhaps another "e" which having a value of 5 could make 726 for a logarithm close to 2.861, (i.e., $\log_{10} 726 = 2.860936621$). We had a nicely separate "E" already in the motto "E PLURIBUS UNUM" but it doesn't appear again distinctly anywhere else in the Seal's letters. At that point I was aware of Davidson's number being used on the dollar in the graphics, so this similarity was very interesting. But I hadn't yet worked out the diagonals based on the thirteenth line and LIR. This number would turn out to be (nominally) 2 ten-thousandths longer than the diagonal PLD2 (see pg 137 et seq.), and as close as one part in fifteen thousand.

We can now see that $\log_{10} 721$ looks similar to:

(1) the shorter of the 13L diagonals a shown in Chapter 4: PLD2 = 2.857745339 and

(2) the reciprocal of 0.0035" (285.7142857) seen in at least three separate places, and

(3) the number-form of the shorter elliptical chord length 0.2857" created in the downward shift of the ellipse.

But, there is a *fourth* type of place. We should look at an interesting fractional relationship occurring between the crucial Latin words NOVUS ORDO (New Order). Notice that their sums as a fraction give

$$52 / 91 = 0.571428571,$$

a faction resembling the above *alternate base* of the little pyramid (LPB4: 0.57220431, see pg 161) on the dollar, half of which is:

$$0.571428571 / 2 = 0.285714285,$$

also resembling the shorter elliptical chord length for size (see pg 59 and 131). The reciprocal is 3.5, making number-form that reminds us of the 0.0035" offset. Another group of curious coincidences.

But the historical chronology seems to put the cart before the horse. So far as we know, nobody but Davidson, all the way up to the 1920's had any knowledge of a number like 286.1. He is thought to be the *source of the information* for the Displacement Factor. Yet here again, as seen above in another problem about a source of Pyramid data on the Pyramid Cubit height question, Smyth (between 1860 and 1880) and John Taylor (ca.1859) could not have been the originators of the Gold grain weight number: 232.2. Smyth *seems* to have been the primary source, yet now it clearly appears to have been in the coinage at the very least 22 years earlier. One becomes suspicious that a body of Pyramid-related math was developed earlier than shown in the literature, an undercurrent among the *monde savant*. When looking at this above $\log_{10} 721$ number, there is a strong temptation to say "coincidence." But that is what it has *seemed like all along, in all the other places where this number-form appears*.

Looking at the 13L diagonal PLD2 vs. the encrypted $\log_{10} 721$, which way does Occam's logic razor cut? On one hand, you can make a good case for believing the 0.0035" offset is a modifier of Davidson's Number as a source for the diagonal PLD2 in the 1.0011x paper shrinkage theory--all of which is supported by evidence and traditional pyramid symbolism. But the Bishop Occam might ask "What if the Designer was using the so-called $\log_{10} 721$ as his *actual source of the diagonal?* Ignoring evidence of Davidson's Number, perhaps $\log_{10} 721$ was a source for a connection to the Great Seal. It came first."

But there's more possible here. Curiously, dividing this number into the ideal D/100,

$$\log_{10} 721 / 2.861022156 = 1.00108,$$ Results in a ratio not very different from

Piazzi Smyth's "1.0011" ratio or David Davidson's "1.00106" value--both as close as one part in fifty thousand. So was *this* the inspiration for the diagonals tableau? Both numbers, the $\log_{10} 415$ and the ratio 1.00108 are mathematically adjacent, much as PLD2 and 0.0011 were. A *lot* of coincidences here.

Maybe D/100 resembles $\log_{10} 721$ only by some coincidence, but it seems *far less likely* that ratio φ^2 resembles $\log_{10} 415$ only by chance. And yet (amazing if true) this encrypted $\log_{10} 415$ must have originated in the 1780's at the time of Revolution. We *do know* that the Founding Fathers were aware of some number-letter ciphers. The use of English ciphers go back to at least the time of Mary Queen of Scots and the espionage antics of that period. Thomas Jefferson had more than one code among his documents at the time of his death, one of which is said to be a Rosicrucian cipher. But of course, we don't know if they actually knew anything about Briggsian logarithms.

On the other hand, the supposed $\log_{10} 415$ and $\log_{10} 721$ seem *much* less likely to be a chance relationship from the Designer's symbolism. I think he must have found them much as I did. Its likely that one or both log-numbers, (415 and 721) together with the ratio within NOVUS ORDO (52/91 = 0.571428571) inspired the major elements of the 1935 dollar design. He may have noticed that the numerological cumulative sum of both groups, the partial 415 *and* the total 721 are "1". Much of the Designer's symbolism revolves around "one". Was he inspired by the NOVUS ORDO fraction, in the alternate base LPB4 at the little pyramid? Do *all* of these relationships appear by chance? Unless it is from the efforts of Charles Thomson, who we will look at next, we might have to ascribe the coincidences in the mathematics to the hand of Providence. None of this looks like chance.

What could have been the Source of the Encrypted Logarithms?

In the wonderful book on the Great Seal, <u>The Eagle and Shield</u> by Richard S. Patterson and Richardson Dougall, we read:

> "On June 20, 1782, Charles Thomson submitted to Congress his report recommending a design for the Great Seal, and Congress adopted the device the same day."

(See pg 83, <u>The Eagle and Shield</u>.)

My guess is that "Mr. Charles Thomson M.A., the Secretary of Congress" is the ultimate source of these numerical symbols. From what is known of Thomson, we have little to go on. Not known to have been a Mason or anything else unusual, I have no corroborative evidence that he had any tables of logs. The other possible author, the collaborator, William Barton, a consultant on heraldry, does not appear to have been the source of the final draft. But we don't know what his total contribution was. He was 28 at the time of the adoption of the Great Seal. His mother was a sister of David Rittenhouse (1732--1796) an instrument maker, astronomer, mathematician, and the first Director of the United States Mint at Philadelphia. (Isn't that odd? Public money and math again. See pg 49, <u>The Eagle and Shield</u>.) So here is a possible connection to a knowledgeable person in technical mathematics, of the kind that would have included logarithms. Other known collaborators are: Hon. Dr. Arthur Lee and Elias Boudinot (members of Congress); Benjamin Franklin, Thomas Jefferson and John Adams; John Rutledge and Arthur Middleton (Delegates of South Carolina, who didn't add much of anything to the Great Seal); Frances Hopkinson and Pierre Eugène Du Simitière a consultant, a naturalized citizen, born in Switzerland. Frances Hopkinson designed a $50 dollar bill in 1778, showing an unfinished pyramid--a pyramid *very* closely resembling an illustration in John Greaves <u>Pyramidographia</u>. Du Simitière and Hopkinson were prime movers in the Great Seal design early on, and both favored a radiant eye of Providence and the unfinished pyramid. From a portrait of that period, the unfinished pyramid also appears on one of George Washington's Masonic Aprons,. So, two if not three of these men promoted these symbols.

Two intriguing groups of statements from <u>The Eagle and Shield</u> seem to stand out:

(I)   The first group of statements involves the *exact spelling of the Latin words* of the Mottos on the Seal. Quoting [Gaillard] Hunt, the passage from line 625 of Book IX of the <u>Aeneid</u> of Virgil, is quoted as being the source of Annuit Cœptus: *Juppiter omnipotens, audacibus annue cœptus.* (All powerful Jupiter favor my undertakings.) Then:

> Thomson changed the imperative *annue* to *annuit*, the third person singular form of the same verb in either the present tense or the perfect tense. In the motto *Annuit Cœptus* the subject of the verb must be supplied, and the translator must also chose the tense. In his 1892 brochure, Hunt suggested that the missing subject was in effect the eye at the apex of the pyramid on the reverse of the Great Seal, and he translated the motto--in the present tense--as "it (the Eye of Providence) is favorable to our undertakings..."

(See pg 89, <u>The Eagle and Shield</u>.)

Interesting. Later, the <u>Eclogue IV</u> of Virgil (through Hunt) is quoted to show the source of *"seclorum"* (*Magnus ab integro seclorum nascitur ordo.* Hunt: "the great series of ages begins anew.") Then:

> The spelling of the word *seclorum* as used in the second motto requires explanation. There are three normal spellings for the word, all permissible--*saeculorum, sæculorum,* and *seculorum*--but the four syllables of the full word would have distorted the meter in Virgil's line quoted above. To preserve the meter, the poet resorted to the device known as syncope, dropping the first *u* from the word. In Latin poetry the use of the syncope--that is the dropping of a

vowel or syllable in the middle of a word so as to fit the word into meter was very common. With the first *u* omitted, the word could be spelled *saeculorum* (now the preferred spelling based on Virgilian manuscripts), *sæculorum*, or *seclorum*--all three spellings are to be found in eighteenth century editions of Virgil--and Thomson chose *seclorum.*

(See pg 90, The Eagle and Shield.)

The Eagle and Shield goes on to say that only one version of Virgil (by a J. Brindley in 1744) is known to contain all of the various Latin spellings seen in the Great Seal's mottos. But I think it is more to the point that there are sufficient variations of acceptable spellings that *Thomson had some latitude in his choice of letters, if had he wanted room secretly to arrange or encrypt a number-letter message within the three mottos and date.* Thomson was also quite familiar with Latin: "...from 1757 to 1760 he was master of a Latin school in Philadelphia." (See pg 71, The Eagle and Shield.) So this spelling was not accidental, nor from ignorance. Notice that Thomson chose the less preferred form *seclorum* for this motto, and a somewhat obscure *Annuit* for another.

The words as chosen, *Annuit Cœptus* are *13* letters, (separating the œ diphthong as two) and the whole sum of motto letters being 52, is equal to 4 · *13*. This is a suggestion of the presence of at least some intentionally crafted number-letter arrangement to make more 13-based symbolism.

(II) The second group of interesting and intriguing statements in The Eagle and Shield is a written indication that *there is in fact a hidden message incorporated within the Great Seal,* right from the source--Charles Thomson himself:

It is perhaps worth noting here that twenty-two years after Congress adopted the seal, Thomson sent Secretary of State James Madison the following letter dated December 1, 1804:

> Enclosed I send you an explanation of the Device for the armorial Atchievement and Reverse of a Great Seal for the United States in Congress assembled.
>
> It was drawn up when I made a report & contains the Sentiments which I had in mind when I was considering the subject, and which I wished to express covertly by the device. It has never been published nor have I ever given a copy of it. If you think it worth preserving you may lodge it in your office, if not, destroy it.

The footnote says it was taken from: "Miscellaneous Letters Aug.--Dec. 1804...The enclosure was not filed with the letter, nor has it been found in the Madison Papers..." (See pg 87, The Eagle and Shield.)

What this appears to say is (1) there was a secret message in the text of Great Seal "Remarks and Explanation" that *no one else knows of, made at that time.* And (2), the person James Madison may *keep it in his office,* but (3) if it is not to be kept there, *it is to be destroyed.* Unfortunately, we don't know what message this was, yet this is a very important statement.

The authors of The Eagle and Shield, cast doubt on the importance of this passage. They wonder if the word "covert" could really be intended to mean "secretly" and whether Thomson might have forgotten that his "Remarks and Explanation" had already been published and made Law. We don't *know,* but it is doubtful that this "explanation" could be the "Remarks and Explanation." He says, "It was drawn up *when* I made a report [the Remarks and Explanation]..." and it was of record. Then he says: "It has *never been published* nor have I ever given *a copy* of it." Never gave a copy? It doesn't seem like something he forgot, nor anything that he was uncertain about--this is something different. To me, the statement is categorical and clear. Exactly the sort of thing we would hope to find. This is a "for your eyes only" communication to a close friend and another student of the Mysteries, offering covert information.

What would have been needed to make the logarithmic message?

Was the encrypting done by Thomson's efforts alone, or with the help of others? What would have been needed was the theory and use Briggsian logs, a knowledge of classical mathematics to know the significance of φ^2 and a thorough knowledge of Latin Classics. The work might have been about as difficult and laborious as an acrostic puzzle or the construction of a magic square, where all the numbers of the letters of all of the various desired mottos must be summed and analyzed. And then, resulting forms, ratios, logarithms and whatnot examined and incorporated into a whole. A time consuming process. Yet there would have been a short list of desired target messages and/or numerical forms. (There is much that still may hide in the Remarks and Explanation or within the above 52 letters of the three mottos and date.) Then a final selection, before taking the proposal to Congress. But from the history given in The Eagle and Shield there is a suggestion that there were unsettled details of Latin mottos from *just the day before its adoption.* June 19, 1782, was apparently the final date of Barton's input. His Latin mottos were *very different* than what we have now. Our present form would have to have been re-arranged in one day's time for this account to be true:

> Thus Thomson's report was a composite of his ideas and the language of Barton's; and they both had borrowed from the designs of the earlier committees whatever they deemed useful and appropriate. The Secretary brought his report to Congress the next day.

> (See pg 82, The Eagle and Shield.)

And yet this statement seems to contradict Thomson's quote above, where he says: "...and which I wished to express covertly by the device." This doesn't sound like something thrown together the day after dealing with Barton; and the personal pronoun "I" doesn't appear to include Barton at all. Also, this wouldn't fit the presence of the 13-letter divisible arrangement, a form that seems unlikely to have been slapped together over night, and not at all evident in any of Barton's creations.[12] Nor, would this fit with the apparently encrypted logarithms from letters. Thomson's report must have been prepared far in advance, and Barton had been cut out of the process. (See pg 75, The Eagle and Shield et seq.)

Am I "reaching" too far for an idea? How likely is it that Thomson's "...Sentiments that I had in mind" are the peculiar number-letter encrypted logarithm idea? Was he saying (hypothetically), "this is a bunch of mystical mathematical ideas that you can toss out if they seem silly to you?" *Even if* all of the above log/encryption ideas are merely a *chance relation* arising out of the numbers and completely nonsense, the fact remains that the Designer apparently found them too, and appears to have built them into the geometry of the 1935 dollar.

****

The Document "Remarks and Explanation" and Preamble adopted by Congress:

Preamble:

> The Secretary of the United States in Congress assembled to whom were referred the several reports of committees on the device for the great seal, to take order, reports That the Device for an Armorial Atchievement & Reverse of the great seal for the United States in Congress Assembled is as follows.--

> Arms

> Paleways of thirteen pieces Argent and Gules: a Chief Azure. The Escutcheon on the breast of the American Bald Eagle displayed, proper Holding in his dexter talon an Olive branch, and on his sinister a bundle of Thirteen arrows, all proper, & in his beak a scroll, inscribed with this Motto. "E Pluribus Unum".---

For the Crest

Over the head of the Eagle which appears above the Escutcheon, A Glory, Or, breaking through a cloud, proper, & surrounding thirteen stars forming a Constellation, Argent, on an Azure field.--

Reverse

A Pyramid unfinished. In the Zenith an Eye in a triangle surrounded With glory proper. Over the Eye these words "Annuit Cœptis". On the Base of the pyramid the numerical letters MDCCLXXVI & underneath the following motto. "Novus ordo seclorum"

Remarks and Explanation:

The Escutcheon is composed of the chief & and pale, the two most honorable ordinaries. The Pieces, paly, represent the several states all Joined in one solid compact entire, supporting the Chief, which unites the whole & represents Congress. The Motto alludes to this union. The palesin arms are kept closely united by the chief and the Chief depends on that union & the strength resulting from it for its support, to denote the Confederacy of the United States of America & the preservation of their union through Congress.

The colours of the pales are those used in the flag of the United States of America; White signifies purity and innocence, Red, hardiness & valour, and Blue. The colour of the Chief signifies vigilance perseverance & justice. The Olive branch and arrows denote the power of peace & war which is exclusively vested in Congress. The Constellation denotes a new State taking its place and rank among other sovereign powers. The Escutcheon is born on the breast of an American Eagle without any other supporters to denote that the United States of America ought to rely on their own virtue.--

Reverse.

The pyramid signifies Strength and Duration: The Eye over it & the Motto allude to the many signal interpositions of Providence in favour of the American cause. The date underneath is that of the Declaration of Independence and the words under it signify the beginning of the New American Æra, which commences from that date.--

(See pgs 83-85, <u>The Eagle and Shield</u>.)

****

Section IV--

Conclusions and Thoughts on the Design of the 1935 Dollar:

What can be said with any certainty about this design on the dollar? I will try to summarize Chapter 2 through Chapter 5 concluding the major portion of my findings for the back face of the 1935 dollar. And with this, some leads, that for various reasons I chose not to develop in this book. But there is more: much of what is said may be seen as interim conclusions. I invite the reader to have a look at the possibilities of the deeper dollar-related connections in the Great Pyramid. (See Appendix A, pg 194.)

The writing style I have tried to follow relies upon "it seems" or "it appears", etc., as a caution against dreaded certainty. Where I have stumbled in this investigation has been due to lazy grasping for simple answers. As a result of private embarrassments, my inclination is to keep from fixing an identification of facts until all shadows of doubt are gone. Here I am more reporting facts as I found them, within a construction that seemed reasonable at the time. A report like this has no strict format, but I will try to give a logical account. Here I will try to prioritize the (I) historic influences and what they mean with respect to the design, followed by (II) some speculation on how Mr. Weeks composed this design and (III) a summation and speculation on the design elements, their development and possible meanings:

(I) Many of the symbols on the dollar can be seen in the iconography of the Church, Rosicrucianism, and as far back as ancient Egypt. In the 1770's, Francis Hopkinson, Pierre Eugène Du Simitière and possibly George Washington, (through the evidence of a Masonic Apron), all seem to have promoted the Eye of Providence and the Unfinished Pyramid as symbols for the Great Seal. (Washington's actual involvement is unknown. But that a person of that stature, a unique and focal individual in the Revolution, would have such a symbol on his Masonic apron extends the possibility that many other people would have understood it, and were in accord with it's meaning.) The other known collaborators on Congressional committees were Hon. Dr. Arthur Lee and Elias Boudinot, who do not seem to have contributed anything to the design. Benjamin Franklin, Thomas Jefferson and John Adams on another committee did not appear to have had any great influence on the theme of the design, but dutifully contributed designs. (The addition of Shekina, however, appears from this first committee.) John Rutledge and Arthur Middleton, Delegates of South Carolina, didn't add anything to the Great Seal and evidently did little but quarrel. It fell to Charles Thomson, Secretary of Congress in 1782 to create the final form of the Great Seal, summarizing previous committees' work, with a certain amount of influence from William Barton.

(II) The origin of the dollar's design is rooted in Edward M. Weeks' special interpretation of the Great Seal. The idea of the dollar as the carrier of the Great Seal was apparently suggested by Henry Wallace the Secretary of State under FDR, his idea being a coin. FDR wanted a new design of the paper dollar instead, an assignment probably given to the Director of the Bureau of Printing and Engraving, Edward M. Weeks. Weeks' work is a much more attractive rendition of the Great Seal than the work of earlier artists. He had various engravers as assistants, each apparently was given elements of the design:

J. Benzing and W. Wells engraved the Reverse face (Pyramid and Eye); J. Eissler is said to have completed the Obverse (Eagle and Stars), both are said to have completed their work in *April 1935.* Whether any of these men were aware of the deeper elements of the design, is unknown. The simplest presumption is that they were not.

Weeks apparently sent his first printed "model" to the President sometime in 1934. The President at first approved it. Then he scratched out his signature, drew rough designs of the Eagle and Triangle in opposite positions and wrote a note on the printed sheet ordering him to reverse the Seal faces, so that the Reverse (with Eye and Pyramid) would appear on the left, and the Obverse (Eagle and Stars) on the right (See Illus. Q2). Although the President was almost certainly unaware of Weeks' careful arrangement of this design, or what it might mean to him, Weeks and company at the Bureau reversed and revised the faces in the final form that was issued in 1935. Strangely enough, since most of the design was symmetrical, this did not have affect his secret design very much. Note above that the "bush" is still on the right, and the "sharp corner" (the WC point, see Illus Q2 and pg 189 et seq.) would have been directly measurable from the *western edge* of LIR in this configuration as would the upper-left corner of the little pyramid. The J3-symbolism of the "edge of the letter O and edge of the Eye" -diagonal would have still been present in the earlier form, with manipulation of both the right- and left-hand SIR rectangles. Curiously, this display has a more pronounced effect after FDR's reversal, (See Note 1) the dual position sliding effect being made to happen together within the left-hand SIR rectangle. But although the "bush" near the EC point and triangle of the Pyramidon are not symmetrical, most of the important symbolism has probably been preserved in the final form (the *design 1935 A*).

Illus. Q2   The Original Model and FDR's
                   instructions for reversing faces

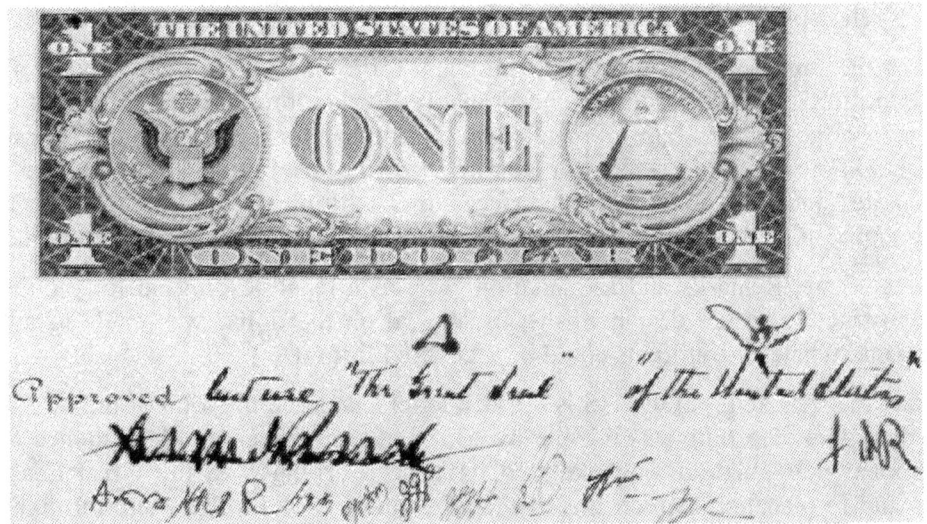

There were printings of the dollar without the IN GOD WE TRUST motto, but it is clear from the geometry shown in Chapter 2 that this graphic element was a part of Weeks' overall plan, even if not an original part. (See Chapter 2, See Illus. D and E where the diagonal touches both serifs on both sides of the letter "I" in this motto.)   The U.S. Bureau of Printing and Engraving says the motto first appeared on some of the 1935 G series Silver Certificates, which were still being printed in 1961. (Note a possible Masonic designation "G" for the God related motto.) The motto appears on all denominations of the 1963 series of Federal Reserve and United States notes issued thereafter. (The suggestion to incorporate "IN GOD WE TRUST" into all currency was sent to the Secretary of the Treasury in November 1953 by Matthew H. Rothert of Camden, Arkansas. Secretary of the Treasury George M. Humphrey liked the idea but believed Congressional approval was required. In March 1955, bills to this effect were introduced to the Senate by Senator Fullbright of Arkansas and to the House by Congressman Harris of Arkansas and Bennett of Florida. This motto was finally prescribed Public Law 84-140, approved July 11, 1955, by President Eisenhower. "IN GOD WE TRUST" was used informally for 93 years, first appearing on a two-cent coin in 1864.)

Speculations on the Development of the Design--Edward M. Weeks' ideas and interconnected elements of the dollar design:

(III)

(1) Edward M. Weeks' design of the dollar is a masterwork of the development of many ideas through *successive approximation,* a theme found elsewhere in the world of sacred geometry. Apparently he followed some method that surfaces in the justifications J1 to J7 given in Chapter 3. After assignment to design the new dollar, Weeks probably began by research on the basis of the Great Seal in the "Remarks and Explanation" as adopted by Congress in 1782, and other lore surrounding the Seal and symbols related to U.S. currency.

(2) Weeks evidently became aware of the gold to silver ratio dating from 1837, from the 17.765-number forms found on the dollar. This was probably from the writings of Piazzi Smyth, in a book on the Great Pyramid written about 1880. It seems reasonable to assume that Weeks' studied most of the lore and related ideas on his own, several years before the design of the dollar.

(3) From research at the time of the design of the dollar, Weeks seems to have found number symbols crafted into the monetary gold to silver ratio by the lawmakers of 1837, and from this he may have gleaned the "pyramid-numbers/Declaration of Independence" concept. These elements were also present together in the legal description of Reverse face of Great Seal. This motif also surfaces in a puzzle formed by the measurements to LIR edges, proposed little pyramid base(s), and a subtraction/reciprocal and squares relationship with the little pyramid first step and base, (See pg 156 et seq.)

(4) Perhaps the presence of the pyramid numbers-forms in the gold and silver values was a start for research into the Great Seal tradition, and probably Weeks drew up some coherent design based on that inspiration. The fact that the ratio of metals gives a number-form similar to the date of the Declaration of Independence, appears to have been (at least) part of the inspiration for the choice of the length of the LOR rectangle, since this number-form surfaces after multiplication by π. This is the basis of the IDT theory (See DFT in 18, below). All the math ratios, dates etc., are variously intertwined in the dollar.

(5) The document "Remarks and Explanation" acts as a legal description for the Great Seal. It appears that the construction of the design of the Great Seal on the dollar was made to adhere closely to that description and whatever traditions could be recovered from it.

(6) The fact that ANNUIT COEPTIS is thirteen letters, and that the total of the three mottoes and Roman Numeral date give 52 a number divisible by 13, would have been ample evidence for Weeks that some kind of lexical manipulation was likely to have been arranged in the Great Seal. This follows the "Remarks and Explanation" which literally requires the use of 13 elsewhere throughout the Great Seal's design. The dollar design implies a "thirteenth line" at the top of the hatching in the Chief; there is a geometrically related association with feather tips on the Eagle (not developed in this writing); a thirteenth arrow point; a thirteenth olive leaf; eyes and beak tip of the Eagle, tip of Shield, etc.

(7) Week's esoteric and graphic use of number and geometry appear to follow (or at least be associated with) a decoding of two number/letter sums found in the "Remarks" by the use of Briggsian logarithms (logs to the base of ten as applied to the partial sum of 415 and total sum of 721.).

(8) Within number/letter sums of the three mottoes and date of the "Remarks and Explanation", Weeks is likely to have found the ratio formed by the words NOVUS ORDO "New Order" produce (52 / 91 = 0.571428571). The use of this number-form, perhaps as an approximation, appears in a LIR subtraction: $(0.826 + (\sqrt{17.76.508197})$ from $\sqrt{31.5} = 0.571621673$. This is a difference of one part in 2900.

(9) The theory that Weeks uncovered the two numerical messages by means of logarithms, shown in Chapter 5 seems evident, since the first of the principle messages $\log_{10} 415$ is practically the same as the primary ratio used in the design for the exterior LOR rectangle, being the square of the golden ratio-- $\log_{10} 415 = 2.618048097$ vs. 2.618033989 (being φ^2). The use of φ and the associated φ^2 are present in a large number of artistic and mathematical schemes, throughout the dollar design. Though the dollar's symbolism takes most of it's mathematical form from the Great Pyramid of Giza, (the π-based ratio as interpreted by David Davidson), the history of φ^2 is only weakly associated with the Great Pyramid lore, (e.g., Herodotus' surface area idea). Davidson says nothing about the φ-ratio, at all. The logarithm connection appears to be the sure source of the dollar's symbolism. Although it seems unlikely, Weeks' independent study may have found the φ-connection within the Great Pyramid math.

(10) The second message based on the $\log_{10}$ provides a number ($\log_{10} 721 = 2.857935265$) that appears to approximate one hundredth of Davidson's Displacement Factor (2.861022156) seen on the dollar. It seems likely that this relationship is coincidental, but the great similarity of numbers makes further speculation on possible mathematical resources at the time of the Revolution difficult to avoid.

(11) It appears as though Weeks' use of the numerical message of $\log_{10} 721$ may have become a matter of fastidious devotion to the traditional mathematics of the original Great Seal legal description, since Weeks used this or similar number-forms in several mathematical connections related to a 1/100th of Davidson's Displacement Factor (2.861022156). This may be the source of the "self-scaling" British inch/Pyramid inch -idea, involving paper shrinkage, seen on Pages 135 to 141.

(12) The above $\log_{10} 721$ is also a comparable to another number-forms in the Great Seal letters, being similar to half of the above NOVUS ORDO ratio, divided by ten: $(52 / 91)/2 = 0.285714285$ vs. $\log_{10} 721$ = 2.857935265 being as close as one part in 3600.

(13) The number-form "2.857" appears in at least four places (1) the shorter of the 13L diagonals a shown in Chapter 4: PLD2 = 2.857745339 and (2) the reciprocal of 0.0035" (285.7142857) and (3) the sum-and-average relationship with number-form 0.2865 to the shorter elliptical chord length 0.2857" created in the theoretical downward shift of the ellipse making an average of 0.2861 and (4) the connected form 52 / 91 = 0.571428571, by way of: 0.571428571 / 2 = 0.285714285.

(14) Both number-letter sums 415 and 721 which gave interesting results by way of logarithms, also separately show a cumulative adding to the number 1, (a numerology detail). It is conceivable that this may have contributed to Weeks' attention to the "one" symbolism found all over the dollar.

(15) It is likely that Weeks was quite familiar with the work of David Davidson' work on the Great Pyramid, and the body of prophecy and speculation from that school of thought. The special date given by Davidson of September 16, 1936, appears encoded in (at least) three places in the dollar design: (1) in the subtraction of BSC and 13L lines taken with the 0.0035 offset; and (2) in the complex arrangement of LDO diagonals a Seal ellipses, which require the use of the correct date of September 16, 1936 (as it's square root, decimally reduced) to solve a puzzle, resulting in the Displacement Factor number-form (0.2861, see 13 above) as a solution. And, (3) the squared form of the length taken from the upper-right corner of the little pyramid frustum to the outside edge of the LIR rectangle to the far right.

(16) The fact that the September 16, 1936 date uses the same square root coding process as the Declaration of independence date, and has an equal weight as a tie to the corners of the little pyramid, appears to imply that Weeks thought this date was of equal or parallel importance. That there is a gap above the frustum to the bottom of the Pyramidon, implies a prefigured interregnum between world Orders. (Not conceived to be an ushering-in of the Messianic Kingdom as Davidson et al. had imagined.)

(17) Weeks' use of Davidson's Displacement Factor in the reduced form of 1/100th (2.861022156) is found on the dollar as: (1) at least three graphic radii, from the center of the dollar and the Eastern and Western Balls (and elsewhere not shown); and (2) as part of mathematical ideas such as the DFT theory LOR length (10 φ/ 2.861022156 = 5.655440261); and (3) in implied reciprocal forms such as the theoretical z dimension 3.495254302; and (4) in the mathematical arrangement of offsets of the Eastern and Western Balls. (0.143" · 2 = 2.861, the curious subtraction and addition ideas 0.143" - 0.136" = 0.007" considered together with 0.143" + 0.136" + 0.007" = 0.286". 0.143" + 0.136" = 0.279". Also: 0.279" + 0.286" = 0.565" There is an odd cross-connection to the IDT-pattern from the height to be computed for a π-pyramid having a side of 0.279" which equals 0.1776".)

(18) The apparent connection of 10 φ/ 2.861022156 = 5.655440261 is a likely source for the average length of the LOR rectangle, (perhaps also a supporting source of inspiration for this length with the date symbol, see above No. 4). Although much good theory underlies this concept of the DFT theory, neither DFT or IDT theory is conclusively distinguishable from the 5.655" length as measured.

(19) Davidson's theory (with Smyth) of the Great Pyramid π-based slope angle, appears in (at least) three implied graphic side profiles, and other mathematical connections using circumferences, lengths and a number of other relations not shown.

(20) There are many other arcane geometric relations on the dollar not shown in this writing, the 3,4,5 triangle, and a possible "$\sqrt{2}$-1 : 1 sacred cut" ratio and others. But these connections appear to be somewhat weaker with respect to evidence, and to Rules 1-3. There are φ- rectangles (a pair that overlap across the center) having a length of 3.495", where one diagonal touches the inner serif on the G in GOD on the line "IN GOD WE TRUST". There are $\sqrt{7}$ length and $\sqrt{6}$ width rectangles hidden in the design. The overall art of the design is framed in a hidden grid of small, derivative φ-based rectangles.

(21) A Briggsian log connection appears on the dollar (3.677" between vertical arms in filigree in the lower part of the design giving $\log_{10} 3.677 = 0.5655$ related to the length of LOR) and elsewhere on the dollar not developed in this writing. (Note 0.565" above in 17.) The Displacement Factor might be said to imply the LOR length by *natural logs*: ln 286.1022156 = 5.656349144. But, either one or the other IDT or DFT ideas above seem to be a better explanation for the source of the length of LOR.

(22) The distance from the westerly edge of LIR to the right corner of the Pyramidon may be a reference to Psalms 118:2, also a reference to a capstone, since the measured number-form is about 1.182". Or it may be a reference to a similar point on the lower structure the frustum at EC which measures about 1.397 which is the *square* of 1.182. Another intriguing possibility is that Weeks drew *both* numbers out of the Great Pyramid's math as shown in the Appendix A from Borchardt's incised line. The pyramid is said to signify "Strength and Duration" from the Remarks and Explanation. This is curiously similar to the Masonic symbolism of the Two Pillars at the door of the Temple of Solomon, called Jachin and Boaz. It seems odd that these two qualities are needed to describe one Pyramid, when this is more similar to the traditional dual pillar symbolism. Although there wouldn't appear to be any connection to the Pyramid to two pillars, later in Appendix A, we will see a possible connection.

Notes:

[1.] Presumably, the Designer readjusted these puzzle problems to his satisfaction after the Chief Executive meddled in the design. The base of the little pyramid is apparently symmetrical to the ellipse of the Seal face so that might not matter if the two sides are switched--but for the Pyramidon it *does* matter--it is not symmetrical at all. This difference is important, and the problem can be seen from the extended edge line of the Pyramidon, shown in Illus. K2 of the last chapter. The opposite edge of the Pyramidon has a different angle, so it would point to some other place than where the Designer intended. But mirror reversal of the Pyramidon might have fixed the problem--if Weeks was willing to do that. How annoying that must have been for the Designer--he certainly couldn't very well have complained. Sort of a J3- turn of events irony for Weeks. In the later design version the measurement problem is a maybe little different than it might have been. (The original Model should be measured.) In either circumstance though, we still have seen a clear form of the imposed uncertainty in the ends at the Eastern and Western balls.)

[2.] It does match something. Through this mathematical route there are some other very interesting things--especially in *the double height* about which we will see more later, relating to the "1776" number and several other things.

[3.] LPB1 was postulated in chapter 2 as B / 5.655 or, 0.555542" a J3 reverse from B · 5.655.

[4.] Backward: September 17, 1789: Day Number 260 divided into 365, plus 1789, divided by 100, (17.89712329), then the square root, (4.230499177) minus $\sqrt{31.5}$, minus LPB3, leaving 0.825699902: which is about three ten-thousandths (0.0003) smaller than the measured 0.826". (Note, that September 16 1936 in a leap year of 366 days *has the same Day Number* as September 17 in the non-leap year of 1789, of 365 days. A curious and intriguing fact. Also, notice that the length of LOR in the DFT theory comes near to the 260 Day Number for the year 1776: 5.655440261 · B = 17.76708958, then, n · 100 - 1776 = 0.70895767, then, 366 · 0.70895767 = Day Number 259.)

[5.] I can't resist noticing that this implied period is practically an even 147 years. The fact that these dates are part of the Designer's scheme may well turn out to have some sort of meaning.

[6.] There were some variant forms of lesser fineness after 1853 and 1873 as half-dollars, at his time and later.

[7.] Pyramidographia, by John Greaves, 1646 pg 69.

[8.] America's Secret Destiny, Robert Heironimus PH.D.

[9.] Pyramidographia, by John Greaves, 1646 pg 94. This original figure of Greaves is a little shy of the above figure from

Smyth, which was determined through measurement and corrected by a least-squares process that used all of the chamber's proportions, done by a Mr. James Simpson.   The Great Pyramid, (Originally, Our Inheritance in the Great Pyramid) by Piazzi Smyth, London, 1880, pg. 196, et seq.

[10.] These two numbers could also be said to represent *two fundamental elements* of the mathematics of the Great Pyramid design, as shown in many equal area schemes with respect to the mathematics of the Great Pyramid ratios. One is tempted to suspect that a symbolic identities are being signaled between the 232.52-based area and *gold*, and the 412.132-based area and *silver*. Maybe copper and nickel should be looked into. See Smyth's table PLATE XXI. These two ideas appear in one diagram as equal areas in Our Inheritance in the Great Pyramid.

[11.] Number by John McLeish, pg 165, 1991. (A good book.)

[12.] Barton's four mottos: *Virtus sola invicta* "Only virtue Unconquered"; *Deo favente* "With God's favor"; *In vindiciam libertatus* "In defense of liberty"; and *Perennis* "Everlasting". (17,10,21,8 letters, a total of 56 letters: Not divisible by 13.) Barton *also* took some liberties with Latin. Although these mottos were not used, perhaps they also deserve some analysis. Barton's last useful idea to be added appears to be to make the stripes on the Chief vertical--but this wouldn't have affected the three mottos and date, that Thomson appears to have settled on before June 19, 1782.

APPENDIX A

OBSERVATIONS AND THEORY ON THE ENTASIS OF THE PYRAMID OF GIZA

This is an analysis of the mathematical "fine structure" of the base course of the Great Pyramid of Giza, showing a subtle mathematical motif. This delicate form is similar to the *entasis* or "stretching" built into temple configurations of ancient Greece. An earlier form of this essay was privately published for friends in 1994[1]. This paper is more of a technical report, different in style from most of the writing in this book. It was written to answer a friend's remark that "there wasn't any validity to Davidson or his Displacement Factor." The hobby research I was doing around that time in 1990 on the oddities of the centerline in the little pyramid on the dollar, inspired me to research "Borchardt's incised centerline" on the northerly pavement of the Great Pyramid. Curiously, I am happier with my theory ideas for the Great Pyramid, than for the results on the centerline of the little pyramid on the Reverse Seal face on the dollar, for which I found no satisfactory solution. This paper has been greatly modified and direction changed for clarity. There are changes, some elements removed, and it is focused on the dollar research project. This analysis is mostly speculation and mathematical experiments based on survey data made by Cole in 1925, done for Professor Ludwig Borchardt[2].

In his book *War as I knew it*, the American general, George S. Patton Jr., gave descriptions of his travels and sights in various places in World War II. He gave particular attention to the sites and lore of ancient history. After the invasion of Sicily, Patton and Major General Hugh J. Gaffey were riding around the Sicilian countryside when Patton spotted a ruined ancient Greek temple. Gaffey and Patton tried a visual experiment related to the Greek architectural form:

> "At a small road junction called Sagesta, Hugh Gaffey and I saw the most beautiful Greek temple and theatre that I have yet encountered. With the exception of the fact that the roof of the temple no longer exists, it is in a perfect state of preservation and has been very little repaired. Since the Greeks were driven from this part of Sicily in 470 B.C.--that is, some twenty-five hundred years ago--the temple must have been built at an earlier date.
>
> "There is one rather peculiar thing about this temple. The columns are not monolithic, composed of two or three blocks, as is usually the case, but are built up of a number of small stones. It is further noteworthy that, after the lapse of two and one-half millenniums, you cannot get a sharp knife edge between the joints of the stones.
>
> "When I was about eight years old, a minister named Mr. Bliss told me that when he visited the Parthenon he had put his silk hat on one end of the steps, and having gone to the other end sighted across and could not see the top of his silk hat, indicating that, in order to cut agreeable lines, the straight lines of the temples were actually curves. Gaffey and I tried the same thing at Sagesta with two steel helmets, one on top of the other, and were unable to see them over the curve of the steps."

War as I Knew it pgs 65-66

This curious building form is typical of ancient Greek temples. It is widely thought by architects that the purpose of this kind of subtlety in ancient building proportions is for the aesthetic effect on the eye, like Patton's "agreeable lines". Yet there is a body of evidence to suggest that this was done for ceremonial reasons having to do with the temple's mathematical proportions rather than art.[3] Indeed, there are, in fact, many designs for objects where small shifts are made only for aesthetic effect. (In the Japanese game of Go, the playing boards appear to be a square regular grid, but their lines are actually *slightly rectangular*. They are longer in the direction between the players, so that at the height of the two seated Go players the appearance will be more like a square grid.) But this doesn't seem to be the motive for the various oddities in the lines and lengths of Greek and Egyptian temples. Probably the most telling piece of evidence is that the northerly lines of nearly all of these structures are shorter than the southerly lines, (in both Greek and Egyptian structures) so this can't have been done for an observer's aesthetic tastes. According to Livio Stecchini, this kind of curvature is a *parabolic arch*, which is a very complex refinement, but completely inappreciable by eye. If the statement is true, it would have been a far more difficult shape to layout and build than a regular arc or a catenary curve. These proportions might only please a Deity or a deep student of temple Mysteries. In any case, the intent doesn't seem to have been for the sake of art.

The entasis for the Great Pyramid is not a curve. Earlier in Chapter 4, we touched on David Davidson's idea that various considerations of the base course of the Great Pyramid can be made to produce the Tropical, Sidereal, and Anomalistic year lengths. His theoretical idea was that the base is a twelve-sided shape. In three dimensions, this pyramid would be thirteen symmetrical planes, in three types of shape above the base. (See pgs 84 and 96.) These planes would have small differences barely distinguishable from a plain, truncated pyramid. Though there are some reasons to suspect Davidson's idea is correct, there is scant evidence for some of the lower elements of his proposed design. His geometric proportions produce close approximations of these year-lengths by implied geometry. Even if these subtleties only existed in the mind of Davidson, this too falls under the category of a type of a hypothetical entasis form. The following ideas developed in this essay only concern an eight-fold symmetry at the base of the Pyramid, for which there is some corroborative photographic evidence. This proposed eight-fold division might be said to be concordant with part of Davidson's system, and doesn't appear to add to, or contradict Davidson's general concept.

This photo was made (accidently) near the moment of sunset at the Autumnal Equinox, taken by British Brigader P.R.C. Groves, about 1920.

Unusual geometry is demonstrated by shadows on upper pyramid (The Great Pyramid).

Note the alignment of southerly edge line of pyramids and shadows. The faintly lit triangular face on the Great Pyramid is suggestive of Davidson's proposed geometry, or at least some slight eight-fold symmetry of faces.

(Photograph faces Northeast.)

Secrets of the Great Pyramid, by Peter Tompkins, pg. 109

Without concerning ourselves with the prophetic scheme by David Davidson, or confusing his religious ideas with his technical skill, as shown in The Great Pyramid: Its Divine Message[4], we should spare another look at his work. Specifically, his **(I)** geodetic metrology, **(II)** his mathematical concept, and **(III)** his idea of a special "Displacement Factor". We need not concern ourselves with his reconstructed pyramid-as-built-concept however, which seems to be greatly in error. Nonetheless, the data he shows and mathematics he uncovered appear to provide solid clues to part of the original intent of the Great Pyramid's builders. This analysis uses independent source of measurement of the Pyramid's base course, as provided by a contemporary, James Humphrey Cole, a professional surveyor from Great Britain. This theory awaits a high precision, thorough replication of Cole's survey.

(I) Geodetic Metrology:

Davidson believed that the Pyramid was designed and laid out in special units of measurement very nearly the same size as the British Inch, which he called the "polar inch or primitive inch", (PI). He concluded that the pyramid builders or someone of their tradition, at or before 2700 BC had measured the earth with great accuracy, and used as the basis of their measurement the polar diameter of the earth, for which they divided the distance into 500 million units. He said that this unit was "about 1.0011 times larger" than the British Inch, yet from the data given, he apparently used[5] a rounded form of this from data giving a factor of 1.00108. (His published value was: 1.00106.) He believed that the British and certain other cultures originally inherited their inch-like units from a common source in ancient times. He also thought that these ancient Egyptian engineers had a special cubit composed of 25 of these units, which would produce a polar radius having an even 10 million of these pyramid cubits.

William Herschel the British astronomer had shown many years earlier, that there are around 500 million British Inches through the earth pole to pole, and suggested that an actual 500 millionth part of the polar diameter might be adopted by British (and the world) as a scientific unit. (See Chapter 4.) Herschel thought this was a more logical standard than the French idea of making a unit of one 10 millionth part along the uneven and elliptical surface distance on the earth between the North Pole to the Equator, (the meter). But, this would have required the British to change their traditional inch to a slightly longer, and inconveniently longer length by somewhat more than a thousandth of an inch.

From recent geodetic data published in the U.S. National Oceanic and Atmospheric Administration's professional Paper NOS 2 (December 1989) in <u>North American Datum of 1983</u> using the ellipsoid for the World Geographic System 1984 (WGS 84), the following table gives the Earth's modern polar radius as the "Semiminor Axis" in Table 22.2. This is a statistical average many observations from satellite telemetry, and is an accurate model irrespective of specific seasonal change of ice, tides etc. This is half the diameter between north and south poles, said to be an accuracy of about one third of a meter, or around one foot:

250   *North American Datum of 1983*

TABLE 22.2.—*Derived geometrical constants*

Parameter	Notation	Units	Ellipsoid	
			GRS 80	WGS 84
Semiminor axis	b	m	6356752.3141	6356752.3142
Eccentricity squared	e^2		0.00669438002290	0.00669437999013
Flattening	f		0.00335281068118	0.00335281066474
Reciprocal flattening	f^{-1}		298.257222101	298.257223563
Polar radius of curvature	c	m	6399593.6259	6399593.6258

Notes on the Semiminor Axis and Inch/Meter Conditions:

(a); I will use World Geographic System 1984 values (WGS 84) and, (b) first translate from meters into inches of the pre-1946 British foot of 0.30479916 meters per foot, the value that Davidson would have used. (This is not the International Foot of "0.3048 meters per foot, exact" nor is this the US Survey Foot of "12.00/39.37 exact", which is 0.30480060901 meters per foot. All three are different standards. In 1946, the Empire Scientific Conference redefined [changed] the yard to 0.9144 (exact) of a meter, which became the above International foot. The U.S. National Geodetic Survey favored the use of the above U.S. Survey Foot having 39.37 inches per one meter, equaling 2 parts per million longer.[6])

This is the conversion Davidson probably would have used with our modern figures, into old British feet* based inches:

Eq. 1

$$6,356,752.3142 \text{ m} / 0.30479916 \text{ m*} = 20,855,544.0711 \text{ pre-1946 British Feet,}$$

using the WGS 84 semi-minor axis shown above;

Eq. 2

$$20,855,544.0711 \text{ ft} \times 12 = 250,266,528.853 \text{ pre-1946 British Inches,}$$

from a pole to the center of the earth;

Eq. 3

$$250,266,528.853 \text{ in} \times 2 = 500,533,057.706 \text{ the earth's polar diameter}$$

in pre-1946 British Inches.

So there are, in fact, *about* 500 million old British Inches pole to pole. Using this data, how big would Davidson's proposed "Primitive inch" be in these pre-1946 British inches?

Eq. 4

$$500,533,057.706 \text{ pre-1946 British inches} / 500,000,000.00 = 1.00106611541$$

(Or about 1.00107 pre-1946 British Inches.)

Davidson's rounded "1.0011" is not bad as a given value for this polar unit, correct to four places. (An accurately computed value from Hayford is 1.00108, giving a difference of about 1 part per hundred thousand. Davidson said he used "U.S.A. Geodetic Survey results of 1906 and 1909" [Hayford 1910].)

If we wanted *a modern metric unit value for Davidson's "primitive inch"* we could divide twice the WGS 84 Semiminor Axis by 500 million. Davidson's published value "1.00106" fits modern estimates:

Eq. 5 $(2 \times 6356752.3142 \text{ m}) / 500,000,000.00 = 0.02542700926 \text{ meters} = \text{"primitive inch" (PI)}$

Eq. 6 per $0.3408 \text{ m} = 1.00106335657 \text{ International Inches per one PI}$

Eq. 7 per $0.30480060901 \text{ m} = 1.00106135444 \text{ US Survey Inches per one PI}$

Since the above number from Eq. 5, (a length of 0.02542700926 m) given for one hypothetical PI was derived directly *without recourse to inches of any period, US or UK,* we need not be concerned with the extra step in conversion, nor which sort of inch should be used in conversion to before comparing with metric measurements. This factor will be used below to convert metric units to ideal Primitive or Polar Inches.

Notes on precision:

The original theory in this essay was inspired by the evidence provided by J.H. Cole who published his data in meters to the thousandth of a meter, 0.001 m, or to one millimeter. His angular headings were given to the arc second. The given decimal value in reported measurements is usually taken to mean that the surveyor was reasonably satisfied with the stated result, and not beyond that level of accuracy. And we may be sure that the smallest unit Cole could have measured was the millimeter using a chainman's gage or card. But he didn't do this--the markings on Cole's tape were to one centimeter:

> "The short distances between these marks and the nearest traverse point were measured with a standardized steel band graduated in centimetres and read to millimeters by estimation."

Pg. 5 Determination of the Exact Size and Orientation of the Great Pyramid of Giza.

It is important that these numbers be used with caution, since he didn't measure to the millimeter. The next decimal, millimeter unit would be called "the doubtful digit" in statistical jargon, due to estimation. Cole's stated "margin of error" at each of the Pyramid's corners was quite a bit larger than a millimeter, but his final stated figures appear to have been the result of a statistical mean of measurements, temperature correction of Invar wire measurements, and possibly other adjustments. Much of science rests on doubtful digit data. Using one form of his data for the north line of the Great Pyramid, a persuasive group of ceremonial proportions emerge through the use of his figures by an exact scaling into modern hypothetical Polar Inches.

This theory is about measurement translated into supposed ancient inch-like units of measure, based on a pro-ration of the measured proportions. Since one thousandth of an inch is a vastly different size than a thousandth of a meter, what decimal place limit should be displayed in translated comparisons?

In this essay where measured meters and inch-like units will appear translated, I will follow a convention similar to that now used in electronic dual-measurement micrometers, on occasion adding one rounded digit for accounting, if it seems reasonable, which is usually not shown.[7] However, in certain cases, extrapolation of simple fractions of the smallest unit of measure into the next decimal place should also be permissible. Where, for example, a reading of 0.01 mm (a hundredth of a millimeter, 0.00001 m or the fifth place) is found in measuring in metric units, the translation to inch equivalent will be 0.0004 inches, (i.e., 0.0003937 inches, rounded at the fourth place.) This would be a difference of one decimal place. But, here I may add another rounded digit. So, a finding of say, 0.053 m (or 53 mm), which is 2.09 inches strictly speaking, I may call this "2.087 inches" showing this extra rounded digit, for accounting. And if I know this measured number to have been divided into say, exactly a half or a third, I will add one more rounded digit, as a special case.

In dividing to produce a ratio between two numbers, having two numbers of three decimal places of precision above or below the decimal place, I will allow both values in computation (six places) plus one more for accounting--but not to exceed 10 floating decimal places. (The primary distance measurement data here will not exceed six floating places in metric units, and if this is multiplied by a ratio or a scale factor, use of any more than the sum of both sets of floating place accuracies is only for accounting.)

The above considerations are very important, since the theory expressed here rests at the edge of practical ability to measure the data, and my evidence and clues are derived from reputable measurement of dimensions and angles. The angular closure of the Pyramid figure supplied by J.H. Cole is of higher accuracy than the dimensions, and the dimensions provided by Cole are the beginning point for extrapolation and mathematical experiments into the realm of ceremonial mathematics. I will try to treat the data with great care.

My sympathies lie with the skeptics of the theory that follows. Inspiring ideas and conclusions may be drawn from data to some extent, but the data must not be made to fit conclusions. We must stay within the reasonable bounds of precision--but then, not ignore telling results if they appear under acceptable conditions of evidence. Where I show greater ranges of precision, of say, 10 to 12 floating places, the reader should understand that the figures shown are either (1) mathematical proportions, (2) geodetic data, (as in the above computations, with a much higher given range of precision than land surveys), (3) data multiplied by scale factors, or by mathematical constants, or (4) a totally theoretical constructions. That is to say, imaginary, an ideal mathematical construction of absolutes, based on natural constants, proportions, geodetic information, or a plausible combination of these. I will attempt to make these distinctions clear where necessary. Much of this theory rests on the recovery of very fine *numerical ratios,* some of which are the result of what appears to be elaborate care on the part of the ancient builders. It should be remembered that this work is an attempt to ferret-out a very old message from this structure and with slightly imperfect data. The chief tool in this process is the presumption of good angular observations by Cole, and the translation and small adjustment of the total base into a plausible ceremonial length. For this theory to be credible, great care must be taken with the manipulation of values, within the restraints of survey precision.

To sum up: the above segment establishes the basis of geodetic values for the Earth's size between poles from the US NOAA data; the actual computed length of a modern polar inch (one PI equals: 0.02542700926 m); and general remarks and constraints on precision and notation in this essay.

****

Ken McGrath

(II) Mathematical Concept--A Pi-Based Pyramid:

David Davidson said that the Great Pyramid was designed so that the total base course of masonry (tbcm), if divided by twice its height (h), would prove to be the important mathematical proportion pi, or 3.141592654...:

Eq. 8   (tbcm) / 2h   = π =   3.141592654...

For this to be true, the slope angle at the base must have an angle of 51°51'14.31", which would be 90 degrees minus the arc tangent of $\pi/4$, a well known mathematical theory, over a hundred years old:

Eq. 9 $arc^{-1}(\pi/4)$ =   38°08'45.69"
           - 90°   =   51°51'14.31" (The ideal "pi angle" of Smyth and John Taylor.)

Davidson believed that the builders chose the above 51°51'14.3", but unfortunately, here we would be conjecturing about this now rough slope. The casement stone face that might prove or disprove this idea was mostly removed in medieval times to build Cairo. But many examinations of surviving casement stones and surveys of this slope come within a few minutes of this angle, and J.H. Cole accepted the generalization that the average slope angle was close to 51°51'. Curiously, some measurements of this slope, notably that of W. Flinders Petrie, argue toward another old idea, the "Golden Ratio" proportion:

Eq. 10 $(\sqrt{5}+1)/2$ =   1.61803398875...
                                   = φ (phi, the Greek letter symbol)

This second angle is derived from 90° minus the arc tangent of the reciprocal of the square root of the golden ratio:

Eq. 11   arc tan $(\sqrt{1/\varphi})$ =   38°10'21.75"
               - 90°   =   51°49'38.25" "phi angle"

Between these two proportions of π and φ angles there is a span of about one minute and thirty-six seconds of arc, or an angular difference of about 87 mm (perpendicularly) to the slope. W. Flinders Petrie's actual observations by transit falls more or less between these two at 51°50'40"(+/- 1'05")[8] for the North slope, an angle of some 14 seconds or so steeper than the average of the two proportions. It is possible that this angle may indicate an important observation, possibly pertinent to later theory developments in this essay.

****

(III) "The Displacement Factor"

Davidson claimed that the base of the Great Pyramid was built to represent the solar year (or the astronomic tropical year, equinox to equinox). He said that the base measured in primitive inches (PI) would give 36524.2465 PI, or a number that was 100 times the average solar year over the passage of several thousands of years as Davidson estimated it, and also from Egyptian lore. We will call this a ceremonial length or "total base course ideal of Davidson", **(tbcid)**. (The modern astronomical data of the tropical year is 365.242191 days. If this difference is important, the difference would amount to about three hundredths of a Polar Inch over the tbcid ceremonial length.)

Davidson wrote of a Displacement Factor (D) of 286.1022156 PI, (7.2747 m). Davidson noted (reproducibly) that (a); the centerline of the Descending Passage, Ascending Passage and Grand Gallery was offset about 286 PI to the East of the true centerline, an observation that had been approximated by several other researchers. And (b) Davidson said there was a *implied* "rhomboid of displacement" (See Illus. 1) having sides of 286.1 PI, in three dimensions, that was a function of the geometry of the South end of the Grand Gallery at the Pyramid's central axis: between the roof line plane of the Grand Gallery and the extension of the roof line plane of the 1st Ascending Passage, the above (a) East-West centerline displacement running Westerly along the East-West plane from the central axis of the Pyramid, (along the Great Step), and an implied definition running North, but on the slope down, along the North-South central geometric plane of the Pyramid and along the slope of 26°18'09.73" which is the angle of the ascent of the Ascending Passage and Grand Gallery. (i.e., this three dimensional shape is rhomboidal facing East-West, and right angled facing North-South.) [9]. Davidson also noted, (c) that the above slope of 26°18'09.73", (his ideal angle for the ascent), was the resulting angle from a right triangle, which has a vertical height of 2861.022156 (10 times D), where it would intercept a slope length of 1/4th of the sum of the diagonals of the square formed of the above length of 36524.2465 PI. (This diagonal length is 6456.635594 PI, and would not physically evident in any particular place, except as a one-fourth part of the base diagonals. So, the three dimensions and angle, are related to the Displacement. (See pg 102. Plate XXXVII: <u>The Great Pyramid: Its Divine Message</u>.)

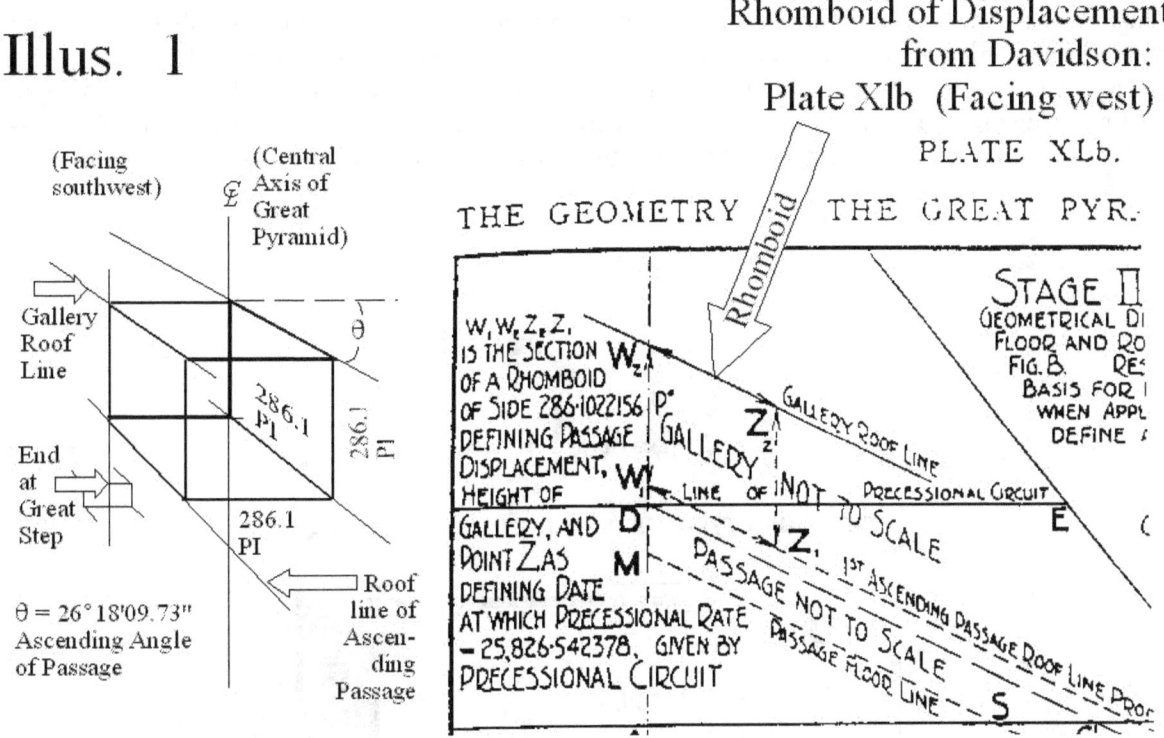

Illus. 1

Rhomboid of Displacement from Davidson:
Plate Xlb (Facing west)

In these physically evident offsets or relationships, there are three distinct types: there is an East-West lateral difference of 286.1 PI in (a); a vertical difference of 286.1 PI between sloping parallel planes in (b), [the North dimension is only implied]; and (c), another sort of vertical difference implied by the trigonometry of the Ascending Passage slope, and an angular representation of this number, though a symbolic form of this number. It must be said of the last two, that the Grand Gallery roof line is somewhat rough, and the slope only is symbolic through trigonometry, where the true slope is within a minute or so of the ideal angle, but on average both (b) and slope (c) relationships clearly exist from physical evidence.

Davidson also claimed that this 286.1 PI length related to the length of the sides of the "missing capstone" though there remains no evidence of this supposed feature, since all of the carefully worked casement stone surface in this area is now missing. There is however, an interesting set of geometric proofs given by Davidson, (See pg 96 and 97. Source: Plates XXIII and XXIV <u>The Great Pyramid: Its Divine Message</u>) using squares of area and equal quadrants using a tenth of the supposed Pyramid year circuit and Pyramid height that provide a mathematical basis for his 286.1022156 number; and Petrie's measured displacement from the centerline of the structure.

Davidson gave no complete analytical reasoning for his choice for this special length beyond the geometric proofs mentioned above, and physical measurement by W. Flinders Petrie. He seems to imply that he arrived at a formula through the mathematics of those proofs. This is the formula he gave for the Displacement Factor: The square root of π divided by 128 multiplied by 25, and then multiplied by the solar year:

Eq. 12 **Formula:**

$$\sqrt{(\pi/128 \cdot 25)} \times 365.242465 = 286.1022156$$

Eq. 13

(Or using the above formula, using modern astronomical figures: x 365.242191 = 286.102001)

This number could pass unnoticed at this point in time, except for the fact that, **(1)** it appears elsewhere in recent mathematical research on the Pyramid unknown to Davidson, and **(2)** also appearing in the Earth's parameters. This seems pertinent, since Davidson's arguments are based upon a supposed special knowledge by the Pyramid's builders of theoretical geodesy, or that there was a formula known to them, that will produce these astronomical constants. (See Sections VIII and XII for a third possible relationship.)

Illus. 2

THE SARCOPHAGUS IN THE KING'S CHAMBER

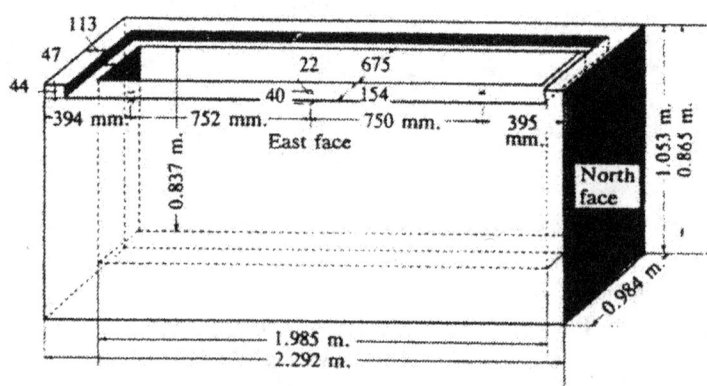

(1) Andre Pochans' data[10] gave measured dimensions of the King's Chamber in cubits, (in cubits of 0.5237 meters*) as:

Eq. 14

height	5 × √5 cubits*	=	5.8551 m	(H)
width	10 × cubits*	=	5.2370 m	(W): H × W × L = volume: 321.1683 cubic meters
length	20 × cubits*	=	10.4740 m	(L)

Later, in a description of the Coffer within the Kings Chamber, Pochans shows a diagram of the Coffer (See Illus. 2) which gives the interior of the Coffer in dimensions in meters, from an unnamed source.

From The Mysteries of the Great Pyramids pg 25, and pg 39 by Andre Pochans, 1978

Of the part of the Coffer that would contain a three dimensionally bounded volume, that is, from the *lower lip* below the stone rail keeper for the original stone lid of the Coffer, (or the actual cubic volume of *the interior* of the stone box). Andre Pochans gave these dimensions in the diagram as:

Eq. 15

height	0.837 m	(H)	
width	0.675 m	(W):	$H \times W \times L$ = Volume: 1.1215 cubic meters
length	1.985 m	(L)	

Dividing the above King's Chamber volume by the above Coffer's volume:

Eq. 16   321.1683 / 1.1215 = 286.380169

In other words, a fraction of about *1 / 286.1th*, as close as one part in a thousand to Davidson's ideal Displacement Factor. Note that the numerical ratio is three digits before the decimal, like Davidson's number.

Again a familiar number-form, but this time as *a volume fraction, and independent of any units of measure.* This apparently would be a fourth and distinct kind of representation of this number. (This stone Coffer should be carefully re-measured, to see if Pochans's results can still be replicated.) It is doubtful that this is coincidental or a mistake. I don't think there is any bias toward this number--it is doubtful that there is any partisan altering of facts on Pochans' part to favor this fraction. Pochans' work is fairly hostile and sarcastic to everything Davidson and his supposed Displacement Factor, and these two pieces of information from different parts of Pochans' book and were not stated by Andre Pochans to have any relationship. Consider however, the elaborate care taken by the ancient builders in the architecture of this chamber and this sacred stone box. This stone box and surrounding chamber is the central focus of literally millions of man-hours of effort and millions of tons of stone built to seal up and enshrine. Mathematically, the Coffer's solid stone volume is said to be double its internal volume as measured from the *top rail*. Also as noted by Petrie, this volume (taken from the top rail) is said to be an aliquot unit of the King's Chamber volume, divisible by five.[11] We should recall the religious emphasis given by the ancient Egyptian religion to measures and volumes. The Coffer as it happens, must have been *built into the structure* at the time of construction. We know this because *it is taller than the entrance to the King's Chamber and could not have been taken through the doorway.* This says something of its precise shape and primal importance to the Builders.

(2) Coincidental Ratios:

As noticed by other researchers, the square root of the Earth's eccentricity, (or the fourth root of e^2-- shown above, in the next entry on Table 22.2 of the NOAA's North American Datum of 1983), shows what may be an other odd coincidence:

Eq. 17

$$(WGS\ 84)\quad (0.00669437999013)^{1/4}\ =\ 0.28604054056$$

Yet another *fractional form*--yet this time one thousand times smaller, here again resembling Davidson's Displacement Factor. (Also, numerically, similar to the largest elliptical chord on the dollar. See Chapter 3.)

It should also be noted that dividing the measured base courses of the Great Pyramid (921.453 m) into the Second Pyramid, (861.046 m)[12] make a number that if squared and divided by four results in a ratio of 0.2863, a number-form reminding one of the Coffer volume/King's Chamber volume, shown above.

Additionally, as for other astronomical proportions, various writers (John Mitchell and Bonnie Gaunt) have proposed an imaginary pyramid having a height made from the diameter of the Moon plus a radius from of the average diameter of the Earth, whose base, using the average diameter of the Earth will have a slope angle approximating the π-based slope +/- of 51° 52'. Bonnie Gaunt[13] also shows that the centered, adjacent Earth and Moon square figures of the above data will yield a 3,4,5 triangle (2160, 2880, 3600 miles) between corners of the two squares. This natural geometry is strongly suggestive of *both known proportions* found on the two largest pyramids on the Giza Plateau. (There is also a possible hint to traditional units angular measure: degrees, minutes and seconds that could conceivably be based on a proportion of the diagonal 3600, or alternatively, half of the imaginary triangle's fundamental unit of 720.)

There is an argument to be made to justify the use of these facts: although our modern science is greatly superior in astronomy and telemetry to anything known to ancient Egyptian science; ancient astronomers might have somehow discovered the Earth's eccentricity and other fundamental dimensions at some level of accuracy. They might have done this through mathematical theory such as above proportions or arrived at it through geodesy, by long celestial observation much after the fashion of Eratosthenes and other ancients. They need not have used anything like decimal convention as we do now, their known fractional system would have sufficed to annotate or handle all of the mathematics shown here, decimal or otherwise. But some form of the knowledge of the Earth's eccentricity and flattening might well be a convergent source of inspiration for the measured Displacement, in addition to Pyramid geometry. And then perhaps, for political, religious, astronomical or purely mathematical reasons, the builders then built these ratios into their Pyramids, perhaps as a statement of holy wisdom.

****

(IV) Starting Points:

My introduction to J.H. Cole's survey data was as a comparison to Davidson's data, taken from <u>The Secrets of the Great Pyramid</u> by Peter Tompkins, given in the appendix by Livio Stecchini. Davidson said his data was derived from the survey work of W. Flinders Petrie, a skeptic of most pyramid theory. I was surprised at the solid reasoning behind Cole's method discussed in <u>The Mystery of the Great Pyramids</u> by Humphrey Evans.[14] According to Humphrey Evans' account:

> "James Humphrey Cole extended the line of the edge of the casing blocks that was still traceable on the pavement on each side of the Great Pyramid. In this way he was able to fix the position of the corners and measure the length of the sides with an error of less than 4 centimetres (1.6 in.)."

If this is in fact how Cole determined the length of the sides, it appears that this would in fact be the most reliable method of re-establishing the original corner points, while *avoiding the century old debate surrounding the character of the sockets and missing corner stones.* This was indeed the purpose of this survey: to settle once and for all the controversy caused by pyramid theories. Cole's survey data was given by most sources as dimensions with headings. Having at that time no independent access to Petrie's data outside of Davidson's work, I started by investigating various ideas using Cole's data [6] from various sources. Eventually a friend located a copy of Cole's original nine page report for me.

The following is Stecchini's arrangement of Cole's data into distances and headings, with original side lengths of the Great Pyramid. (Here Stecchini's millimeters are shown in this essay as meters for uniformity.):

TABLE A (L. STECCHINI)

North side:	230.253*m	2'28" South of West
South side:	230.454 m	1'57" South of West
East side :	230.391 m	5'30" West of North
West side :	230.357 m	2'30" West of North

(921.455 m = Cole's original sum of base lengths.)

MINISTRY OF FINANCE, EGYPT.

Survey of Egypt.

Determination of the Exact Size and Orientation of the Great Pyramid of Gîza.

By

J. H. COLE B.A., (Cantab) F.R.G.S
Inspector, Computation Office.

SURVEY OF EGYPT PAPER No. 39.

Government Press, Cairo, 1925.

To be obtained, either directly or through any Bookseller, from the GOVERNMENT PUBLICATIONS OFFICE, Ministry of Finance (Dawâwîn P.O.), Cairo.

Price - - - - - - - P.T. 10.

As Livio Stecchini showed, [15] these angles "...deviate from as follows from a right angle:"

TABLE B (L. STECCHINI)

Northwest corner:	- 0°00'02"
Northeast corner:	+0°03'02"
Southeast corner:	- 0°03'33"
Southwest corner:	+0°00'33"

(If heading angles are subtracted, the northwest corner is practically a right angle, internally, 90° minus 2 seconds. See Fig 1 below.)

Eq. 18

This is the positive sum of the above headings, as the headings deviate from a right angle:

Sum: 0°07'10"   (Compare: π / 26°18'09.73" = 0°07'09.98")

The angle given here is a sum of residuals, all converted to the positive sign: i.e., 0°07'10" = 430 total arc seconds of residual degrees against a square. Curiously, if Davidson's slope angle for the Ascending Passage (26°18'09.73") is divided by π, practically the same angle value will result: 0°07'09.98". This is a probable connecting theme. We will return to this sum later. These angles, plus or minus a few seconds, Stecchini continues, "deviate from a right angle according to the following pattern", which is a rough but intriguing generalization:

TABLE C   (L. STECCHINI)

Northwest corner:	0 minutes
Northeast corner:	+3 minutes
Southeast corner:	-3 1/2 minutes
Southwest corner:	+1/2 minute

This was interesting since there appeared to be a regular pattern of some kind. Each is in 1/2 minute values. I concluded that the differences in lengths of sides and headings were not accidental, as builder's error, as some have said--but carry a sacerdotal form of some sort. This form, slightly diamond or rhomboid in shape was clearly artificial in character--especially so, with what appeared to be a Right Angle at the Northwest corner. (This is the *second* rhomboid shape we have seen thus far, in connection with the Pyramid, but of a more subtle distinction.) This *biased* rhomboid brings to mind the question of what the Pyramid base's "natural shape" might be. What would its *symmetrical conformation* be, using the same dimensions? (Later we will look at these dimensions in an "unbiased" configuration, relaxed into a double bilateral symmetry in Traverse Experiment B.) It also appeared as though the Great Pyramid's designer demonstrated a knowledge of the "three hundred and sixty degree--sixty minutes--sixty seconds" angular system that we still use today, or perhaps some similar system dividing evenly into our system.

TRAVERSE EXPERIMENT A

The above data of distances and headings of TABLE A enables us to compute a traverse in bearings to provide a mathematical check on the above account of Cole's survey:

Traverse Notes: (a) the mathematical origin of these coordinates (the northings and eastings) is an arbitrary North 10,000 m by East 10,000 m, at the intersection of centerlines of the Pyramid, i.e., at a point of intersection between the mid points of the four lengths of the sides. (b) See discussion of the North line below. (c) No scale factor or rotation was used. (d) Headings were converted to Bearings. (e) This traverse does not represent a physical survey procedure, and is a mathematical experiment after the fact, solely using the above data taken from Cole.

(Note: A "traverse" is a commonplace land survey computation, ordinarily done today by computer programs. This converts survey information into an accounting process on a grid that allows one to inspect how closely the measured angles and distances actually return the beginning point of a survey, and to compute the precision of the work. The test of relative precision is found in the closure distance divided into the perimeter, giving a fraction, similar to tests of precision given elsewhere in this book. The resulting North and East coordinates (points on a grid) are used to create an exact mathematical map, and simplify other computations.)

Fig. 1: Plan of Traverse Experiment A

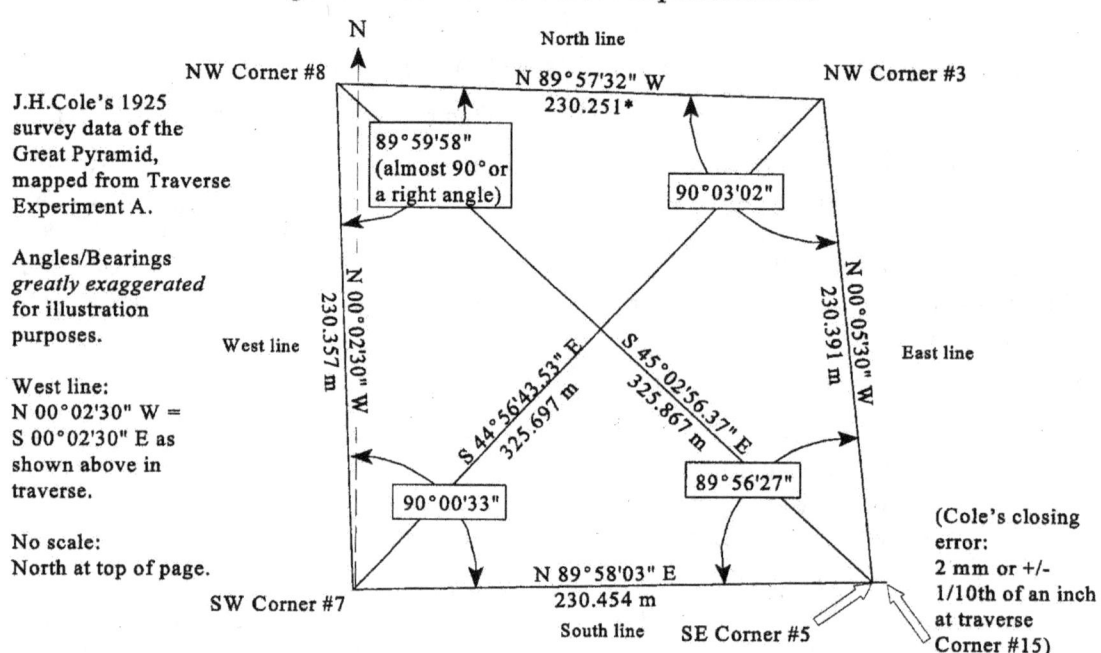

TRAVERSE EXPERIMENT A

course origin	distance(m)	deg min sec (bearing)	northings	eastings (m)	description
			9884.805	10115.227	PT# 5 SE COR
1 (EL)	230.391	N 00°05'30.00" W	10115.195	10114.848	PT# 3 NE COR
2 (NL)	230.251*	S 89°57'32.00" W	10115.030	9884.607	PT# 8 NW COR
3 (WL)	230.357	S 00°02'30.00" E	9884.673	9884.775	PT# 7 SW COR
4 (SL)	230.454	N 89°58'03.00" E	9884.804	10115.229	PT# 15 SE COR 5
(N/A	0.002	N 69°34'32.91" W	9884.805	10115.227	CLOSURE ON #5)

	921.453 m				

(921.453 m is the sum of lines of the perimeter, or total base course of masonry by Cole, or "**tbcmc**" This closure is an error of 2 millimeters over the total length. See Fig 1)

Eq. 19

error of closure: 921.453 m / 0.002 m   =   1:460,726.5   (About 2 parts per million)

Eq. 20

Final angular error: $\tan^{-1}(921.453 / 0.002)$ =   0.000124359°; (decimal degrees)

0.000124359° / 360°   =   1 : 2,894,829.9

                         or about 0.35 parts per million (ppm)

Were we to accept this data at face value, we might say that this work by its "closure error" as shown here is every bit as good as very high quality survey work done today. But this data should be approached with some caution. Cole himself states: [16]

> "The distances were measured with the Base Line Apparatus designed by M.M. Benoit and Guillaume using 24 metre standardized invar wires. Each length was measured twice with an accuracy of about 1 in 500,000.

> "The angles were measured on 4 arcs with a 6 inch Troughton and Simms micrometer theodolite.

> "The top of the flagstaff on the top of the Pyramid was visible at points 1, 6, and 7 and was included in the round of angles at these stations.

> "The closure angle of this traverse was found to be 9 · 6" which was adjusted by adding 1·2" to each angle.

> "Using these adjusted angles, the co-ordinates were computed and a closing error of 8 millimetres in North direction and 0 millimetre in East direction was found. These co-ordinates were adjusted by the normal traverse method for closure."

These statements mean: an *angular closure* of one part in 135,000 (9.6" / 360°) and a final *distance closure* of about one part in 115,182, (0.008 m / 921.453 m) which is fairly good--being an error of about 1/2 an inch over a mile, a combined average of +/- 10" angularly. (But, we still don't know which closure form he used. The "normal traverse method" could have meant what Americans call the "traverse method" which favors holding dimensions over angles; or "compass method" favoring angles to be held over less certain dimensions, a common procedure; or even the "least squares method". This last is a very laborious scientific closure route, used by engineers as far back as the time of Smyth in the 1860's, now common due to the use of computers. These techniques distribute error to produce a balanced picture of the mapping of evidence. With a closure as high about one part in 125,000 it may not greatly matter which method was used.)

Livio Stecchini points out that the general figure given in Cole's totals for the length of the North side (230.253) is two millimeters *longer* than the sum of two measurements given for the same side elsewhere in his report (230.251). This information was the result of two given measurements to a thin line that ancients builders cut into the pavement at the base of the North side of the Pyramid. Cole stated that this line is "...probably the original line of the axis."[17] Cole said he measured 115.090 m from the Northwest corner to this line, and then 115.161 m to the Northeast corner, there being a difference of 71 mm (or 0.071 m).

of the original extension.

Prof. Borchardt pointed out a small line on the pavement which projected a few centimetres from the edge of the casing-block about the middle of the North side.   This line was neither a joint in the pavement nor in the line with the joint of the casing block.   The measurements from the two northern corners to this line are as follows: —

to N.E. corner=115·161 metres.
to N.W. corner=115·090   „

Diff.—   71 millimetres.

Thus this line is probably the original line of the axis.

MAXIMUM ERRORS TO BE EXPECTED IN THE NEW DETERMINATION.

ile

See FIG 1-A.   What this means, is that this line is 35.5 mm (half of 0.071 m) Westerly of the midpoint which would be at 115.1255 m from either side.   See FIG 1-B.   (This small offset will prove important later.)   Since we are presented with both sets of data, with no special reason to choose either, I chose to reject the longer value--much as Stecchini did--but for different reasons, as will be seen in Sections VI and VII.   Ordinarily, we might ignore a two millimeter difference over such a large survey, as I suspect Cole did.

Yet through mathematical trials using this *smaller version* of the north line length, I was able to uncover what I think is evidence of a careful proportioning scheme involving the multiple use of the pure mathematical functions π and φ and Davidson's Displacement number-form.   My first use of this sum was inspired by Stecchini's attempt to uncover a metrological explanation for the four sides of the Pyramid using the same data provided by Cole's survey.

Stecchini chose to accept the smaller figure due to the fact that it was acceptable evidence from the stated measurements, and since it made a lot of sense with respect to his elaborate metrological theory.   I have formed no opinion about his theory, but I have also come to accept the shorter dimension 230.251 for different reasons.   One of these reasons is that this scribed line can have no correct or certain position were one to use the longer length.   (i.e., If your going to use any value at all for this line, it would then have to be this one, derived from these two lengths, since no division is provided for the stated length.)   Of all the other lengths, only the North line is supported by two stated dimensions, which seems to increase its likelihood of being a truer length, and we have only these stated dimensions to go from.   Beyond this, there is good reason to suspect the excess two millimeters in the total may have been *an arbitrary addition* on the part of Cole.

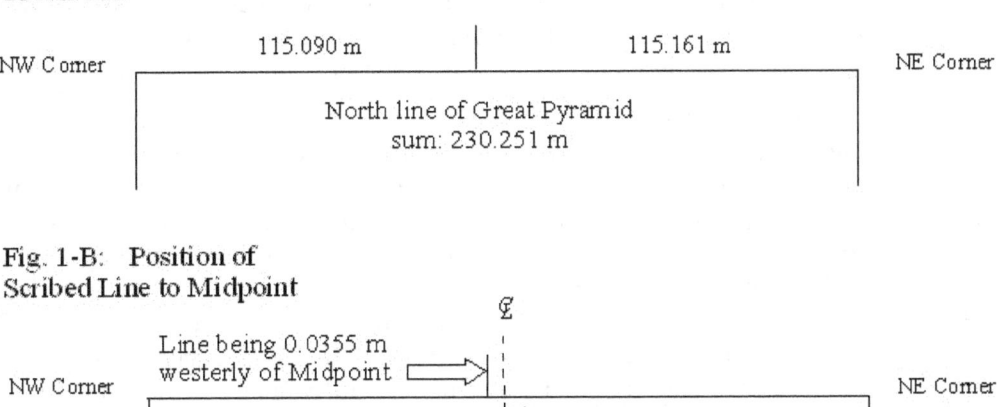

Fig. 1-A: Position of Borchart's
Scribed line using the shorter sum
230.251 m

NW Corner ——— 115.090 m ——— | ——— 115.161 m ——— NE Corner

North line of Great Pyramid
sum: 230.251 m

Fig. 1-B: Position of
Scribed Line to Midpoint

NW Corner

Line being 0.0355 m
westerly of Midpoint

Midpoint

115.1255 m, from either
corner to Midpoint of the
Great Pyramid

NE Corner

(Proportions greatly
exaggerated.)

If, for example, we use the longer North line dimension of 230.253 m in the same type of traverse (not shown) it will improve Cole's closure precision dramatically, *doubling* the above Traverse figure's precision to just under one part in one million. It may be that the two millimeters *were arbitrarily added* to the North line, since this would have the result of making the angular closure *practically perfect.* And this would be one of many semi-legitimate methods of distributing error, especially if it were his closing course.   Otherwise, without the two added millimeters he would have to consider sub-millimeter adjustment of the measurements at all four corners. Cole may have been avoiding this, which he may have seen as outside of his own rules of precision of one millimeter. We must remember, this is around the range of a tenth of an inch--and Cole (and the surveying profession) might have felt it would be hubris to average into the next decimal place of tenths of a millimeter--especially when the millimeter had been estimated. So, the excess of 2 mm is suspect.

Given Cole's statements above, we might say he closed to an average of about 1:125,000 which is much more believable. This is not unusual for good, modern survey work, though not typical for the time and technique of the period around 1925.   Many surveyors of that period would have been really quite happy with closures of one in forty- or fifty-thousand. Cole's work is extensively quoted, and highly thought of by many authors, notably Livio Stecchini, the metrologist, Humphrey Evans, and many others who regarded the Cole Survey as the definitive survey of the Great Pyramid.

In the determination of bearings Cole used the Eastern or Western Elongation of Polaris [12] (as Petrie is said to have done), which is still the most accurate method, even today. He lists results to the tenth of an arc second (nominally, 1:12,960,000); not unusual for astronomical work, but which is work of a very different character than the survey on the ground. During that period some really good survey work, even from a modern standpoint, was being done all around the world. (It is said that Clifford Holland, in his construction of the Holland Tunnel during the 1920's, engineered the two intersecting underground tube alignments under the Hudson River so closely they missed meeting by about 3/4ths of an inch, which (in three dimensions over a distance of more than a mile) would have a closure value of at the very least about one part in 300,000.)

Cole's requiring as a condition of the survey, that the base of the Pyramid be completely cleaned and cleared of rubble, immediately suggests that he intended to use a "supported tape", (e.g., a tape or wire flat on the ground or pavement) and that he wanted to find and use the base edges of the Pyramid. This is also suggestive of substantial money, personnel and time being invested in the survey. No doubt, he must have used the a supported Invar wire apparatus, ("Base Line Apparatus designed by M.M. Benoit and Guillaume using 24 metre standardized invar wires") and paid close attention to ground temperature corrections; double readings; plus an excellent theodolite. If he had an experienced survey crew, a possible closure of about one to one-hundred thousand or two-hundred thousand might be expected.

During the 1920's, and much later, angular measures were comparably better than distance measurement--where in many ways the reverse may be said to be true of work today. Global Positioning System (GPS) techniques at this time are still only plus or minus 10 mm locally at every point, but over very large distances one can get a precision of one part in 20 million. There are electronic instruments in use today for local situations, (e.g., Kern, and others, for the alignment of hair-thin cyclotron beams and factory floor alignments) that can produce final results of +/-200 microns, (0.0002 m, or two tenths of a millimeter). Error of 3 mm per one km (0.003/1000= 1:333,000) is more typical today. Quite old instruments (Wild T3) can read to better than a tenth of an arc second. Cole's headings, are given to the single second of arc, a much finer value than his given metric-unit values and his "possible margin of error" (TABLE C). This shows a greater faith in the instrument work than that of the his Invar wire. I do not know his method for adjusting his final values, though the figures have "a millimeter" limit, which is suggestive of statistical reduction of data.

TABLE C (L. STECCHINI)

West side:	30	millimeters at either end
North side:	6	millimeters at either end
East side:	6	millimeters at either end
South side:	10	millimeters at the west end, 30 millimeters at the east end

Having accepted these caveats, Cole's work in all probability remains as good as any present survey of the Great Pyramid's base course.

****

(V) Comparisons and Translation of Cole's data:

In the most conservative sense, if we took Cole's survey data of the base course of masonry (tbcmc), and translated the data into Polar Inches using the above modern value for the computed Polar inch of 0.2542700926 m (from Eq. 5, above) we will get a figure very close to Davidson's estimate of the 100x year circuit minus his Displacement Factor:

Eq. 21

Davidson's theoretical 100x Year circuit is: 36524.2456 PI (tbcid)

36524.2456 PI (**tbcid**) *minus* the theoretical 286.1022 PI (four places)	= 36238.1443 PI
Pyramid base (**tbcmc**) as measured by Cole:	= 36239.14 PI

The above Davidson number is quite similar to Cole's survey data. I think this evidence and the Passage way's offset (286.1 PI) argue for the ceremonial year length idea. I propose what I will call the "virtual pyramid theory"--an imaginary, mathematical pyramid outside of the real one. Suppose then, we directly added the hypothetical Displacement Factor to Cole's total base course of masonry (tbcmc), converted to Polar Inches; how much difference is that from Davidson's ideal base course (tbcid)?

Eq. 22

(a)   (921.453 m / 0.2542700906 m)   = 36239.1424 PI   then,

(b)   36239.1424 PI + 286.1022 PI   = 36525.2446 PI

(c)   36525.2446 PI   (b, above)

      36524.2465 PI   (Davidson's ideal base: tbcid)

         0.9981 PI   (or about one Polar inch if rounded at two places.)

In this approximate value, the two totals are not very different: one part in 36,594. Cole's total only needs to be shrunk by about one PI to be the same length that Davidson said it would be.

The "virtual pyramid theory" is the idea that, first; the Pyramid's builders may well have built a physical pyramid with a base very much as Cole and Petrie found. Second; the base secretly served ancient initiates of an arcane system as the hallowed physical base for a virtual mathematical construction of the 100x Tropical Year that could be said to *surround the real structure.* This would only be possible to know at such time as the secret Displacement Factor is added by the technically able student of the Mysteries.

This is not unlike the design intent and method apparent in other architecture of the Pyramid. It is replete with hidden construction lines, large and small offsets, implied factors, chambers where the implied wall height below the floor gives a second special room volume different than the floor level, etc. Sacred knowledge of this kind may well have been totally secret, and the 100x Tropical Year length may have been forbidden to display to the outsider in any plain, physical form.

How do we deal with the "extra" amount of length of about one Pyramid inch?[18] We could easily call this an accuracy error in taping of about 1:36,500 and adjust proportionally. Perhaps the Egyptian Units are a little off in length. Or perhaps not: the ancient Egyptian Manetho coded the number "*36,525*" into his year lists according to Davidson and Aldersmith. There are a large number of possible questions and answers that may be posed here about this little excess. I propose the following solution (for now), so that the four lengths can be examined in the hypothetical Polar inch units, *within the side proportions that Cole found.*

In survey practice when the originally intended distance is discovered by a resurvey to be longer or shorter than it is supposed to be, the assumption can be made, (without prejudice to either the new or old surveyors' measuring efforts), that a *scale factor* can be used to make both the new and old measuring data translatable. The idea we will be getting to here, is that we are going to assume, in theory, we know just what this total, imaginary, secret base length is *supposed* to be: that is, Davidson's tbcid number, or the astronomical tropical year times 100. Then using these four lengths as measured, we can find out *just how long they would be* if measured in Davidson's theoretical Polar inch base. Here we are making an assumption for the sake of argument, that the ancient builders may have actually intended a polar inch unit but accidentally made them slightly too big, intended for an imaginary, virtual base surrounding the stone masonry of tbcmc. These numbers are too close to ignore the symbolic 100x year-length possibility.

If hypothetically, we thought that Cole's measuring wire was a little too short, over the sum of all his measurements to fit the 100x year length concept, (which would give longer readings than would be correct, let's say about an inch longer as seen above), we can find a tape correction by division:

Eq. 24

(Davidson's tbcid minus the Displacement Factor:)   36238.1443 PI

                                                ------------------   =

(Cole's measured tbcmc:)   36239.1424 PI

Scale factor=   (tbcid-D) / tbcmc   = 0.999972457

Multiplying this scale factor by each side length of the Pyramid converted to PI, plus 1/4 of the Displacement Factor, will produce a length which would be exactly *as if* Davidson's 100x year length was the correct length for the total; and will give four virtual sides of correctly scaled lengths with respect to Davidson's ideal 100x year length. Then conceivably, any recognizable sacerdotal values might stand out in a way that would not appear through metric units. This conversion will be done by Eqs. 25 through 27:

Eq. 25 (Conversion to Polar inches)

		(Conversion factor: Eq. 5)		(Cole's data translated into PI)
(EL)	230.391 m	/ 0.02542700926 m	=	9060.876960 PI
(NL)	230.251 m	/ 0.02542700926 m	=	9055.371003 PI
(WL)	230.357 m	/ 0.02542700926 m	=	9059.539799 PI
(SL)	230.454 m	/ 0.02542700926 m	=	9063.354640 PI
sum:	---------------			---------------
	921.453 m			36239.1424 PI
	Actual length m			Actual length in PI

The next step is to scale each length for each side:

Eq. 26 (Scaling to Davidson's Ideal base length)

	(lengths in PI, above)		(scale factor)		(scaled sides)
(EL)	9060.876960 PI	×	0.999972457	=	9060.627401 PI
(NL)	9055.371003 PI	×	0.999972457	=	9055.121596 PI
(WL)	9059.539799 PI	×	0.999972457	=	9059.290277 PI
(SL)	9063.354640 PI	×	0.999972457	=	9063.105013 PI
sum:	---------------				---------------
	(921.453 m or				(Scaled down to
	36239.1424 PI)				36238.14429
					[36238.1443])

The last column in Eq. 25 represents Cole's measured lengths in Polar inches. Since some discoveries as ratios appear out in the region of the seventh place past the decimal, I will carry these conversions from the ratios to ten floating places. (Later on, with respect to predicted observation and of measurement, we can return to three or four place precision, if needed.) Then,

Eq. 27 (Adding 1/4th Displacement Factor 71.5255539 PI to each side)

	(Cole's data in PI, scaled)	(Adding D/4 or a quarter of Disp. F)		(Cole's survey data proportioned to Davidson's Ideal Base)
(EL)	9060.627401 PI	+ 71.5255539 PI	=	**9132.152955 PI**
(NL)	9055.121596 PI	+ 71.5255539 PI	=	**9126.647150 PI**
(WL)	9059.290277 PI	+ 71.5255539 PI	=	**9130.815831 PI**
(SL)	9063.105013 PI	+ 71.5255539 PI	=	**9134.630567 PI**
sum:	---------------------			---------------------
	36238.14429 PI [36238.1443]			36524.2465 PI

This is Cole's data converted: The final column at right of Eq. 27 represents: (a) Cole's survey data from meters; (b) converted to modern-value Polar inches; (c) scaled to Cole's data to make lengths equivalent proportionally to Davidson's Ideal base course length and (d) with virtual Displacement Factor added in Polar inches. This is Davidson's base with the sides proportioned to Cole's survey, so that possible ceremonial proportions or other values might become more easily visible.

This conversion may allow us to create the invisible pyramid of mathematics around the real stone structure, and perhaps see what sacred, mathematical or astronomical symbols are present, if any.

****

(VI) A Connecting Clue to The Dollar and to the Pyramid's proportions:

While experimenting with Stecchini's metrology theory, (using the scaled dimensions shown above) something surfaced that later appeared to be an important clue to the Pyramid's proportions--as well as that of the dollar design. The sum of the North and South lines when subtracted from the sum of the East and West lines gave a difference of 43 mm, or 1.69 PI if translated into Polar inches. This becomes important in my theoretical reconstruction.

Eq. 28   (In Cole's original measurements. Due the small size of this residual dimension, there will be no appreciable difference using the scaled values above. This is in the same range of as Borchardt's difference.)

(EL: 230.391 m + WL: 230.357 m)	= 460.748 m	(East and West sum is larger)
(NL: 230.251 m + SL : 230.454 m)	= 460.705 m	(North and South sum is smaller)
	(-) ---------	
	0.043 m = 1.69 PI	

The difference is 0.043 m. The number 1.69 in PI is for all practical purposes the *square root of 2.86 PI*. My initial reaction was to ignore this, but I later realized that this might well be one of the possible small offsets or number symbols to be looked for to get a mathematical starting place to analyze the base lengths. The other small difference, to Borchardt's line, as discussed by Livio Stecchini, amounts to an offset of 35.5 mm to the West of the North line midpoint, described above. This length translates to 1.396 PI.

Here (because of the pairs of sides) I was cued to mathematically investigate a *half* of 1.69 PI, since, (in a similar way), half of this number would be a number being *a quantity plus or minus the half of the total base course* defined as the sums of North + South lines and East + West lines. (The pair-sums are *minus* for N+S and *plus* for E+W.) Soon I found that the square of the reciprocal of half of 1.69 PI was practically the same as half of Stecchini's noted difference, or about 1.40 PI, if we only accept two digits past the decimal. (If instead, using the theoretical square root of 2.861, being 1.691; half of this 0.8455, which has a reciprocal of 1.183, and if squared gives a value of 1.398. This is practically the same as 1.396, also being nearly identical to 35.5 mm.) The intended relationships are not perfectly clear yet, but we can say:

Eq. 29 **Hypothesis:**

$(1.691 \text{ PI} / 2)^2 / x = 1.398 \text{ PI}$

(Or $1.691 / 2 = 0.8455$, then, $0.8455 / x = 1.183$, $1.183^2 = 1.398 \text{ PI}$)

The most important part of this train of reason, is that there could be a well thought out relationship between these small differences, suggestive of a geometric relationship--with a special connection to Borchardt's line.

Surprisingly, these numbers *surface in the dollar bill design, as the same lengths in Polar inches.* It is important to notice that this information could have been available to the dollar's Designer Edward Weeks by way of Professor Borchardt's work sometime after 1927, and Weeks would have had no

215

trouble converting 0.0355 m into PI (i.e., pre-1946 British inches, then into Davidson's PI) to find the number 1.398 PI. This would have been close enough for him to draw some conclusions and similar computations as were done here. The Designer of the dollar need not have gone any further than here in his research. The dimensions 1.182 PI and 1.398 PI on the dollar are found from the westerly outer edge of the LIR rectangle (see Illus 4 pg 230) to the easterly corners of the Pyramidon and a parallel theoretical point under the bush (near the EC point) at the base of the little pyramid, respectively. (See Chapter 5. Notice that the Designer made the 1.398 PI *also at the base line of a pyramid, like an offset.*) We will look at these figures again later, where the above 1.691 plays a role in the theoretical reconstruction of the four sides.

****

(VII) Analyzing the Four Sides:

In Davidson's mathematics, a given side of the Great Pyramid is treated as though it was exactly one quarter of the total tbcid or 9131.061625 PI, (which·is also a one quarter *average* of the total of the new values of Eq. 27, but not strictly one of the actual measured sides.). His proposed height of the ideal pyramid was 5813.014373 PI, is based on the above formula in Eq. 8. (A simpler procedure which I will alternatively use here is: the height 5813.014... divided into half of 1/4 tbcid, [i.e., 1/8th tbcid], will give the tangent value of the "pi angle slope" of 51°51'14.31")

My first efforts were to analyze and compare the re-computed Cole data of Eq. 27 with the above 1/4 tbcid, through various methods. I was initially unsuccessful in finding any satisfactory patterns. Later, I tried some algebraic schemes for lengths of adjacent sides, and this, too, was mostly unsuccessful. In modern pyramid literature there is an idea that the North side of the Pyramid is somehow related to an angle based on the arc tangent of the square root of φ, and that the West side relates to in some way to the arc tangent of $4/\pi$. There also is a notion that the ancient Egyptians conceived of an equivalency between these ratios, by means of the square root of φ and $4/\pi$. Although there may not be a lot of clear evidence to completely support either idea, I started investigating in that direction.[19]

My attention was drawn to the North and East lines since they were at almost 90° to each other. The ideal North line was supposed to relate to φ. This (adjusted) length of (NL) 9126.647150 PI when divided in half, and divided into the height 5813.014, gives neither a close approximation of the φ or the π proportions. But, it does, as it happens, place them in *proximity to an average of these two functions.* The translated West line however, (WL) 9130.815831 PI, is not really far from the π function, which would be a length of 1/4 tbcid, or 9131.014373 PI: close to a quarter of a PI short of the ideal π related length.

When I chose to think of these side lengths as a *sum of two lengths: the φ and π math proportions*, a conceptual break occurred. Whatever difference we might find from regular fractions of the ceremonial functions for one side, was something that was important to know. I proposed that a 1/4th of the total, or a quarter of the average length, 9131.06 PI, (Davidson's ideal side of the Pyramid) was *half* of a side--or an eighth of the total tbcid base. To then subtract this half a side, or an eighth of the total, might provide a clue or a hint of what the rest of that side was all about, by proportional multiplication by eight. (Shortly before this, I had been thinking about the faint mid-side seams, or an apparent indentation noted in solstice photos of the Pyramid. These acted as a clue.) This is approximately the line of reasoning and steps that I took:

THE NORTH LINE:

Eq. 30   Subtracting half of a ideal side:

(NL) 9126.647150 PI - (9131.0616... PI /2) = 4561.116338 PI

What happens if we considered this figure as one eighth part of a new virtual base? (i.e., multiplying by 8):

Eq. 31

$$8 \times 4561.116334 \text{ PI} = 36488.9307 \text{ PI}$$

This is pseudo-base a, it is 35.3158 PI smaller than tbcid. Since this could be regarded as another imaginary base, mathematically, what height based on π would this smaller virtual pyramid have?

Eq. 32

$$(36488.93067 \text{ PI} / \pi) / 2 = 5807.393689 \text{ PI (new height, a)}$$

(This new height a, is 5.6206841 PI less that the ideal height, of 5813.014..PI). How then would this new height relate to the ideal base of 36524.24...PI ? (The formula here: a / (1/8 tbcid), and the result squared):

Eq. 33

$$5807.393689 \text{ PI} / 4565.630813 \text{ PI} = 1.272008432 \text{ (a ratio, } \alpha) \quad [\tan^{-1} \alpha = 51°49'37.37'']$$

Eq. 34

$$1.272008432^2 = 1.61800545$$

This number has a very close resemblance to the golden ratio or φ. So close, in fact, that one might say "we need look no farther". This is a difference from the ideal golden ratio of one part in fifty-seven thousand. But look what happens if we compare by subtraction, the actual golden ratio from the above result:

Eq. 36

(a) Error Computation, from Difference: $(\varphi / \alpha^2 - 1) / x$

$(1.618033989.../ 1.618005548 - 1) / x = 1:56,697$

(b) (Difference):

   1.618005450

   1.618033989... (φ, or the Golden Ratio)

   (-) --------------------

   - 0.000028538

   (or what looks to be *a form of the Displacement Factor*, a familiar number but in the minus sign.)

It appears that the North Line was designed to have the proportions of (1) one part π length (2) plus one part φ length, and (3) minus one 10 millionth of the Displacement Factor.

The first two elements do not seem too hard to accept as a mathematical undercurrent within the π based Pyramid. But the third part having such a small representation of the Displacement Factor is difficult to accept--until an analogous procedure is tried on the West line.

THE WEST LINE:

When I tried the same procedure on the west line, the result was not a function of φ, but was near the function π when the same 1/8 tbcid length was subtracted out:

Ken McGrath

Eq. 37

$$(WL)\ 9130.815831\ PI\ -\ (9131.0616...\ PI\ /\ 2)\ =\ 4565.285019\ PI$$

Again, what happens if this is considered as one eighth part of a new virtual base? (again, multiplying by 8):

Eq. 38

$$8 \times 4565.285016\ PI = 36522.28015\ PI$$

This is a pseudo-base b, 1.96635 PI smaller than tbcid. And again, what height, based on π, would this second, smaller virtual pyramid have?

Eq. 39

$$(36522.28012\ PI\ /\ \pi)\ /\ 2\ =\ 5812.701419\ PI\ \ (new\ height,\ b)$$

(This new height b, would only be 0.3129546 PI less that the ideal height, of 5813.014..PI, and 5.30772955 PI taller than new height a.) How then would this new height relate to the ideal base of 36524.24...PI as before with the North side? The formula given here is b / (1/8 tbcid) x 4:

Eq. 40

$$5812.701419\ PI\ /\ 4565.530813\ PI = 1.273170997\ (a\ ratio,\ \beta)\ \ \ \ [tan^{-1}\ \beta = 51°51'08.91"]$$

The next step of squaring this similar ratio β however, does not produce an immediately recognizable ratio, certainly not φ. However, as noted above about ancient Egyptian sacred mathematical concepts, the square root of the φ-ratio is supposed to have been held by the ancients a parallel equivalent to value to 4/π. It just so happens that the ratio of Eq. 40 does closely resemble this function, and we will exchange it *quid pro quo*, "something for a thing in return" for the other ancient ceremonial function--the proportion's own reciprocal multiplied by four as though the intent was π:

Eq. 41

$$1\ /\ 1.273170997 \times 4\ =\ 3.141761796$$

Note how similar this to the function π. What happens if we (likewise) subtract π from this number, or 3.141592654? In the equations below, the one one-hundred thousandth of the square root of the Displacement factor is divided into the found difference in a formula for the error ratio, and then π is subtracted from the result of Eq. 41 above--the comparison and difference as before:

Eq. 42

(a) Error Computation from Difference, similar to the above Eq. 36-a: (π / [1/β x 4] - 1) / x

(3.141592654 / 3.141761797 - 1) / x = 1: 18,575

(b) (Difference, as above, Eq. 36-a):

3.141761797

3.141592654... (the circle's function, π)

(-) --------------------

(+) 0.000169142

(or the *other* now familiar number. In this case, approximately the *square root* fractional form of the Displacement Factor, now in the positive sign.)

218

We see the "1.69" number again. Here however, we see a number that is almost exactly one hundred thousandth of the square root of the Displacement Factor, good to five places at the front of the decimal value, (i.e., through eight places past the decimal in mathematical precision). This residual, from several things we have seen earlier, does not appear to be a thing to have appeared by chance.

Are these small differences meaningful?

Do these tag-end residual fractions have a real significance? Might they only be mathematical "noise" occurring as they do out in the fourth through eighth decimal places? One could argue that the clearly intended ratios were just a simple, equivalent mix of π and φ, on the West and North lines, seemingly well demonstrated, and these residuals are only random error on top of the builders noble attempt to build at those proportions. And indeed, were this idea of proportion correct, we would need only add one millimeter the North line, and add about 9 and 2/3rds millimeters at the West line--all well inside Cole's possible error scheme as seen in TABLE C. Many people will be uncomfortable with such fine or ephemeral ratios. Yet professionals commonly compute state plane coordinates using delicate twelve-place ratios, all the time.

We should note, that if all of these equations were done in metric the original units, (the same scaling to tbcid, and adding a metric Displacement factor), all of these signature ratios would still be present. We are seeing *the relative line proportions* from Cole rather than measurements. Here, *the proportions* are used to reveal possible ceremonial dimensions from a scaled base, found at nearly the ideal size. The above counter-argument wouldn't explain the improbable and clear result of our answer at the *North side--being only being explainable as a subtracted result from the sum of *two* easily identifiable ceremonial lengths, and not as an average of the two. *If the two sides were simple proportions,* the West side would then only be related to π, which wouldn't make any sense looking back at the clear pair of contrasting ratios on the North side. Plus, we have *already seen a pattern like this above,* (Section VI) where the square root the 2.861 number-form has apparently been *added* to the sum of East +West sides, also appearing as a *subtraction* on the sum of North + South sides. This might call out for something of a parallel sort on the North side, where we do, in fact, see an excess in another form again *subtracted* from that side (- 0.0000285). Then on the West side, another--and related--form of excess is *added,* (+ 0.0001691) again like the pattern seen in Section VI. So, these numbers: - 0.0000285 and +0.0001691 act like a signature, and apparently follow a previous pattern.

This theory is from observations and evidence from Cole shown above, with nothing of an arbitrary nature or modification of the figures given above, beyond the addition of the Displacement Factor, scaling and translation into Polar inches. These relationships are complex and closely symmetrical in character-- and it is the equity or symmetry in these equations that argue strongly for accepting the observations above. Were it not for these small distinguishing residuals, small as they may be, and with the evidence of the well known mid-side seam indentations, the original proportioning scheme might not be known.

As a curious side note, the polar flattening number, (see fourth entry of the NOAA chart on page 196, "Reciprocal flattening") can be produced, from the sum of subtracting both pseudo-bases from Eq. 31 and Eq. 38 from Davidson's tbcid base of 36524.2465, and multiplying by eight:

(36524.2465 - 36488.93070 = 35.31580)
+   = 37.28215   × 8   = 298.2572
(36524.2465 - 36522.28015 = 1.96635)

The WGS 84 value is 298.257223563. Isn't that amazing. This kind of discovery is near and dear to the heart of the pyramidologist who believes that all of the Earth constants and astronomical ratios are to be uncovered in the dimensions of the Great Pyramid. Livio Stecchini believed that the ancient Egyptians were aware of this Earth flattening, and as a logical consequence, they might have computed the Precession of the Equinoxes. Perhaps this is evidence that they did. Later, in my reconstruction, the eight-fold pseudo-base sum that can be drawn from Eq. 44 and Eq. 45 will not match the above result-- these will give: 298.2639328. Close, but not as close as above. My reconstruction scheme is probably still missing something.

****

(VIII) A Theory of Mathematical and Religious Symbolism:

It seems at first as though these two sides show a general, over all proportionality pertaining to π, but, on closer examination this idea is found to be more complex by degrees. This is *a successive approximation* process as three regular steps:

Step I: (Eqs. 5 through 27) The general π proportion is given by the builders having made the total stone base course and height closely based on a π-ratio to height, and that they also made half of all the sides the π-based length of 1/8th of the base circuit.   But, within the other half of a side is hidden a deeper Mystery, perhaps as a formal question posed to an ancient initiate of sacred lore. One can not get beyond this point without a knowledge of the year length in Polar inches, and a willingness to consider proportional relationships from measured dimensions through mathematics.

Step II: (Eqs. 28 through 34 and Eqs 37 through 41.)   From here inwards, mathematically, it seems that the builders intended to enshrine a close representation of a parallel importance of the two sacred functions of π and φ. *Two virtual pyramids* must be created (the pseudo-bases), probably connected to the dual pillar symbolism of ancient temples of the Mideast. These are given to appear as only a close proportionality (see Eq. 34 and Eq. 41), since we are only entering the second step of understanding, or inner court of the ceremonial forms. These are correct π and φ values to four decimal places at this stage.   We have seen in Chapter 4, from John Greaves, that there is a definite equivalence of *pillars* and *pyramids* in ancient Egypt and Biblical lore, and that the relationship is said to be of an astronomical nature. Curiously, there is a trace of a dual symbolism given for the U.S. Great Seal's Pyramid seen in the legal description Remarks and Explanation (See pgs 186 and 187), stated as "Strength and Duration". This appears to exactly parallel symbols for the Pillars of the Porch, Jachin and Boas of Masonic lore. These are at the entrance to Temple of Solomon, called in reverse "Establishment and Strength".

Step III: (Eqs. 36 and 42.)   Then, in the third step, the earlier general forms are more completely demonstrated, finally revealed accurately through subtraction. But as before, only by the work of the investigator of the Mysteries.   This is from the fact that only those who technically aware of the sacred functions (π and φ) may correctly divide or subtract them. This is like a password or key needed to unlock the hidden information. This reveals on the principle North and West sides, two regular but different forms of the same number-form, the sacred fraction 286.1 and it's root, 16.9. Or this reveals some very similar numbers, (e.g., "285") correct to at least five decimal places. It is also interesting, that none of the sacred functions are ever actually revealed: they are only known from a *fractional relationship between two measured numbers.*   They only appear by mathematical discovery. This is much like the mathematics of the Coffer to King's Clamber in Section III, where the same ratio is also found, but in a different decimal position. It is possible the above fraction 0.00002861 is a *symbolic identity.* It may be connected to the meaning of the open tomb symbol of the Coffer. Several writers have made a connection to the symbol of the open tomb or coffer and spiritual resurrection. It was originally found open and empty, and its position on the floor of the King's Chamber is said to be centered 286.1 PI south of the east-west axial line of symmetry of the Great Pyramid.

(Later in Sections X and XI, we will see that even the bearing of the West line, and the division between the original and symmetrical configurations of the Pyramid's base follow forms of the 286.1- number ratio. With regard to the number-form symbolism of Step III, it is possible that another mathematical value for the Displacement factor could be implied--that of being the diagonal of: x = the fourth root of π, and y = the square of φ, giving a length of 2.86102033... a relationship connecting all three ideas as a simple form. See Eq. 12 and 13 pg 202 and Illus. 3, pg 228.)

****

(IX) Reconstruction of the Four Sides:

With the above information, it will not be difficult to closely reconstruct Cole's base course lengths, from a completely theoretical basis, using only theoretical mathematical proportions and ideal theoretical lengths. Note that in the first two equations below, (a) the signs change in subtracting and adding the small forms of the Displacement factor, (b) the difference between the factors is 100 fold, one with the factor in the square root form, $\sqrt{\varphi}$ is exchanged for $4/\pi$, and (c) all other operations the same. In the next two equations the Displacement factor is a 1/10th of the square root form, where half of it is *subtracted* from half of the ceremonial year length in one equation, and *added* in the next.

A theoretical construction in four equations to create this base using only four givens:

Eq. 44

Given: π, φ, 286.1022156 PI and 36524.2465 PI.

The North side, the above equations with ideal factors, but in reverse:

$\sqrt{[\varphi - (286.1022156 / 10,000,000)]}$ = ratio α' [1.272008404]   then,
$\alpha' \times (36524.2465 \text{ PI} / 8) \times 2 \times \pi / 8$ then,
$+ (36524.2465 \text{ PI} / 8) =$ NL'= **9126.647049 PI**
$\qquad\qquad\qquad\qquad\qquad\qquad\qquad\qquad\qquad$ [$\tan^{-1} \alpha'$ = 51°49'37.37"]

Eq. 45

The West side, also using above equations with ideal factors in reverse:

$[\pi + (\sqrt{286.1022156} \text{ PI} / 100,000)] / 4 / x$ = ratio β' [1.273170996]   then,
$\beta' \times (36524.2465 / 8) \times 2 \times \pi / 8$ then,
$+ (36524.2465 / 8) =$ WL'= **9130.815827 PI**
$\qquad\qquad\qquad\qquad\qquad\qquad\qquad\qquad\qquad$ [$\tan^{-1} \beta'$ = 51°51'08.91"]

Average:$(\alpha'_\measuredangle + \beta'_\measuredangle) / 2$ = 51°50'23.14 See Chapter 2, pg 78 (dollar) and pg 242 (W.F. Petrie).

Eq. 46

The South side, (See Eqs. 28 and 29, above.) Here we adopt an assumption that the discovered difference of "1.69 PI" between sums of parallel sides is actually a tenth of the square root of the 1/100th Displacement Factor (1.691455632 PI). This small difference has the effect of *defining the sum* of the N+S and E+W lines. When this is subtracted or added to *exactly half* the tbcid base of 36524.2465, the correct sums of two parallel sides are found (see pg 215, Eq. 28). Since this appears to be the intended proportion, and since we now have North and West line lengths, we can now uncover what the South and East lines must be by subtraction from the half-sums plus or minus 1.6914. (We should note at this point, that the ten millionth of the Displacement Factor that we found subtracted from the North line, is in harmony with the fact that a parallel factor is required to also be subtracted, but from half of the ceremonial year length to produce the South line.) The 1.6914 addition/subtraction clue simplifies and anchors a determination of the South and East sides:

221

[18261.27752 = theoretical sum of N+S line lengths]

(36524.2465 PI - √286.1022156 / 10) / 2 = 18261.27752 = c   then,
c - NL' =   SL'= **9134.630473** PI

Eq. 47

The East side, as above for the South line, though *reversing the sign to plus*, where the effect is to define the sum of East and West lines:

[18262.96898 = theoretical sum of E+W line lengths]

(36524.2465 PI + √286.1022156 / 10) / 2 = 18262.96898 = d   then,
d - WL' =   EL' = **9132.153151** PI

Eq. 48   A check sum of the above theoretical lengths, Eq. 44 to Eq. 47, showing that they sum to Davison's ideal tbcid base course:

$$
\begin{array}{ll}
\text{NL} & 9126.647049 \text{ PI} \\
\text{WL} & 9130.815827 \text{ PI} \\
\text{SL} & 9134.630473 \text{ PI} \\
\text{EL} & 9132.153151 \text{ PI} \\
& (+)\text{--------------------} \\
3 & 6524.246500 \text{ PI}
\end{array}
$$

These lengths may have been the original intent of the Builders from proportional considerations. This is the best solution I have found. But much about this proportioning is unknown in a strict sense, as noted above in the flattening discussion on page 204. Having found an apparently exact reciprocal flattening for the Earth of 298.2572 in Cole's data, is not suggested or required in any part of my theory. But since it happens that the flattening form is not perfect as computed in the proposed reconstruction, suggests that something could be missing. It is almost as though these calculations are still missing some small coefficient, requiring another fundamental step beyond the three described above. Maybe the Builders knew something we don't.

This is not the end of surprises for this base. Later, in Section XI, we will see that these four length-proportions of Cole's survey have some unexpected and apparently unique properties not visible here. This suggests that the reconstruction is only a small part of a much larger and unknown design scheme.

****

Comparison of Measured to Theory:

Here below, are the measured, converted lengths subtracted from the theoretical lengths from above. Note that there is no difference at the rounded third place for all sides, and practically no difference at the fourth place for the West and South lines:

Eq. 49 NL - NL' =

9126.647150 PI as converted from Cole's measurements, Eq. 27 (NL)

9126.647049 PI the theoretical construction Eq. 44 (NL')

(-) -------------------

   0.000101 PI   (0.0026 mm)

Eq. 50 WL - WL' =

9130.815831 PI as converted from Cole's measurements, Eq. 27 (WL)

9130.815827 PI the theoretical construction Eq. 45 (WL')

(-) -------------------

   0.000004 PI   (0.00001 mm)

Eq. 51 SL - SL' =

9134.630567 PI as converted from Cole's measurements, Eq. 27 (SL)

9134.630474 PI the theoretical construction Eq. 46 (SL')

(-) -------------------

   0.000093 PI   (0.0024 mm)

Eq. 52 EL - EL' =

9132.152955 PI as converted from Cole's measurements, Eq. 27 (EL)

9132.153151 PI the theoretical construction Eq. 47 (EL)

(-) ----------------

  - 0.000196 PI   (- 0.005 mm)

(The small differences are due to the small change in number-forms from - 0.000028538 to - 0.000002861 and from + 0.000169142 to + 0.000169145 in the formulas above.) The interesting thing about these results, is that if one substituted the ideal numbers, and turned all of the equations around in reverse, and ran these computations backwards, inverse scaling, removing all the fractions of the Displacement Factor, and finally converting back to meters at the millimeter level--there would no difference whatever at the rounded millimeter. The largest difference, shown just above, of 0.000196 PI is very small indeed being 0.005 mm or about one two hundredth of a millimeter.   From the point of view of lengths, this is a satisfactory result.

Since we scaled out the excess 0.9981 PI shown in Eq. 22 to preserve the lines' relative proportions-- we still remain in the dark as to *the accurate length* of the base course. An accurate survey, having an error of two tenths of a millimeter at each corner, would be needed to verify J.H. Cole's work. It should follow Cole's method of *intersecting edge lines,* rather than getting involved in the socket points. Two tenths of a millimeter is about the practical limit, at one part in a million for each line. A survey such as this would not, however, distinguish between the above proportioning theory and these Cole-based dimensions--all the differences would be much too small to detect in any survey. It would be interesting if a scheme could be devised to find exact evidence for the seam-line positions at the mid-points of the four lines.

****

223

Ken McGrath

(X) The Angular Relationships of the Four Sides to Astronomical North:

Livio Stecchini and others have speculated that the West line's bearing of N 00°02'30.00"W, (See Traverse Experiment A, Course 4 above), relates to the angular change in the heavens over three years caused by sidereal motion. I don't know why it must be three years, but it is not far off--if we accept this hypothesis. The Encyclopedia Britannica gives 50.2" arc seconds as the as the angular change for one year, which will give an acceptable value of 0°2'30.60", or a figure of just over a half a second Westerly of the bearing measured at the West line. More interesting with respect to the discovered mathematical proportion motif, is this bearing's relationship to the positive sum of the internal angular differences from a right angle as shown above of 0°07'10", at Eq. 18. Using the stated West line bearing:

Eq. 53

0°07'10" / 0°02'30" = 2.8666667

Here again, we find the familiar number-form. If instead, we used 0°02'30.60" or the supposed "three year" astronomical figure, we get:

Eq. 54

0°07'10" / 0°02'30.60" = 2.8552456

Now, suppose we worked backwards from one thousandth of the ideal Displacement Factor and the astronomical value. We could have a look at what the sum or the bearing might look like, if the other two elements were possibly intended:

Eq. 55

(1) 0°02'30.60" × 2.861022156 = 0°07'10.87" or

(2) 0°07'10" / 2.861022156 = 2'30.30"

This being a difference of about one arc second in the first case, and about a third of an arc second in the second case. Although much might be extrapolated from this, the clear message may be the that the Builders were quite taken with the Displacement Factor in almost every step of the design of the Great Pyramid. This is only a continuation of the motif. If this is true, the N 00°02'30.00"W bearing as measured by Cole may not have an astronomical or time-based significance, but a geometric one.

Also, in addition to the obviously artificial configuration of the base lengths, we can now see that the whole structure was canted to the west by 2'30.00" so that the signature 2861-number could appear, with respect to true North. But to find the 2861-number, one must first have computed the positive sum of side angles with respect to a square (See TABLE B and Eq. 18, pg 207.) This bearing is the closest of the four sides to being a cardinal north bearing, and it is often quoted as an example of how good the astronomical work was for this period. But this idea is ill-informed: from the point of view even primitive astronomers and surveyors, astronomical north is not all that difficult to find. One sights a line from a point to a known star at the Eastern Elongation with a plumb line and hand signals to have a mark made on a point on that line on the ground at a distance. Then, with the same star at the Western Elongation at the same distance to the horizon, another point is marked. Splitting the difference between these two marks provides a perfect north bearing to the sighting point. (In ancient India there were simple procedures for temple and town street alignments using Sun shadows that gave eastern orientations of better than a half minute of arc.)   From the point of view of the ancient engineers who built the Great Pyramid, this west line bearing was to be a clue, for the initiate of the Mysteries. After a number of 2861-number coincidences, the student will be greatly encouraged to dig deeper into these relationships. Or, perhaps the bearing value is only praise to the Supreme Being.

****

224

(XI) A Confirmation of the Pattern found per Eq. 29 in Bearing patterns:

Earlier, after Eq. 18, it was noted that the configuration of the sides seemed to be *artificial,* thus raising the question of what its "natural shape" might be. What would this shape look like if it wasn't tweaked so that the Northwest corner was close to a right angle? Is there a more normal, or symmetrical conformation of these lines? In the traverse experiment below, two motions will be made to manipulate this shape angularly with no change to Cole's original dimensions. In this trial (1) the bearing angles will be balanced to mathematical symmetry while retaining original lengths, and (2) the resulting form is rotated so that it's bearings are made symmetrical with respect to the cardinal directions of the compass. (Suppose by analogy, that these four distances are rigid boards with their corners loosely hinged. If this was a hinged wooden frame, starting from the first configuration in Figure 1 of traverse experiment A, we would press together the opposite corners at Northwest and Southeast somewhat, so that the bearings would become symmetrical to each other about a centerline, then the whole frame rotated so that the convergence of symmetrical bearings would be mirrored on either side of cardinal North-South lines and East-West lines.) A comparison between the original form and this transformation is surprising and revealing with respect to the ancient Builders' intent and mathematical skill.

By experimentation, I learned that only certain quadrangles' lengths can be arranged so that their sides are "relaxed" to *a mirror-image symmetry to cardinal directions.* This turns out to be very a small and unusual class of quadrangles. This can only happen when both pairs of opposite lines are potentially bilaterally symmetrical about the North-South axis as well as the East-West axis. As an example, one can imagine two very long V shapes whose points intersect at 90° crossing centerlines--where the resulting enclosed shape is such a balanced quadrangle. However, not just any four line lengths will be able to do this, since there are sets of four lengths that can not have a bilateral symmetry on central axi separated by 90°. But the four lengths given by Cole *will do this,* and provide other curious properties besides. This shape turns out to be practically unique for a number of reasons--this is not a garden variety quadrangle.

My procedure to find these bearings, (somewhat simplified and not shown here) amounted to a regular distance to distance mathematical intercept. Going in cardinal directions North and East from a center, I made mid-points *at half of the average length* for the sides parallel of the Pyramid for North+South and East+West sides. From these points, an intersection was made for a beginning point at the Northeast corner using halves of the stated dimensions for the North and East sides. Later, I ran a traverse using the two resulting bearings to the intersection point, using mirror-opposite bearings for opposite sides. This strangely produced an exact closure, with practically no residual.[20] After computing the diagonals and their bearings and finding both *exactly* symmetrical in length *and* in bearings, I shrugged and assumed that probably any such quadrangle made symmetrical in this way would have similar symmetrical bearings and averaged diagonal lengths. But trying this elsewhere, I found this was not at all the case. I tried experiments with several different sets quadrangles, reasoning that overall balanced proportions would result from oblique triangle intercepts made from an average of original diagonals and two sides. Yet, I have not found any other quadrangle proportion than the one in Experiment B below, that have these symmetrical characteristics. Apparently some complex algorithm was used to make these four lengths through proportion from the overall base length of 921.453 meters. My above reconstruction theory of the sides lengths doesn't explain the character of this shape.

With this, an ancient riddle is uncovered: delicate trigonometry over a stone surface covering about 13 acres. The internal order of this shape is complex, perhaps as a result of a quadratic or some other equation.   Strangely, the (1) *diagonal bearings are exactly opposite and symmetrical,* (2) the internal diagonals *are of exactly equal length,* and (3) the *internal angles of opposite corners add to exactly 180°.* Additionally, the distances between mid-points of sides of different lengths, (4) both of these lines *have a center at precisely the same point,* which need not be true of quadrangles of lesser classes of symmetry. The difference between mid-point lengths (logically) must be half of 0.043 m seen above, the difference of N+S and E+W sides (or half of the theoretical 1.691 PI being 0.8455 as seen above). Since the existing configuration of the base of the Pyramid seems to relate by π and the Ascending Passage, angle, (see

225

above Eq. 18) and the other elements of the fine structure display φ with π, together with a 2861 number-form: my suspicion is that φ somehow plays a parallel role in this *angular form* and that "2861" will again be part of it. The proportioning ideas above in IX Eq. 44 to Eq. 48 are probably deep within a larger mathematical form. But I have no proof, and have not solved any rigorous formula behind this elegant shape. This is a beautiful and delicately proportioned shape: we are left to deduce that *this* was the ancient Builders' intended secret form of these four line lengths.

TRAVERSE BEARINGS ADJUSTMENT, EXPERIMENT B

course	distance (m)	deg min sec (bearing)	northings	eastings (m)	description
			500115.1955	2000115.1255	# 103 NE COR
1 (EL)	230.391	S 00°01'30.87" E	499884.8045	2000115.2270	# 105 SE COR
2 (SL)	230.454	N 89°59'44.78" W	499884.8215	1999884.7730	# 107 SW COR
3 (WL)	230.357	N 00°01'30.79" E	500115.1785	1999884.8745	# 108 NW COR
4 (NL)	230.251	N 89°59'44.77" E	500115.1955	2000115.1255	# 103 SE COR

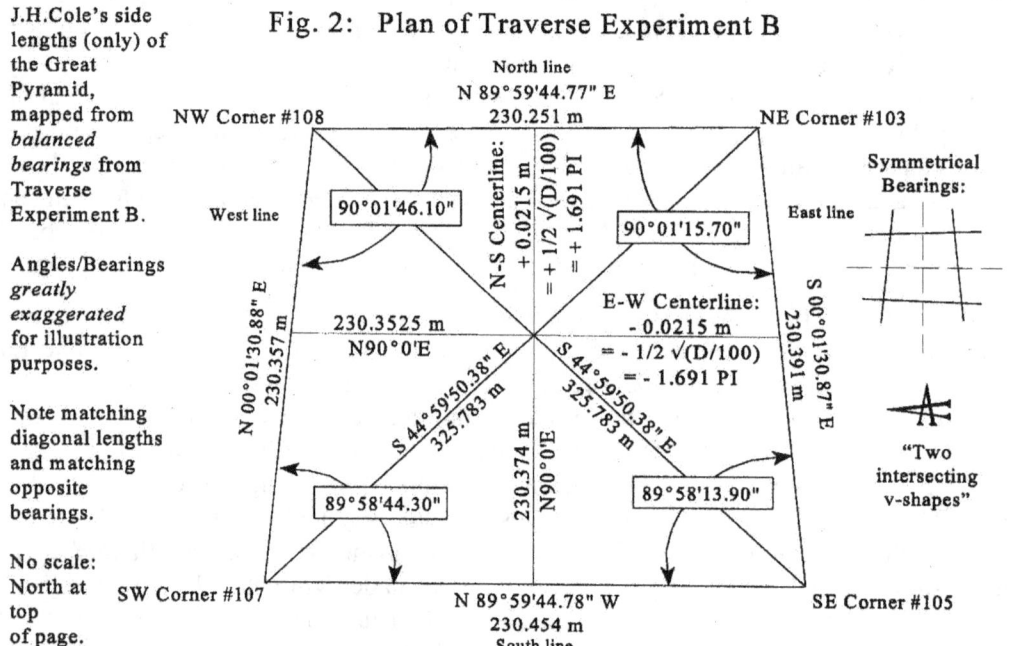

Fig. 2: Plan of Traverse Experiment B

J.H.Cole's side lengths (only) of the Great Pyramid, mapped from *balanced bearings* from Traverse Experiment B.

Angles/Bearings *greatly exaggerated* for illustration purposes.

Note matching diagonal lengths and matching opposite bearings.

No scale: North at top of page.

Cole's three place data is preserved, but is extended here into the fourth decimal place in coordinates. There is no closing course, since there is practically no residual closing distance. (This shape was centered with respect to an imaginary coordinate origin N 500000 m, E 2000000 m at Point #14.[21]) The East and West line bearings are symmetrically opposite, as are the North and South lines. These sides are like two overlapping symmetrical V shapes, one with its apex far (hundreds of miles) to the North, and the other much farther to the West. Note that the North and South lines are only 15.22" in divergence from being straight East-West bearings. The two diagonals if translated (in the above proportioning scheme Eq. 24 to Eq. 27) are 12913.27 PI, half the sum of Davidson's ideal precessional number, 25826.54 PI of Pyramid diagonals. These diagonal lengths, strangely, *are identical* to Davidson's diagonals made from *a square base* having the same perimeter. *The above shape shares the same diagonal lengths with a square of the same circumference.*

Following the method of Stecchini, we arrive at a table of the differences from a right angle, and then as before sum as a positive residual angular difference from a square:

Eq. 60	TABLE D
Northwest corner:	+0°01'46.10"
Northeast corner:	+0°01'15.70"
Southeast corner:	- 0°01'46.10"
Southwest corner:	- 0°01'15.70"

(+)----------------

0°06'03.60" (made into a positive sum)

Curious Coincidences--both within the Pyramid and in the dollar:

If we take the positive sum of the internal differences from a right angle from the original survey data, shown at Eq. 18, as 7'10" and divide into it the new residual sum in Eq. 60, based on symmetrical bearings, several familiar relationships appear, in the form of *fractions*--but not from distance measurements:

Eq. 61   Here we will look at two paths following different directions in division:

(1a) 6'03.60" / 7'10.00"   = 0.845581395;   (1b) 7'10.00" / 6'03.60"   = 1.182618262

(2a) 0.845581395×2 = 1.691162791;   (2b) 1.182618262^2 = 1.398585953

(3a) 1.691162791^2 = 2.860031585

Once again, all familiar numbers, where the accuracy is high--correct to about four floating places, at an error of one part in about twenty-nine hundred. Only 7 hundredths of an arc second need to be subtracted from 7'10.00" (7'09.93") for the relationship to be an exact division of the Displacement Factor number-form. Notice the astounding fact that four out of five elements of this group of fractional ratio numbers *also appear on the Pyramid as small, measurable dimensions in Polar inches.* In other words-- *these abstract fractional relationships,* which are the arithmetic result of the division of angular data, appear again represented *by small, physical linear dimensions.* Even (2a) could be said to be marked by the difference between two lines computed between mid-points. (See Fig 2 Traverse Experiment B, above.) We should also notice, stranger still, that four elements above (2a, 3a, 1b and 2b) appear as parts *of the dollar design of 1935.*

If we subtracted from TABLE D, the Northwest corner difference from the Northeast corner difference, and divided the positive sum of Cole's original bearings by this factor, the result looks close to the square root of 200, suggesting a proportionality. This same number divided into the symmetrical sum also produces some interesting results:

Eq. 62

(1)   Northwest corner: +0°01'46.10"

Northeast corner: +0°01'15.70"

(-) ------------

30.40";

(2)   7'10" / 30.40" = 14.144736;

(3) 14.1447^2 = 200.073   Looks like the square root of 200.

227

Suppose that we try this backwards with the actual square root of 200:

Eq. 63

$$14.14213562... \times 30.40" = 7'09.92"$$

Not only pretty close to 7'10", but as seen above in the remarks under Eq. 61, extraordinarily close to 7'09.93", which relates to the symmetrical sum by exactly 2.861022156. And if we divide this factor 30.40" into the symmetrical sum, square and multiply by 2:

Eq. 64

(a) $6'03.60" / 30.40" = 11.96052632$;

(b) $11.96052632^2 = 143.0541898$;

(c) $2 \times 143.0541898 = 286.1083795$

Once again, a familiar number-form good to about one part in forty-six thousand. This is all speculation, but the above result may call into question what the true value of the Displacement factor is, or what proportions it may actually be based upon. Many of the relations show a three sided relation between π, φ where the 2861 number-form is somehow involved. Is there any exact relationship between them? As noted earlier, in Step III Section XIII respecting the symbolism of π and φ, another mathematical value for the Displacement Factor that may be considered is the diagonal of the eighth root of π and the square of φ. I discovered this relationship accidentally in 1990, amid similar work on the dollar's diagonals (see Chapter 4) and it is included here as a possible explanation for the connection between π, φ and the 2861 number-form, and as a possible source for ancient inspiration for an astronomical relationship. Note that the resulting theta angle is reminiscent of the astronomical angle of the Ecliptic. It is given by the <u>Astronomical Companion</u> as 23.442° or 23°26'31.20", a difference of 20'32.37". (Compare with the result of Eq. 13 above):

Eq. 65

(1) $(\pi^{1/8} + \varphi^2)^{1/2}$ = diagonal
$(1.153835068... + 2.618033989...)^{1/2}$ = 2.86102033
(2) $\tan^{-1}(\pi^{1/8} + \varphi^2)^{1/2}$ = θ
$\tan^{-1}(1.153835068. / 2.618033989...)$ = 23°47'03.57"

Illus. 3

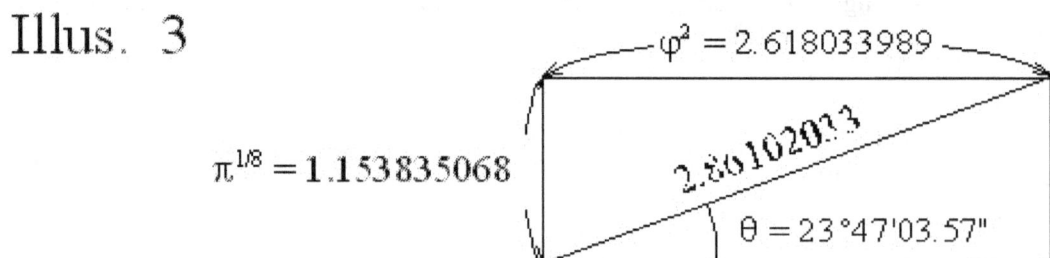

So, with the relationship in the above equation, we should stop, since the value of the hard evidence is small, even though some proportions appear plausible. But the intent behind the choice of these proportions is still not completely unveiled, and there are still multitudes of unanswered questions.

****

(XII) Conclusions--Pyramid theory and the apparent dollar connection:

(1) There is a clear case for an unusually complex and self-referential group of proportions, in both the Pyramid side lengths and in the conformation of it angles. These appear in two forms— possibly a π-

form with the Ascending Passage angle as the original configuration of the base, and a balanced implied form; and an overall π and ϕ-based eight-fold symmetry--together with the 2861 number-form as a very minute fraction; all in addition to Davidson's much larger displays of the 2861 number-form and π-ratio slopes.

(2) I think the likelihood of the use of the Polar inch has been convincingly demonstrated, as well as

(3) the 286-number form in at least seven separate forms of dimensions, fractions, astronomical elements, volumes, angles (of passage ways); geodetic azimuths (the West line of the Base of the Pyramid to true North) and interrelationship of original to symmetrical base line configurations. How the ancients could have arrived at this is unknown, but there is evidence for these.

(4) I think that there is evidence for sophisticated mathematical thought and possibly celestial mechanics by the ancient builders of the Great Pyramid, who clearly intended to demonstrate their sacred forms geometrically. I believe that eventually someone will tie the whole scheme together, and that the building and its open and hidden configurations will be found to be a large scale, multi-layered mathematical stage rather like the original quadratic diagrams of Arabic mathematician al-Khuwarizmi.

(5) From the point of view of the dollar's design and possible inspiration of it's Designer Edward M. Weeks, there seems to be enough circumstantial evidence to connect Weeks' design and the published work of J.H. Cole for Professor Borchardt after 1926, and the exact reference to the angular relations and dimensions shown in Eq. 61 and elsewhere. This could have been discovered by Weeks at our Eq. 29 step and he would need go no further along the trail that we followed. We should also note in passing, how similar the Borchardt's line is in different units, greatly resembling the 0.0035 PI offset at the base of the LIR on the dollar. Also, the 1.398 PI length, appearing as it does in a similar way, along the base line of the little pyramid on the dollar.

There is no reason to believe that Weeks solved the balanced configuration of the Pyramid's lines as seen above, or the elements of the internal eight-fold symmetry. But on the other hand, there is the possibility that he went even beyond these efforts. My feeling after working with these numbers for quite some time is that I have stumbled onto the edges of the "mother-lode" of ancient mathematics, and I am fully prepared to believe that Weeks had a complete picture of the this whole mathematical form. I suspect that evidence will emerge that he had gone even further down this trail than I have. (Why, for instance, if he had gotten to my Eq. 29 step, would he have stopped there? These numbers have no importance until there is some larger connection to the whole. Why did he think them important enough to put them on those fundamental places on the dollar? (See Illus. 4 pg 230.) There is some reason to suspect that he didn't stop. Not discussed earlier, is the interesting fact that the Thirteenth Line has a slightly skewed bearing from 90°:

$$\tan^{-1}(0.0035 \text{ [vertical offset at right Seal Face]} / 3.39448608 \text{ [BSC2]}) = S\ 0°03'32.68''\ E$$
Compare to Southeast corner, TABLE B on page 207: $= -0°03'33''$

(See pg 75-77 and Illus G5, pg 55) Is this coincidence? Weeks appears to have been a deep student of the Mysteries. The dollar is a small window into the deeper world of the Builders' art.

****

Illus. 4

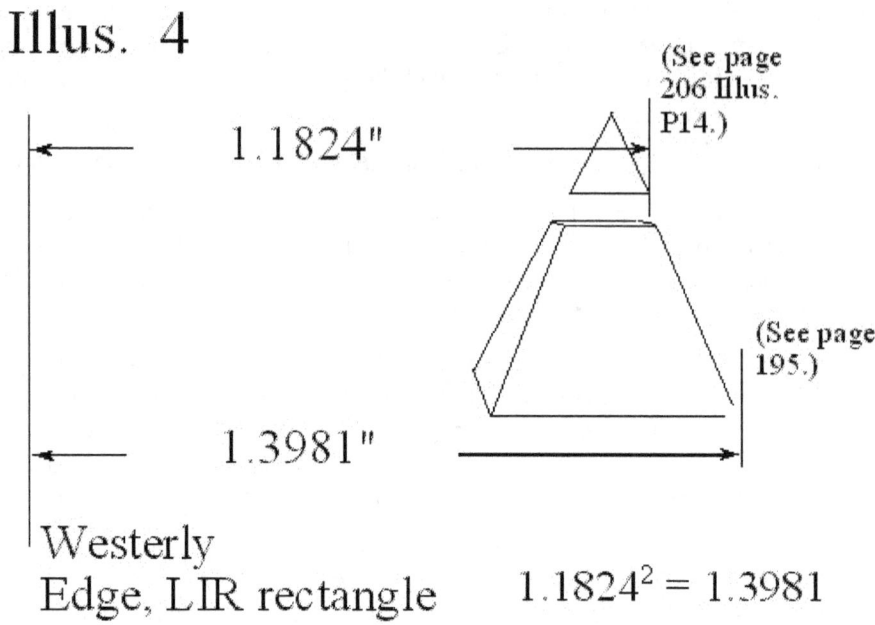

1.1824"

1.3981"

Westerly
Edge, LIR rectangle

(See page 206 Illus. P14.)

(See page 195.)

$1.1824^2 = 1.3981$

Notes:

[1] The Entasis of the Pyramid of Giza, 1994, Ken McGrath

[2] Längan und Richtungen der vier Grundakten der Grossen Pyramide bei Gise By Professor Ludwig Borchardt, 1926

[3] Pg. 358 et seq., The Secrets of the Great Pyramid by Peter Tompkins, in the appendix by Livio Catullo Stecchini.

[4] The Great Pyramid: Its Divine Message by D. Davidson and H. Aldersmith, Second Edition, 1925.

[5] The Great Pyramid: Its Divine Message "U.S.A. Geodetic Survey results of 1906 and 1909" Pg 236

[6] Pg 30, Basic Geodesy by J.R. Smith, 1988

[7] Statistical Methods in Scientific Measurement and Surveying by Bro. Barry B Austin

[8] Pg. 6, Determination of the Exact Size and Orientation of the Great Pyramid of Giza. J.H. Cole.

[9] Pg. 274 The Great Pyramid: Its Divine Message.

[10] Pg. 25, and pg 39 The Mysteries of the Great Pyramids by Andre Pochans, 1978

[11] Pg. 258, The Great Pyramid by C. Piazzi Smyth, 1880

[12] Pg. 366 and 378, The Secrets of the Great Pyramid by Peter Tompkins, in the appendix by Livio Catullo Stecchini.

[13] Pg. 80, Stonehenge and the Great Pyramid: Window on the Universe, by Bonnie Gaunt 1993

[14.] Pg. 175, <u>The Mystery of the Great Pyramids</u> by Humphrey Evans, 1979.

[15.] Pg 368, <u>The Secrets of the Great Pyramid</u> by Peter Tompkins, with appendix by Livio Catullo Stecchini.

[16.] Pg 2, <u>Determination of the Exact Size and Orientation of the Great Pyramid of Giza.</u> J.H. Cole

[17.] Pg. 8, <u>Determination of the Exact Size and Orientation of the Great Pyramid of Giza.</u>

[18.] What about this "one PI" difference between the ideal ceremonial length and the measured length? With the evident precision of the Pyramid's builders, was this an intentionally added length or error--or how precise were they? What about small differences caused by the propagation of random error found everywhere in measurement? Or is this difference really a measurement error giving an apparently longer length of the base course? In the arguments above in section IV, there appears to be reason to accept a possible precision of plus or minus two centimeters over the total length of tbcmc, (1:125,000) a value about half as small a Polar inch. A relative *scale factor* is a quite reasonable method of adjustment this error if you wanted to learn the original intent of these proportions.

For some, there will be a temptation to see this excess as an obviously "additional one PI", added by the builders' intent, as though it is an other one of many other small, symbolic offsets, like those to be discussed later, such as the 0.0355 m Westerly offset and a 0.043 m difference; the theoretical 1.398 PI. This could be true. It is not unreasonable to ask this, due to the smallness of these other offsets, but these offsets are of another character, found by differences of more direct measurements rather than a balancing of a whole base length.

But, suppose as a tempting opposite argument; there is indeed a "Boss"as described by Davidson and others, a semicircle escutcheon on the "Granite Leaf" structure in the Antechamber of the Pyramid, said by some to *define the length of a single Polar inch* in it's profile. (In Egyptian hieroglyphics, the symbol of a semicircle with the flat side down could mean "bread" or "father." It measures 5" x 8", a φ-proportion.) Indeed, in a similar way to the Boss, one is tempted to see this "extra PI" in the tbcmc as a symbolic fraction of one over 36524.24, (or 1/36524.24th), thus elegantly defining the Polar inch measurement standard and the 100x year length in yet another, new way. And looked at in that light, Davidson and Aldersmith's Egyptian "King Lists" as taken from Manetho out of Egyptian-Greek classical literature, not only will show the right number and digits in thousands for Davidson's 100x year length, but shows this number as having *one extra unit*, 36525, suggesting in yet another way, this extra ceremonial unit added to the 100x year circuit. This, perhaps, might demonstrate Manetho's knowledge of precise Pyramid lore, to even a greater extent than Davidson thought. And beyond this, the possible connection to the huge, ancient Sothic Year Chronology Calendar (of 365.25 days, exact) into the Pyramid design, which is an Egyptian calendar requiring a very long epicycle, based on the rising of the star Sirius.

But, this is unfair to the data, since, strictly speaking, Cole's limit of the accuracy *is not enough* to clearly define something as small as an inch over the total base course tbcmc of four corners. Even if my proposed *proportioning* uncovers possible details correctly by means of theory, we must unfortunately wait until a truly precise physical survey result replicates Cole's work to have any useful information that might pertain to this question. W.F. Petrie's work is similar, but neither are sufficiently accurate or of high enough precision to prove or disprove any hypothesis like this.

[19.] Pg. 359, <u>The Secrets of the Great Pyramid</u> by Peter Tompkins, in the appendix by Livio Catullo Stecchini. Note the apparent π and φ connection to the Parthenon's entasis.

[20.] It seems strange to me now that I wasn't surprised then, by the fact that these dimensions just snapped into place from these bearings with next to nothing leftover. I expected something of the kind as a hunch, but it took some time to for me realize just how odd and unlikely a result this was. In survey work, few if any relations have arithmetically clean closures. (The residual in traverse B is negligible, about 0.00002 m.) I was ready to make closure adjustments for this figure, but there were none needed--and the resulting *exact balance of the diagonal bearings and lengths seems to show that this is not accidental.* This has left me with the feeling that some of the discoveries and trial use of earlier ideas (such as the use of the shorter form of the North line of 230.251 m) were fortuitous. If the same procedure is tried using the longer 230.253 m (stated) length, mathematical trials will show an excess and imbalance in the 2 mm range in the diagonals, which also appears in an average of diagonals in Figure 1 for Traverse Experiment A.

This is clearly a carefully contrived shape. What formula could have been used in ancient times to create this shape? What formula would we use today? Required: a shape needed to (1) have four symmetrically opposite bearings about two cardinal axi, having (2) interior angles at opposite corners that sum to 180°, and (3) the diagonals have symmetrically opposite bearings, where (4) the diagonals are the same length as the diagonals of a square of the same perimeter. All of these requirements seem to describe a *square,* but this is not a square. Some important differences are: the side lengths are all different, and the diagonals do not intersect on both cardinal axi. Notice, that the only other shape that would normally have had this pair of diagonals is a square of the same perimeter. I am not knowledgeable enough in mathematics to answer this question.

If this hidden shape actually pertains to geodetic and astronomical formulae as imagined by the ancient engineers, they used a small number of values to make their definitions, at many scales of size. Some writers have speculated that φ and π stood for abstract male and female procreative functions in the universe. We know that the symbolism that the Egyptians used for a Right Triangle had the god Osiris as the x-axis and goddess Isis as the y-axis. (Their offspring god Horus, was the shape's hypotenuse). In Egyptian, these names are *Ausar* and *Auset.* "Isis" was not a distinct character, she had only the female word ending *"et"* to distinguish her from Osiris. Perhaps these are simply archetypical of male and female. (Together with Horus, this was the "divine family" of ancient Egyptian religion.) My intuitive guess is that φ is the masculine function and π is the feminine function. With this in mind, the two sides and 2861-number form diagonal seen in Eq. 65, Illus. 3, might show ancient religious identification for these mathematical functions.

[21.] Even three place precision demonstrates the sharp symmetry present in unadjusted figures of length, as originally provided by Cole. Here are some other coordinates associated with work on Traverse Experiment B, omitted from the material above, provided for experimenters:

North:	East:
500000.0000,	2000000.0000 PT#14 the center of cardinal axi;
500115.1870,	2000000.0000 PT#109 the Northerly mid-point;
500000.0000,	2000115.1762 PT#110 the Easterly mid-point;
499884.8130,	2000000.0000 PT#111 the Southerly mid-point;
500000.0000,	1999884.8238 PT#112 the Westerly mid-point;
500000.0000,	1999999.9915 PT#113 the intersection of diagonals.

Rather than a simple symmetry this is a dynamic symmetry. (Note that point #113, due to the nature of the diagonals, is slightly off center.) A new experiment based on the original intersection procedure for bearings given above, using the theoretical reconstructed lengths might prove revealing. Since this would be developed out of proportions rather than the rougher survey data, the new elements of the math behind this extraordinary shape might be made to surface through some means. This is by no means the end of the story

How did the ancient engineers develop this kind of precision in their work? Traverse Experiment B was implied by Traverse Experiment A, and presumably the first arrangement representing the structure was so crafted to intrigue an investigator into making an analysis like this. The internal angle of North and West sides are two seconds shy of a right-angle (89°59'58"), or the sacred Builder's Square. The difference at the Northeast and Southwest corners would be a total of 2 mm, or perhaps 1 mm at each corner. This would be good work anywhere today on two lines of 230 m in length. But if they could get *that* close, why didn't they make it perfect? From the evidence of Traverses A and B, we now know that they did "make it perfect" since every arc second is important in the mathematical shift to the symmetrical form. We know the Egyptians sighted with plumb-lines and it is not likely that they had telescopic devices of any kind. They probably made repeated measurements over the prepared, flat stone base surface--found by Cole to be perfectly flat to the millimeter. How did they measure? The oldest records of surveyors (in ancient Europe) describe the use of special hardwood rods with bronze caps at the ends. In Britain, these would have been 16.5 feet in length. One end was centered at a point, and laid flat in the direction of the objective. Then the next rod was carefully butted in front of the first. Then, the first is butted in front of the second, etc., until the end point is reached. Perhaps the Egyptians had rods like these, possibly marked off in units. My guess is the delicate work of very small units were read by means of vernier scale, a small marked rod to be aligned to the end point of measurement. (The vernier scale is an simple, easily made device like a slide rule, that can read hundredths or thousandths of an inch by means of moreé effects. If two scales of the same length differing by one unit, are slipped near each other, the fractional unit difference can be measured by counting which marks line up.) Some kind of temperature correction for measurement would have been necessary, and I also suspect something like logarithm tables for trigonometry functions were used for creating angular relationships. We do not know how, but they managed to lay cut rock blocks with paper thin layers of mortar to a precision of one hundredth of an inch over hundreds of feet.

APPENDIX B

Biography of Edward M. Weeks
from the National Cyclopedia of American Biography
Vol. XLIV, pg. 344:

WEEKS, Edward Mitchell, engraver, was born in Riverside, N.J., Aug. 20, 1866, son of Robert Mitchell and Caroline (Berner) Weeks. His first paternal American ancestor was George Weeks, who came to this country from England and settled in Dorchester Mass., in 1635. From George and his wife, Jane Clapp, the descent was through Ammiel and Elizabeth Aspinwall, Ebenezer and Deliverance Sumner, William and Sarah Tukekee, Lemuel and Peggy Gooding, Joseph and Lois Freeman, and Joshua Freeman and Elizabeth Ingersoll Mitchell, the grandparents of Edward M. Weeks. His father was a plant superintendent. The son received his preliminary education at the Riverside public schools and his art training at the Spring Garden Institute and the Pennsylvania Museum and School of Industrial Art, both in Philadelphia, and the Corcoran Art School, in Washington D.C. He later studied law at George Washington University, where he graduated LL.B. in 1907. He was admitted to the District of Columbia bar and for several years maintained an independent practice of patent law in Washington. Meanwhile, he served as script letter engraver with Bailey, Banks and Biddle, Philadelphia Pa., and in 1887 became engraver for the stationary department of the department store founded in Philadelphia by John Wanamaker (q.v.). During 1887-89 he served in a similar capacity with the Samuel Ward Co., Boston, Mass., and in 1891 with Dempsey & Carroll, New York City. From 1890 to 1900 he was square letter engraver in the Philadelphia office of the American Bank Note Co., following which he went to Washington as engraver with the United States Bureau of Engraving and Printing. He became foreman of letter engravers in 1925, assistant to the superintendent of the engraving division in 1932, and superintendent of engraving in 1933. He relinquished that position shortly thereafter to return to the practice of engraving, in which he was engaged until his retirement in 1940. Weeks outstanding achievement with the bureau was his steel engraved facsimile of the Declaration of Independence, completed in 1939, which he copied from the first engraving of that document, made in 1828, and which at the time of his retirement was the largest work produced for the bureau. He was also responsible for the design adopted for the reverse of the one-dollar denomination silver certificate, the dollar bill, in 1935, and for the frames and lettering on a number of postage stamps. Among these were the three-cent stamp of 1902, a ten-cent special delivery stamp of 1917, eight- and sixteen-cent airmail stamps of 1923, a fifteen-cent special delivery stamp of 1931, and a five-cent Virginia Dare commemorative issue of 1937. He did the entire design for a New York World's Fair commemorative in 1939 and for several revenue, snuff, and tobacco stamps of various years. Weeks was the author of "Letters Analyzed and Spaced" (1952), in which he presented his system of typographical analysis based upon what he called spacing tables. The first of the two tables given for each of the seven different types of Roman, Gothic, and Uncial alphabets presented the proportions of the various elements of the letters; the second, the spaces to be used between any combination of two letters. Along with an explanation of the principles of topography used by him, the book contained more than sixty illustrations. Weeks was a member of the Masonic order. His religious affiliation was with St. Albans Episcopal Church, Washington, where he served as vestryman and superintendent of the Sunday school. Politically he was a Republican. Playing chess was his favorite leisure activity and he was a member of the Capitol City Chess Club of Washington and the Saint Petersburg (Fla.) Chess Club. He was married twice: (1) in Riverside, N.J., June 12, 1889, to Mary, daughter of Charles S. Walcott of Delanco N.J., his first wife died in 1932; (2) in Riverside N.J., Apr. 2, 1940 to Thekla, daughter of Stanley Horton Fundenberg of Cumberland Md., a physician. By his first marriage he had three children: Robert Walcott, Dorothy Walcott, and Ruth Walcott, who married Laurence Carlton Staples. Edward M. Weeks died in Washington D.C., Feb. 20 1959.

Edward Mitchell Weeks

THE NATIONAL CYCLOPEDIA OF AMERICAN BIOGRAPHY

Courtesy of University Microfilms

Commentary and Speculation from the Biography:

Edward Weeks appears to have had two important themes in his life: a strong technical talent, and the other an artistic bent of an extremely technical sort. After going to technical and art schools, he (surprisingly) got a law degree and was admitted to the bar in Washington D.C. But, this either didn't hold his interest or business was tough, and he starts engraving bank notes. After working his way up to Superintendent of the Engraving Division at United States Bureau of Engraving and Printing after 33 years, he gave up this top position (somewhat after the design of the dollar) to return to what must have been his first love, the actual practice of engraving. Perhaps a close look at his designs for stamps is called for. There he remained until about the eve of the second World War in 1940, having worked for the Bureau for 40 years. Now at 74 years of age, he marries for a second time to Thekla Fundenberg. In retirement (at the age of 86) he wrote a book on his theory of lettering[1], with the help of one Esther A. Richards (see below). It should come as no surprise to the student of the dollar that Weeks was among the Masonic brethren, and that there was a strong element of religious idealism, as seen in the more than passing involvement with St. Albans Episcopal Church. (I do not know what Masonic lodge he belonged to, but there are many in the Washington D.C. area. A "vestryman" is like a trustee position at the church.) His interest in chess may come as no surprise, considering the very cerebral, riddle and game-like character of his work in the dollar's design. My guess is that he was a ferocious chess player. We should notice the devotion or patriotic sentiment in his pursuit of the large engraving project of the Declaration of Independence. It seems that this large, career-end engraving project is a final salute to the Declaration of Independence concept. (Note the IDT-1776 connection, the many-fold theme seen throughout the dollar's design.) Considering how far away this would be from Bureau's regular business of currency and stamps, it seems likely that he would have had to talk them into this unusual project. The implication from the biography is that Weeks retired in St. Petersburg, Florida, a popular retirement spot, then as now. But, it seems that he may have been a commuter to D.C., since he died there in 1959.

The book <u>Letters Analyzed and Spaced</u>:

This curious book is evidently a labor of love, with Weeks' personally devised tables for various types of lettering. It lists a collaboration with Esther A. Richards, "Instructor in Art and Lettering, Moore Institute of Art, Science and Industry of the Philadelphia School of Design for Women". The book's dedication reads:

To

T. C. F. W.

the helpmeet of my senior years

This is undoubtedly Thekla Fundenberg, probably abbreviating "Thekla C. Fundenberg-Weeks". This is a small book (8 1/2" x 5 3/8", 109 pages) made sideways as a rectangle in *landscape* format, opening to a wide 17 inches. This has an obvious practicality of being a book that you can open and lay on the table beyond your work, easy to copy from. Indeed, the copy I located showed occasional small signs of India ink, evidently from a student of lettering. But, to us, studying the mysteries of the dollar, this format is suspiciously familiar, oddly reminiscent of the joined right and left SIR/SOR rectangles in the middle area of the dollar--something only to be noticed by those researching the life of Designer of the dollar.

Is there a hidden meaning here? (The opened-book-ratio of length to height turns out to be nearly the square root of 10 to 1: $17"/5.375" = 3.1627...$ For those interested in the J3-theme, we should note the odd properties of $\sqrt{10}$: $\sqrt{10}$ is $3.16227766...$ and it's reciprocal is the *same number, reduced by exactly one decimal place:* $1 / \sqrt{10}$ is $0.316227766...$ Isn't that curious? we have seen this sort of thing before.)

After our non-stop and many faceted exposure to Weeks' use of serifs and edges of letters as fundamental parts of his design of the dollar, one cannot resist a laconic chuckle in reading from the Introduction where he tells us about *serifs*:

> "The strokes of Roman letters are terminated by serifs. They give grace, character and finish to the letters, and add to their legibility." Pg 15.

On the dollar, Weeks appears to have used his "Blackface Bank-Note Roman" font, and elsewhere in his book he gave examples of it's use in stamps. It would be interesting to analyze how rigidly he stuck to his spacing scheme on the dollar, since he obviously located certain serifs and letter strokes solely for design geometry on the dollar. (e.g, the geometric framing of the word "HE" from the word THE in Illus. D and E, as shown in Chapter 2 and elsewhere.) Weeks' spacing scheme is only on the horizontal axis, and he shows that any height may be used without disturbing the letter proportions. This being the case, an examination of Weeks' choice of *heights* of the letters on the dollar also seems warranted. His book is set up as six lessons. In his Lesson 2 pg. 33, the first lesson to demonstrate the use of his exact numerical spacing, we should note of the odd coincidence in spacing regimen for the first letter P of the example word "PARKING":

"Parking"

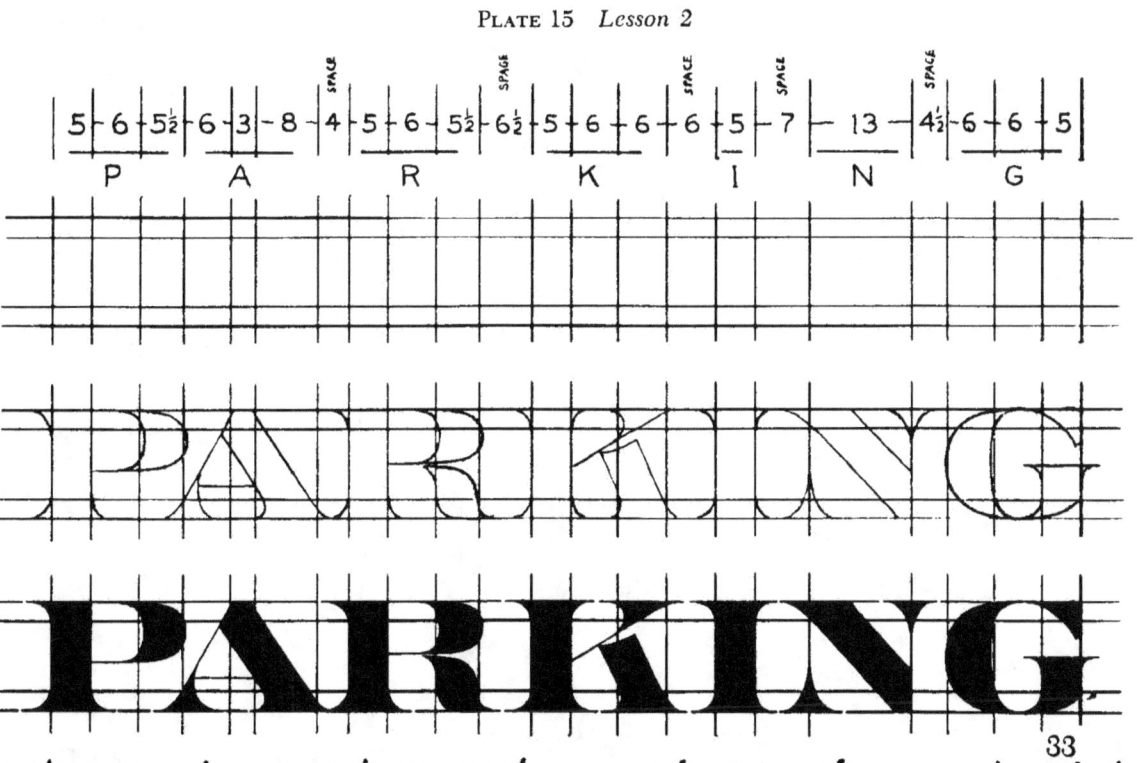

PLATE 15 *Lesson 2*

Notice the beginning of this letter spacing at upper left shows "5, 6, 5 1/2." This exactly *codes* the four decimal digits of the measured length of his masterwork, the dollar. When I first measured the length of the dollar with a one-hundredth inch scale, I read 5.65 and, estimated a half to the next hundredth. (i.e., 5.655") My guess is this was done only to make a smile at a distant point in the future-- but, who knows? he may have hidden all sorts of things in this odd little book. For those who are interested in lettering, this is still a valuable book even in the days of computer supplied fonts, and otherwise provides an insight into the old time bank-note engraver's art.

Notes:

[1]. <u>Letters Analyzed and Spaced</u> by Edward M. Weeks, 1952 Exposition Press, New York.

ABOUT THE AUTHOR

Kenneth James McGrath (1952--   ) Born in Evanston Illinois.   Began work in land surveying in 1973.   Twelve years in Illinois, (lot and block –type surveys, hydrography, ALTA surveys.) Arizona: 1985—1989, public and private sector survey work; control, traverses and construction. From 1989, public and private sector work in southern California; GPS control and traverses on Santa Fe rail right of way, deed writing, survey mapping, acquisition and relinquishment of highway right of way. Interests: science, history, math and language. Master Mason (Aztlan Lodge No. 1 Prescott AZ). Member Southern California Epigraphic Society. Go player (2 Dan).